NATURE'S PALETTE

Nature's Palette

The Science of Plant Color

David Lee

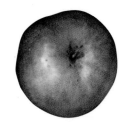

THE UNIVERSITY OF CHICAGO PRESS

CHICAGO AND LONDON

DAVID LEE is professor in the Department of Biological Sciences at Florida International University and Director of the Kampong of the National Tropical Botanical Garden, in Miami.

The University of Chicago Press, Chicago 60637
The University of Chicago Press, Ltd., London
© 2007 by The University of Chicago
All rights reserved. Published 2007
Printed in China

16 15 14 13 12 11 10 09 08 07   1 2 3 4 5

ISBN-13: 978-0-226-47052-8 (cloth)
ISBN-10: 0-226-47052-0 (cloth)

Library of Congress Cataloging-in-Publication Data

Lee, David Webster, 1942–.
   Nature's palette : the science of plant color /
David Lee.
     p. cm.
Includes bibliographical references and index.
ISBN-13: 978-0-226-47052-8 (cloth : alk. paper)
ISBN-10: 0-226-47052-0 (cloth : alk. paper)
1. Plants—Color. I. Title.
QK669.L44 2007
580—dc22

                    2007004229

⊗ The paper used in this publication meets the minimum requirements of the American National Standard for Information Sciences — Permanence of Paper for Printed Library Materials, ANSI Z39.48-1992.

For Carol, Sylvan, Katy, and Shaun   For helping me become a botanist with a small b

....

CONTENTS

....

....

PREFACE

....

About birds and trees and flowers and water-craft; a certain free margin, and even vagueness—perhaps ignorance, credulity—helps your enjoyment of these things, and of the sentiment of feather'd, wooded, river, or marine Nature generally. I repeat it—don't want to know too exactly, or the reasons why.

WALT WHITMAN

....

This book was personally satisfying to write because it enabled me to combine experiences from my life, from childhood to adulthood, with my scientific background as a botanist. I have always had a strong connection to plants and their colors in the various landscapes of my life. As a toddler I watched my mother garden, adding flashes of bright color to our yard. My emotional attachment to the landscapes of my childhood, and my fascination with the color produced by their plants, inspired my training and career as a botanist. Part of my research during that career, particularly the last thirty years of study in the tropics, focused on the colors produced by plants, particularly colors produced by foliage—and their adaptive significance. So I write this book from the perspectives of my emotional attachment to those landscapes, amazement at the intricacy and beauty of plants, and curiosity about how they function.

My life, in the tropics and then on the edge of the tropics in Miami, has combined a scientific fascination with nature and an emotional attachment to the landscapes I studied. These interests have made transparent for me an issue we all face in our lives, a reconciliation of the intellectual with the emotional and intuitive. So my approach in this book is to write about the science *and* culture of plant color. By the science, I mean the physical and chemical explanations of colors, and the selective advantages they provide the plants in their natural environments. By the culture, I mean our aesthetic and emotional responses to plant colors, which have become parts of the diversity of ways by which people live on our planet.

Nature's Palette: The Science of Plant Color explains the science behind the production of color by plants: the nature of light, optical principles relevant to color production, vision, and the perception of color. It describes the production of color on chemical and biological levels, and discusses the biological importance of plant colors. It surveys the remarkable diversity of color production in leaves, stems, flowers, and fruits. Two chapters describe two particularly remarkable color phenomena: iridescent metallic colors produced through physical effects and not by pigments, and the phenomenon of red leaf color and autumn coloration in temperate forests.

This book also explores our fascination with plant color and its importance in our lives. Since plants were the principal materials for the fabrication of tools, clothing, and shelter, and were the main source of food from our beginnings, their colors became culturally important as well. The colors of plants, particularly of flowers and fruits, were incorporated into culture and language, and they continue to be important today by giving us a connection to nature and adding aesthetic diversity to our lives.

Plants provided fibers for the production of fabrics, and these natural fabrics are still very important today. Plants also provided agents for the curing of leather. Dyes, derived principally from plants, were applied to leather and fabrics to make them more attractive. The collection and cultivation of dye materials along with the dyeing of fabrics became an important economic activity, part of the mix of commodity trade that fed the Renaissance and justified the colonization of the tropics by the European countries. The value of dyes stimulated the development of organic chemistry and the discovery of synthetic substitutes.

In some way, poorly understood, plants soothe us and may be therapeutically beneficial to us. Perhaps we are attracted to the greenness of plants for deep reasons that relate to our origins as forest primates and then as humans. I hope that the cultural descriptions will add a dimension to the science, and reveal the broad reach of plant color in human life.

My target in writing this book is an informed nonscientist, probably with an interest in gardening or natural history. The text is, I hope, straightforward. It is illustrated with a variety of photographs and diagrams. It is not cluttered with footnotes, references, or detailed information on illustrations. At the same time, I provide the additional information in separate sections so that the interested reader can go much deeper. I have added proper scientific names of every illustrated plant, including authorities and families, in either the chapter or illustration notes, and have mainly used common names in the text, with few exceptions. I have standardized varietal names, which traditionally vary among different plants, to provide the name of the genus followed by the variety name, capitalized and enclosed in single quotes. Common names are always uncapitalized, unless they include a geographical location or other proper noun. At the end of the book, the notes for each chapter document the sources of information I used, written in order of the ideas in the chapter. A section at the back of the book provides full information about the identification and location of the plant in each figure, and acknowledges the individuals who provided the photographs. Providing this information certainly satisfies the scientific side of my personality. It may also make this book detailed enough to satisfy some of the demands of a professional biologist. Thus I am trying to find the balance that Whitman wrote about (but I am certainly erring on the side of the scientist!).

Thus this story about plant colors combines elements of the scientific along with the aesthetic and emotional. These traits are common to all of us, in some combination or another. For me, understanding the scientific basis for a phenomenon, such as plant color, adds to my aesthetic appreciation of it. I hope that this book fulfills the same purpose.

ACKNOWLEDGMENTS

As this book has been a long time coming, I have many people to thank for helping me along the way with information relating to plant colors, stored away for use at the right time. My high school chemistry and physics teacher, Leo McIntee, introduced me to the chemistry and the physics of color. A college professor, Jens Knudsen, inspired me with his mixture of love for and scientific fascination with nature. My graduate school mentor, David Fairbrothers, added scientific discipline to the other qualities and emphasized the importance of integrity and compassion. While in graduate school I was also fortunate to become acquainted with Oved Shifriss, who opened my eyes to the genetics and domestication of plants, and Carl Price, who taught me plant biochemistry. My first academic position was at the University of Malaya, and I benefited greatly from my friendships with Brian Lowry and Benjamin Stone, and other colleagues at the university, including Engki Soepadmo. While at the University of Malaya, I became friends with Francis Hallé and later worked with him at the University of Montpellier. Like Ben and Brian, he has an enormous wealth of knowledge about plants and tropical biology.

Since moving to Miami, I have benefited from my colleagues at Florida International University, many of whom have helped on various aspects of this book. They include Jennifer Richards (aquatic plants and botany); Steve Oberbauer (plant physiology and tropical ecology); Kelsey Downum (plant natural-products chemistry); Brad Bennett (ethnobiology); Kevin O'Shea, John Landrum, and Martin Quirke (pigment structure and chemistry); Suzanne Koptur (pollination and dispersal biology); William Vickers (cultural ecology and the Secoya people); Tom Philippi (plant ecology and lore); and Tim Collins (phylogenetics and evolution). Stacy West beautifully crafted the figures illustrating various scientific principles, and Roberto Roa completed the molecular diagrams. George Taylor produced most of the transmission-electron-microscope photographs, and Barbara Maloney of the Florida Center for Analytical Electron Microscopy provided the scanning-electron-microscope images. Members of the Instructional Media Photography Laboratory helped with preparing transparencies for publication. Two of its photographers, Michael Upright and George Valcarce (who recently passed away and will be missed by many), inspired me to take better photographs.

I also received help from two botanical institutions in Miami. Larry Schokman, director of the Kampong of the National Tropical Botanical Garden, provided much information and access to plants for photography. An eminent botanist and associate of the Kampong, Barry Tomlinson, also provided much useful information. At Fairchild Tropical Botanic Garden, several scientists provided information and photographs, particularly Jack Fisher, and Carl Lewis, Scott Zona, David Bar-Zvi, and Mary Collins. FTBG was the source of the most plants photographed for the book. My friend at the University of Miami, Ted Fleming, was a good role model for getting the book completed and a provider of helpful information about pollination and dispersal biology. Bill Stern, emeritus professor at the University of Florida, provided the microscope I used for all of the recent anatomical photographs. Also in Miami, two nurserymen and judges of the American Orchid Society, Bill Peters and Robert Fuchs, provided me with very helpful information on orchids.

The final impetus to write this book came from my friendship and collaboration with Kevin Gould of the University of Otago, whose research is featured in two chapters. My writing about red leaves, and ability to complete the entire manuscript in a timely fashion, is due to my receiving a Bullard Fellowship at Harvard University, which I spent at the Harvard Forest, in central Massachusetts, in the autumns of 1998 and 2004, with the support of its director, David Foster. There I collaborated closely with John O'Keefe, a scientist at the Forest, and with Missy Holbrook, a professor at Harvard. In 1998, we also worked with Taylor Feild, who was Missy's Ph.D. student at that time.

Much of what I have written in this book is informed by my own research, and was partly published with other people. I would like to acknowledge their help as well, if not already mentioned above: Charles Hébant,

Richard Bone, John Norman, Rita Graham, Susan Brammeier, Alan Smith, Sarah Tarsis, David Storch, Knut Norstog, David Kuhn, Marzalina Mansor, Haris Mohamed, S. K. Yap, Daniel Baskaran, and Tony Irvine.

Many other individuals provided smaller bits of information or photographs used at specific points in the book, and I add their names here in no particular order but alphabetical: Jeff Abbas, Neha Batra and Neema Batra, Paul Beardsley, Mahadev Bhat, Cynthea Bogel, Stephanie Bohlman, Susan Brown, B. L. Burtt, Kunfang Cao, Linda Chalker-Scott, Enrico Coen, Robin Currey, Elizabeth Daynes, Deena Decker-Walters, Dale and Pam Easley, Adrienne Edwards, Jim Erdman, Priscilla Fawcett, Harry Frank, Paul Gepts, Robert Griesbach, Richard Grotefendt, Anita Hendrickson, D. M. Joel, Gernot Katzer, Catherine Kleier, Helga Kolb, Bob Korn, Philippe Latour, Katherine Lee, Diane and Mark Littler, Rodrigo Medel, Allan Meerow, Cecilia Mendes, Nancy Morin, Jerry Palleschi, Ken Robertson, Anne-Gaëlle Rolland-Lagan, Austin Roorda, Gene Rosenberg, Oliver Sacks, Lora Lee Saracino, Doug Schemske, Ed Schofield, John Stevenson, Philip Stoddard, John Stommel, Sharon Strauss, Guoda Tao, Georgia Tasker, John Thomas, Michael Tlusty, Virginia Walbot, Chris Winefield, and Alicia Zuniga. Cary Pirone read through the entire manuscript and found many mistakes.

Publication of this book required financial support to subsidize the use of many color photographs and keep the price affordable. The Ball Horticultural Company, of Chicago (and Jeff Gibson in particular), provided financial support and supplied a number of high-quality flower images. These were particularly helpful because of the precise varietal information that came with each flower image. Important financial support was also provided by Florida International University through the Provost's Office (Doug Wartzok and Ron Berkman), the Office of Sponsored Research Adminstration (George Dambach and Kelsey Downum), the College of Arts and Sciences (Mark Szuchman), and my own Department of Biological Sciences (Jim Fourqurean).

Now that I have gone through the exhausting process of writing this manuscript, looked through hundreds of references, and reviewed thousands of images, I know I must have messed up somewhere. So I must own up to my own responsibility and not blame any of the people who have helped me. I deeply appreciate their generosity and openness, which still is very much a part of the botanical disciplines where I do my research. However, there would be more errors if not for those pointed out to me by two anonymous reviewers (who read and reread the manuscript) and by Christie Henry, senior science editor at the University of Chicago Press (whose grace and enthusiasm kept this project on track). Joann Hoy, both a professional botanist and an editor, remarkably improved the manuscript during the production phase. Family members, wife Carol (who has shared many of my forest sojourns over the decades) and kids Sylvan and Katy, read and critiqued different parts of the manuscript. They, along with grandson Shaun, kept things in balance. My mother, Mary Lee of Ephrata, Washington, still tending her roses at ninety-three years of age, continues to be an inspiration for all that I do.

Chapter One Coloring Our Bodies with Plants

....

To climb these coming crests
one word to you, to
you and your children:

stay together
learn the flowers
go light

GARY SNYDER

The world in which we live is teeming with color: the sky,
earth, water, and fire all have distinct colors. From time
immemorial, we who delight in such perceptions have tried
to reproduce these colors in our day-to-day surroundings.
What could be more normal? For color is the child of light,
the source of all life on earth. The challenge of finding
materials capable of producing lasting colors in the world
around us has preoccupied humankind from prehistory
to the present day.

FRANÇOIS DELAMARE AND BERNARD GUINEAU

....

Rogelio Piaguaje (fig. 1.1), a member of the Secoya nation in Peru, has decorated himself in a spectacular
fashion, using plant and other colors available to him in his
rain-forest home. Plants are important to all of us in many
ways. They provide our food (or the food of the animals we
eat) and a surprising number of our medicines. They also
give us beauty, of color and fragrance. This book is about
the visual properties of plants. Plant colors are certainly
important to Rogelio, and his manipulation of plants for
their colors will be discussed in detail later in this chapter.

Figure 1.1 Rogelio Piaguaje, a member of the Secoya nation of Amazonian Peru, 1979. Photograph courtesy of William Vickers.

The story of our relationships with plants and their colors begins in antiquity, and Rogelio's story spans that past and moves into the present. Perhaps the best way to trace the history of this relationship is to look at individuals and the importance of plant colors in their cultures. Their lives may seem very distant (both in time and space) from our own experiences, or from the human condition in the twenty-first century. Yet we can see similar basic human needs met by plants, as they are used and processed to extend the values of color in our lives. Much of human culture, even today, concerns plants, and their characteristics (particularly their colors) are important parts of all languages. So we begin with some examples, with the stories of three individuals, representing differing stages of cultural evolution and different periods in and outside of recorded history.

The Shanidar Neanderthal

His photograph does not seem very revealing: a skeleton embedded in the soil of a cave set in a ridge slope above a river valley. Today this valley is part of what some call Kurdistan, and others northeastern Iraq. Where the Great Zab River flows through the Zagros Mountains, caves have formed in the limestone formations above the river plain. One such cave, near the village of Shanidar, was first excavated almost fifty years ago as a likely location of habitation by the humans we today call Neanderthal. Their name comes from the location of the first excavations of their skeletons in a valley (Neandertal) in Germany. The name Neanderthal calls to mind some primitive, hunched-over, hairy half-ape, as depicted by many artists. However, these beings were not so very different from us. Excavations in the Shanidar cave have revealed a paleolithic culture that existed some 60,000 years ago. Its inhabitants collected plants, hunted game, produced stone tools, and used fire to cook their food.

Seven skeletons were discovered deep in the soil layers of this cave. Some had accidentally been killed by rocks falling from the ceiling, and were left in place. Three of the people, one an infant, had died of natural causes and were buried in the recesses of this cave, presumably by members of their community. Two of the skeletons were elderly men (worn teeth and joints), around forty-five years old. Presumably they had been ailing prior to their deaths. They also were traumatically injured much earlier in their lives (fractured and healed bones), so much that they may have needed help to survive in the dangerous conditions where they lived. Body number IV was buried in one of these niches, and other skeletons were excavated below and above him. Ralph Solecki, one of the archaeologists working at this site, excavated this grave and carefully removed samples of soil from around the skeleton. Solecki's research revealed the position of the skeleton and the location of some of the soil samples. These were later very carefully examined by Arlette Leroi-Gourhan, a French botanist. Leroi-Gourhan was trained in the esoteric science of palynology, the description and identification of pollen. She examined the soil and found clumps of pollen in regions around the skull of the skeleton. She was able to identify pollen produced by flowering plants presently growing in the areas near the cave: grape hyacinth, ragwort, yarrow, hollyhock, and cornflower. She also identified pollen from the gymnosperm shrub *Ephedra*, similar to the Mormon tea of the U.S. western mountains and the ma huang of China.

From the identification of the pollen grains, coupled with their location along the skeleton, Leroi-Gourhan and Solecki were able to piece together some remarkable details of the burial of this Neanderthal man. The flowers are common in this region today in the month of June, the likely time of death of this man some 60,000 years ago. They were placed around his head as a wreath (the clumps of pollen were found adjacent to the skull).

Figure 1.2 Three flowers used to decorate the grave of the Shanidar Neanderthal IV. *Left*, an unknown species of tassel hyacinth (*Muscari comosum*); *center*, yarrow (*Achillea millefolium*); *right*, hollyhock (*Althaea rosea*).

The pollen could not have arrived by wind; the burial site is far from the cave mouth. These are attractive and colorful flowers (fig. 1.2). *Achillea*, or yarrow, is a handsome herb with flowers generally whitish in color, known for thousands of years for its medicinal properties. *Centaurea* (perhaps the *C. solstitialis* that grows near the cave today) produces long spikes of colorful flowers, some a brilliant blue (see fig. 7.20). This plant looks a bit like a thistle, with white hairy leaves, and has medicinal uses. *Senecio* (several species grow today near the cave) produces large bright yellow flowers, and some have been used medicinally. *Muscari*, grape hyacinth, is a spectacular small herb with very dark blue flowers. Its bulbs are poisonous and have been used medicinally. Again, several species of *Muscari* grow today near the cave. Hollyhock is a robust annual plant with large flowers produced continually on its tall spike. It is a garden or roadside plant widespread in temperate regions. These flowers were most likely red or purple, although different color forms are found today. Again, these plants are used medicinally by the people living near the cave today.

Interspersed with the flower pollen were grains from a nonflowering plant, *Ephedra*. This shrublike gymnosperm does not produce flowers, but its species produce various quantities of the alkaloid ephedrine. In central Asia it has been used medicinally for as long as we know. It was also a sacred plant, used ritually with other plants for its vision-producing trances. *Ephedra* was an important plant in the early religions of this region; for example, Zoroastrian priests used it. Some have argued (perhaps not very successfully) that *Ephedra* may be the legendary soma of the Indo-Europeans, who used it in central Asia, later moved to India, and celebrated its properties in the Rig Veda. We know of this plant today because of its illegal use in herbal ecstasy and in herbal diet treatments; we also use a derivative of this molecule in Sudafed cough syrup (used as a raw material in the illicit production of methamphetamine). Pollen grains from this plant were found near the skull and along the length of the body.

What emerges from this analysis is the death of an elderly man, laid to rest with care by members of his group. They collected wildflowers from the nearby countryside on an early summer day. They may have formed a wreath by twining the smaller flowers (like the *Muscari* and *Achillea*) among the flexible branches of the *Ephedra*. The body may have been laid to rest on a cushion of the latter plants. We can easily imagine that this was a man of some importance in the group, the beautiful flowers celebrating his life, and their medicinal importance possibly commemorating his status and knowledge. Perhaps he was a shaman, wise in the ways of the spirits and of healing.

Not everyone agrees that the distribution of pollen around the skeleton is evidence of such a burial. Perhaps it was deposited by burrowing rodents, far from the cave entrance. The flower-laying behavior seems at variance from the simple culture of this group (*Homo neanderthalensis*): humanoid but not truly human. Its members used fire, but their camps were not very well organized. Their stone tools were simply made, and they left no evidence of jewelry, or of art on the walls of the caves they inhabited. Perhaps they wore garments from the animals they killed, but they left no bone needles as evidence of sewing. Their teeth could have become prematurely worn from the processing of those skins. We can only speculate that they spoke to each other in a formal language; analysis of their skulls suggests that they could not make all of the sounds we use in modern languages. So we don't know if they had names for the vivid colors of the pretty flowers collected for the burial. Yet they surely knew the plants and animals they used for nourishment and healing. Other apes, such as chimpanzees and gorillas, have similar knowledge and select plant food items using color to indicate their ripeness and palatability. They also recognize and use certain plants for medicinal purposes. Our local newspaper, the *Miami Herald*, recently reported on the behavior of monkeys at Monkey Jungle, a local tourist attraction. These monkeys discovered that rubbing a secretion from an exotic millipede repelled mosquitoes, and that licking the secretion put them into an alleged state of euphoria. Perhaps it is an exaggeration to say that the Neanderthal man was a shaman (maybe the flowers were pretty and available, but not collected because of their medicinal values), but apparently his friends cared enough for the old man in their group to place colorful flowers in his grave.

I conclude that this burial is evidence of the importance of plants in early human culture. Furthermore, it is evidence that colorful flowers of plants were valued, and selected for the burial rite. Is such a burial unique? Some pollen grains were discovered in the soil that surrounded another skeleton at Shanidar (VII). Perhaps other deaths occurred at times of the year when flowers were not available. Perhaps we should look more carefully at the soil around other skeletons, in other locations. Knowing of their burial rites, it is hard to accept the cartoons depicting primitive Neanderthals, looking as similar to us as a chimpanzee. Solecki had this to say:

> Interestingly enough, it was originally a Frenchman, Boule, who is credited with the bestial characterization of the Neanderthals. And it was a Frenchwoman, Mme. Leroi-Gourhan, who gave us the soft touch. The observation has been made that the Neanderthal has been ridiculed and rejected, but despite this, according to all the proofs that can be mustered, he is still our ancestor. I feel that, in the face of the growing evidence, especially in the light of the recent findings at Shanidar, while we may still have the privilege of ridiculing him, we are not actually rejecting him. For what person will mind having as an ancestor one of such good character, one who laid his dead to rest with flowers?

Solecki held the mistaken view that the Neanderthals were our ancestors. We now know that they were more like our cousins. At the time of the shaman's death, modern humans (*Homo sapiens*) had already spread into southern Europe from their African home. From this time on, and after the disappearance of the Neanderthals some 30,000 years ago, they spread farther north into Europe. We have ample archaeological evidence of the increasing complexity of human culture: organized hunting and gathering, structured camp sites, more diverse and finely made stone and bone tools, art objects and cave paintings, and elaborate burials. We can only speculate about the relations between these two groups. They probably did interact with each other. Plants, and the perception of plant colors, seem to have been important in the evolving culture of the Neanderthal, but much more so in early modern humans. I prefer the image of the Neanderthal child created by Elizabeth Daynes, reconstructed from a skull taken from a burial site in Spain. Perhaps the description of such a child by Michael and Kathleen O'Neal Gear in their novel *Raising Abel* is closer to the truth:

capable of speech but struggling with certain sounds, acute senses of sight and smell, physical strength, and emotionally vulnerable in a dangerous and unpredictable world.

Rogelio Piaguaje

Rogelio Piaguaje is alive today, residing along the Taricaya River in the Amazon lowlands of Peru (fig. 1.1). Rogelio is a member of a preliterate society, the Siona-Secoya nation of Amazonian Peru and parts of Ecuador and Colombia. The Secoya are identified by the language they speak, which also describes a unique culture. The photograph of Rogelio was taken in 1979 by William Vickers, a friend and colleague who has studied and later supported the Secoya in Peru and Ecuador for about thirty years. On the day William took that photo, Rogelio was dressed up to visit a neighboring household in his village.

For this visit Rogelio decorated himself with facial designs. The reddish lines on his face were made from a paste of seeds of the annatto plant (fig. 1.3). This plant is used for decoration by Native Americans throughout the

Figure 1.3 The annatto plant (*Bixa orellana*), used by Rogelio Piaguaje for body decoration. *Left*, flowers; *right*, open capsule, showing the waxy pigment covering the seeds.

New World tropics, and is still used in our society for food coloring and flavoring, and in some cosmetics. When I was a child, and the margarine manufacturers were battling the dairy lobby for a share of the market, the cellophane bags of white margarine had an orange annatto button that the consumer mixed in to make the margarine look like butter.

The patterns of the annatto design were inspired by Rogelio's visions after taking the hallucinogenic ayahuasca (mainly extracts from a vine, also called *caapi*). Perhaps the choice of color for the body decoration was inspired by the same visions under the influence of ayahuasca shared with his fellow villagers. Taking this plant is an important religious practice among the Secoya (and by other peoples in the Amazonian rain forests), and shamans use it to obtain intuitions about the illnesses of their villager patients.

Rogelio has dyed his lips purple by chewing on leaves of a species of *Justicia*, in the family Acanthaceae (some eighty shrubs in the New World tropics are closely related to this species, and this one has not been identified or perhaps even scientifically described yet). On other days, he might create black designs on his arms using the juice from fruits of *genip*, geometric designs inspired by ayahuasca visions. These designs, along with the lip color, wear away in a few days.

Rogelio also has a colorful headdress, decorated with paint and bits of cloth fabric on rings of the light pith wood of the cecropia tree. His ear ornaments are made from tubular sections of a cane grass and are decorated with colorful macaw tail feathers. Bright feathers of rain-forest birds are a prized item for decoration and not so easily obtained. These decorations are completed by a necklace of glass beads around his neck. On this day Rogelio is wearing a blue cotton tunic. The cloth was woven and dyed commercially. The macaw feathers were bound to the ear decorations with cotton twine, spun by the Secoya from the tree cotton raised in their home gardens. The Secoya weave hammocks and armbands, but traditionally their tunics were made from fabric derived from the inner bark of nearby fig trees. Spinning and weaving are laborious tasks, and the Secoya take advantage of inexpensive, brightly dyed, commercial cotton fabrics to make their own clothes.

Today the Secoya are a people in cultural transition. They value metal pots, shotguns, outboard engines for their canoes, even chain saws. Although much has changed for the Secoya, and you might see a child with a Mickey Mouse T-shirt in a traditional village, the Secoya still dress the same today for special occasions. Rogelio still practices shifting cultivation, a sustainable agricultural system in his tropical rain-forest home, and he hunts and fishes. He knows the names and uses of hundreds of plants in the forest, and cultivates dozens more.

Color, particularly derived from plants, is important in decoration throughout Secoya culture. The loose clusters of houses form villages along rivers in the rain forest. Houses are decorated with flowering herbs and shrubs; many are brightly colored or fragrant. On a typical day, flowers and fragrant twigs might be tucked into Rogelio's body piercings for decoration and scent. The Secoya like to use bright colors. Their body decorations are not limited to plants; Rogelio uses brightly colored feathers and glass beads. However, plants are a reliable and available source for color and body decoration in Secoya culture.

In traditional Secoya burials, the corpse was dressed up in splendid clothes, its face decorated with annatto, and its arms rubbed with annatto mixed with the grated sacred *nuni* plant, or joint sedge. If not already there, the corpse was placed in a hammock. Soon the body and hammock were placed in a grave underneath the house, some of the person's possessions— but not the most precious—were added, and the body was covered with earth. There was much wailing and solemn oration. A potion was prepared from a few locks of hair from the body and special plants in the rain forest. It was boiled and the pot shattered; the poison vapor was believed to fly to the heart of the sorcerer who caused the death. After the burial, the shaman, with the aid of the ayahuasca, communicated with the spirit of the deceased. He could tell the family members that the spirit moved to the east into the heavenly realms. These funerary rituals are changing with cultural contacts and the influence of missionaries, but perhaps Rogelio will be dressed for his funeral in similar fine clothing and splendid decorations.

The Secoya people describe the range of plants and other objects in their lives with only six colors. Rogelio called his purple lips *nea*, or black. He has no name for blue (merged with green) or orange (merged with yellow). He describes some colors, such as pink, in relationship to the six colors; pink is red with less color.

The culture of the Secoya people formed in a tropical rain-forest environment. In addition to the blue of sky, white of clouds, and brown of the soil, the principal sources of color in their culture were, and still are, plants. The background of their existence is green, which has the connotation of tenderness and youth, attractive for its promise of nutrition. Detection of subtle shades of green in the rain-forest canopy and understory is vital to identifying the plants that yield medicines, provide succulent fruits, and attract game. However, all of the colors used in decorating their bodies and dwellings, most of plant origin, are attractive and enjoyable to them. It would be difficult to overestimate the importance of plants—and their colors—in their culture. Vickers, with the help of Tim Plowman (a renowned tropical botanist who worked at the Field Museum in Chicago), documented the plants used by the Secoya, the largest part of their material culture. Some 244 species are commonly used, and they have names for hundreds more.

From time to time the Secoya move their villages to other locations along the rivers; most of their material objects come from the forest and can be refashioned (but they are beginning to accumulate more expensive Western goods). They move their fields of cultivation frequently. Thus they don't accumulate a lot of material goods, and those can be replaced. Yet their culture is rich in songs, stories, and knowledge about their environment. And they know their flowers.

Perhaps Rogelio would look quite strange strolling in a suburban mall in the United States, certainly in Dubuque, but I see some similarities among young people on a balmy South Beach evening in Miami, on their way to the clubs (some of them may even be my students!). Their decorations include piercings on various body parts, tattoos (permanently composed or temporarily applied with extracts from leaves of the henna plant [fig. 1.4]),

Figure 1.4 The henna plant (*Lawsonia inermis*) is popularly used by young people to create intricate designs on exposed parts of the body. Leaf extracts were used traditionally for such designs in India (as for the special visit of the daughter of an Indian colleague) and for tinting hair. *Left*, henna flowers; *right*, an Indian hand design.

hair died shockingly bright colors, blue lipstick, painted nails, colorful attire in a rainbow of colors, perhaps even flowers pressed behind an ear. They are dressed for the occasion, not all that different from Rogelio.

King Tut

The most famous pharaoh is known to us not for his accomplishments, but for the remarkable discovery of his tomb, in an Egyptian valley across the Nile River from Luxor (then the holy city of Thebes). The contents of his tomb have revealed much about the culture and use of plants in Egypt some 3,300 years ago. King Tut is, of course, the pharaoh Tutankhamen, who ruled at the end of the Eighteenth Dynasty of the Middle Kingdom of Egypt. In 1977 my wife and I discovered a late winter refuge in Luxor, having lived in tropical Asia for several years and finding the northern route into Europe too cold for us. We visited the Valley of the Kings several times, by bicycle, and entered King Tut's tomb. The valley is a desolate place, and no plants grow in this desert except by the river and where irrigated. Howard Carter's discovery of the rich and colorful furnishings of King Tut's tomb must have seemed all the more remarkable in contrast to that desert valley.

Tutankhamen was pharaoh for only nine years—around 1336–1327 BCE. He lived and reigned at the time Moses lived with the Jews in exile in Egypt. His predecessor Akhenaton, later briefly replaced by Smenkhare, had changed the form of worship to a single deity, Aton the Sun. He also moved the site of government to the north, to a new town he had constructed on the east bank of the Nile. This was Akhetaton, today known as Tell el-Amarna. Many, particularly the priests based in the old royal capital of Thebes, were profoundly upset by these changes, and Akhenaton's reign of sixteen years was marked by rebellion.

We can only speculate on Tutankhamen's early life, based on items left in the tomb and a few sculptures in temples at Thebes and elsewhere. We are not absolutely certain of his parentage, partly because of the complex relationships (including incest) among the royal family. His mother was probably Nefertiti, the early wife of Akhenaton, and his father was possibly Amenhotep III. As a member of the royal family during the religious heresy fostered by Akhenaton, Tutankhamen would have grown up in the palatial residence of his mother on the northern side of the new city, surrounded by gardens. He would have received instruction in the arts of Egyptian civilization: language and writing, mathematics, geography and natural history, religion and theology. Surely he became acquainted with the flowers growing in the gardens (the same ones that decorated his tomb). He

Figure 1.5 The pharaoh Tutankhamen and his wife Ankhesenpaaten portrayed on the back of the golden throne in his tomb. Note the plants used to decorate the couple and their surroundings. © AKG Images.

also learned about animals and the art of hunting, but may have been too frail as a child to participate in the hunts depicted on some of the artwork placed in the tomb.

He married Akhenaton's daughter, Ankhesenpaaten, as a child. This meant that he may have been both a half-brother as well as son-in-law to the sun king. From the works of art that commemorate their marriage (such as the back of the golden throne placed in the tomb), they appear to have become very attached to each other (fig. 1.5).

Probably under pressure from his advisers during those unsettled times (he became pharaoh at the ripe age of ten years), he moved the capital back to Thebes and reinstated the traditional worship of many deities, particularly Osiris. Whether Tutankhamen was the youthful agent of this change, guided by his revered adviser Aye, or forced by the powerful general Horemheb, can only be guessed. Tutankhamen was responsible for the construction of temples and the erection of new statues in the existing temples (some were later defaced by Horemheb). How did he die? A mystery (some have suggested murder), but perhaps he was frail and prone to illness. The autopsy of the mummy revealed no signs of foul play. The young pharaoh was replaced by his two advisers, first Aye and then Horemheb, and their reigns completed the Eighteenth Dynasty.

At around nineteen years old, Tutankhamen was buried in a tomb in a secret place across the river from Thebes. It was just luck that prevented this tomb from being plundered by generations of well-organized thieves.

Some had found the tomb, but took little and did not share their secret. The tomb was rediscovered in 1922 by the archaeologist Howard Carter along with his patron Lord Carnarvon, one of the most astounding discoveries in the history of archaeology. Carter wrote about the moment of discovery:

> At first I could see nothing, the hot air escaping from the chamber causing the candle flame to flicker, but presently, as my eyes grew accustomed to the light, details of the room within emerged slowly from the mist, strange animals, statues and gold—everywhere the glint of gold. For the moment—an eternity it must have seemed to the others standing by—I was struck dumb with amazement, and when Lord Carnarvon, unable to stand the suspense any longer, inquired anxiously, "can you see anything?" it was all I could do to get out the words, "yes, wonderful things."

The outer room, or antechamber, leading to the tomb was filled with all of the material needs for the afterlife of Tutankhamen: furniture, chariots, religious items, jewelry, perfumes, kitchen utensils, clothing, food, and decorations. This material became the object of decades of study by specialists, including botanists. The contents of King Tut's tomb tell us much about the sensibilities of the king and the culture about 3,300 years ago.

The furniture, layers of the coffin, and fabrics were decorated in bold color. Most of the items were constructed from the raw materials of plants: timber, hardwood, papyrus, fragrant oils, fiber, resins, and so forth. The botanical artifacts were analyzed by Percy Newbury, who had worked on the identification of Egyptian artifacts for years at the universities of Liverpool and Cairo and collaborated with Carter. Many of the motifs of decoration were derived from plants, such as the lotus and the papyrus. The photograph of the back of the Tutankhamen's golden throne shows the young pharaoh and his wife with their shoulders covered by, and both of them surrounded by, floral decorations. Similar decorations were found on the tombs and in the antechamber, and the plants were identified (fig. 1.6): blue water lily, white lotus water lily, corn poppy with scarlet flowers, oxtongue with yellow composite flowers, cornflower with blue thistle flowers (see fig. 7.20), and the yellow daisy flowers of chamomile. These flowers were tied together in a highly disciplined fashion, using stems of papyrus, leaves of date palm, and other materials. Foliage and fruits, as of the mandrake plant, were also incorporated into these designs. Similar wreaths had been discovered in much older tombs, so the details of their construction followed ancient tradition.

The goal in mummifying the corpse, assembling the furniture and ve-

Figure 1.6 Left, the red poppy (*Papaver rhoeas*), native to the Mediterranean region, was cultivated in Luxor and was used in decorating King Tut's tomb. *Right*, the water lily (*Nymphaea nouchali*) was used extensively in decorating the tomb.

hicles, providing the food, incense, and flowers, was to prepare for the celestial voyage to the eternal abode of the gods. This was a perilous and complicated voyage. Various scriptures, notably the *Book of the Dead*, provided information and the correct incantations for different stages of these voyages. Animals and plants were important symbols in these incantations, as in the reference to the lotus (white water lily) in this passage:

> O you two fighters, tell the Noble One, whoever he may be, that I am this lotus-flower which sprang up from the earth. Pure is he who received me and prepared my place at the nostril of the Great Power. I have come into the Island of Fire, I have set right in the place of wrong, and I am he who guards the linen garments which the Cobra guarded on the night of the great flood. I have appeared as Nefertum, the lotus at the nostril of Re; he issues from the horizon daily, and the gods will be cleansed at the sight of him.

Most remarkable are the wreaths that decorated the mummy, similar to those illustrated in the artwork of the tomb. Inside the outer coffin, a miniature wreath of olive leaves, cornflowers, and lotus petals was wrapped around the cobra on the golden head of the second coffin. A second wreath was arranged around the coffin chest, consisting of cornflowers, water lily petals, and olive and wild celery leaves. Most spectacular was the floral collar on the inside coffin. It was arranged in nine rows, using date palm, pomegranate leaves, and berries of ashwagandha. Similar floral displays, and bunches of foliage, were also discovered in the tomb. Based on the flowers that were cultivated and used in the tomb, Newbury estimated that Tutankhamen died sometime in March or April.

Furthermore, flowers, fruits, and whole-plant motifs appeared in the artwork throughout the tomb. Notable were motifs based on the lotus flower, the papyrus reed, and fruits of the mandrake. The mandrake is a plant steeped in mystery, known for its medicinal properties and its stimulation of vivid hallucinogenic dreams. It was also revered as an aphrodisiac and promoter of fertility, as shown by this passage in Genesis (30:14–17):

> In the time of the wheat harvest Reuben went out and found some mandrakes in the open country and brought them to his mother Leah. Then Rachel asked Leah for some of her son's mandrakes, but Leah said, "Is it so small a thing to have taken away my husband, that you should take away my son's mandrakes as well?" But Rachel said, "Very well, let him sleep with you tonight in exchange for your son's mandrakes." So when Jacob came in from the country in the evening, Leah went out to meet him and said, "You are to sleep with me tonight; I have hired you with my son's mandrakes." That night he slept with her, and God heard Leah's prayer, and she conceived and bore a fifth son.

Its yellow fruits and distinct foliage are depicted in tomb designs. The chemistry of the mandrake is similar to that of the *Datura* of the southwestern United States (both members of the same family, Solanaceae), its secrets revealed by Carlos Castaneda in the *Teachings of Don Juan*. The lotus is also known for its narcotic properties, and may have been used ceremonially in ancient Egypt.

The color green had an important place in the decoration of the tomb. The floral garlands included foliage of various plants. Green bouquets of foliage were prominently placed among the furniture in the tomb: branches of the olive, the aromatic wild celery, and persea. There was also a remarkable silhouette of Osiris, an outline in compact soil with a dense stand of emerging grain sprouts. This young tender green statue was wrapped in linen and placed in the tomb. Water, and the life-sustaining growth of crops associated with it, was not taken for granted. Failure of rains in the south, and the lack of the yearly floods on the Nile, meant a failure of much of the agriculture in the Nile valley. Such droughts meant widespread famine among the peasants in ancient Egypt. Thus green was associated with growth, abundance, and water. Egyptians developed several bright and durable green pigments, used in the artwork in King Tut's tomb.

Baskets of woven and colorfully dyed fabrics were placed in the tomb and surrounded the mummies. The fabrics were exclusively linens, woven from fibers of the flax plant. Flax was domesticated in the Near East, and records of its use long predate Egyptian civilization. Cotton came much later, from India. The fabrics were dyed bright reds, blues, yellows, and greens. Madder was used to dye fabrics yellow. This plant almost certainly did not grow near Thebes, and the roots were imported from the Mediterranean region where winter rains support this crop. Safflower was used for red, a royal color, and yellow. A blue dye was obtained from the pods of white acacia. After their discovery and study thousands of years later, these fabrics still held much of their color. The artwork in the tomb was bright and colorful, very much like those spring flowers.

The Egyptians developed pigments to make many colors, also employed in glass and ceramics production, and some were used in Europe 3,000 years later. These pigments were developed from minerals, even derived from some simple chemical reactions, but not from plants.

The tomb of King Tut, with the items selected for his afterlife and their images of the young couple and their society, are a window into a civilization rather late in its development. There is considerable but less vivid evidence of similar plant uses in much earlier times, going back some 4,500 years ago into the Old Kingdom.

King Tut probably knew the flowers; those attending him certainly did. It is hard to deduce how much the preparation of bodies, the burial, the furnishing of the tomb, was an expression of love for this young man, so new to his marriage and reign. Aye, his revered adviser, was probably involved. Certainly duty and tradition were very much in play. Of course, the tomb stored an excess of material wealth in preparation for the afterlife—almost unmatched in history. We can only guess if Tutankhamen preferred the flowers and animals to the trappings of kingship. So perhaps we have only two out of three of Gary Snyder's words to the wise.

Shanidar IV, King Tut, and Rogelio Piaguaje represent humanity at different stages in our cultural evolution. At each of these stages, plants were an important part of the material culture of society. They were sources of building materials, including timber and thatching for waterproof roofs. They were the principal source of food. King Tut's tomb was well-stocked with food for the afterlife, including bread. Neolithic village sites, going back 8,500 years, reveal the foods that were consumed, the woven fibers, and even the dye plants used to color the fabrics. Plants were the medicines, and the psychoactive plants were the means of diagnosis of the illness in the patient. They also yielded fiber for clothing and tanning agents to make leather durable and pliable. Plants were the sources of fiber (along with wool) for the production of fabrics used in clothing, tents, and blankets—all of these were protection against the harsh winters of temperate climates.

Anointing Our Bodies with Plants

Plants became important for their aesthetic qualities—in the celebratory rites of the diversity of human culture—and plants were often chosen for their colors. Consuming certain plants also produced the deep hallucinogenic interior states used to give insights about the use of color in the external world.

Color words are found in every language, although the number of colors and the spectral ranges that the words describe vary among languages. Much of human experience of color was associated with plants, the beauty of blossoms, the ripeness of fruits, the colors of plant extracts, and the green backdrop of vegetation. Color from plants was obtained and used in different ways. Blossoms used to decorate the individual could be picked from wild plants, or from plants grown near the house. The different colors and shapes of flowers took on different meanings in various cultures, some with creation stories about their origins, of how the gods sent them down to earth. The foliage is a source of subtle shades of green, colors that could be differentiated by people living in a forest environment. Leaves of certain plants are brilliantly colored, and so they provided a range of permanent color and are still planted adjacent to village huts, suburban houses, and urban parks and gardens throughout the world.

Juices and extracts of plants had many uses, determined experimentally over time. Some plants and their extracts were medicinal. Many colored the skin, and dyed fabrics or leathers used in clothing. Compared to plants and other animals, the diversity of colors produced by the human body is paltry. Take humanity in all its racial diversity, at all latitudes, and we have neutral shades of black to brown, to tan, to pink-white. Most of this color variation is produced by differences in a single pigment: melanin. Its color is influenced by the flow of hemoglobin pigments in blood vessels near the surface. Some animals, birds, reptiles, amphibians, fish, and many invertebrates (think of butterflies alone) produce amazing combinations of bright colors. Plants, in addition to shades of green in their vegetative organs, produce every type of visual color in their flowers and fruits. Plants, being more abundant and constantly part of ancient human environments, must have been an inspiration for colorful decoration of the human body.

In experimenting with plants, humans undoubtedly found that most bright colors, as in flowers, were ephemeral. If ground or extracted, the pigments producing these colors did not generally adhere to the body and faded rapidly in sunlight. Although a few plant pigments, such as the annatto dye ornamenting the Secoya, were relatively lightfast and adhered to the body through multiple washings, many of the early coloring agents were derived from minerals.

Using plant extracts to dye skin and fabric was a major technological accomplishment. In the discovery of plant-derived medicines, the effects are detectable from consuming the plants directly or by taking dilute extractions. In dyes, the extracts are usually not useful unless combined with other materials to change the dye molecules chemically. The discovery of a range of color-producing dye plants in many cultures attests to the importance of using plants for color.

From what we know of traditional cultures presently living in some isolation from Western society, as well as historical records of earlier cultures (including those in regions presently occupied by "sophisticated" ones), the use of coloring material to permanently decorate our bodies was widespread. Mummies of European ancestors, starting with the Ötzi man uncovered from a glacier in northern Italy (who lived some 5,300 years ago) and then the Scythians, were decorated with simple tattoos. Tattooing, increasing in popularity today in the United States and Europe, is an ancient practice.

The Romans encountered fierce Celtic warriors of strange color, as written by Julius Caesar in the *Gallic Wars*: "All the Britons, indeed, dye themselves with woad, which produces a blue colour, and makes their appearance in battle more terrible." This is reminiscent of the strikingly decorated warriors in the movie *Braveheart*. The Picts probably did not dye their bodies blue with woad (about which we'll learn later), but with a copper mineral still found on the bones in a few graves. The color blue is used in Hindu culture to depict certain gods, particularly Vishnu and his avatar

Figure 1.7 The birth of Krishna, seen as the blue infant with his mother and in the pond, from a nineteenth-century Indian miniature painting.

Krishna (his reappearance in another form; fig. 1.7). Different flowering plants are symbolically associated with these gods, and are used in worship and decorative arts.

Later, European explorers encountered varieties of color, applied externally as pigments or beneath the skin as tattoos in many of the traditional groups they visited in their voyages around the world. Captain Cook's sailors encountered tattooed natives in Polynesia, and adapted these practices to their own bodies (fig. 1.8). Tattooing was raised to a high art at different epochs in Japanese history, and was practiced at the most sophisticated level when American sailors first visited in the mid–nineteenth century (fig. 1.9). The tattoos they encountered were amazingly complex and multicolored, and their creators were revered as cultural treasures.

The Japanese employed a variety of pigments in tattooing, similar to those used in making wood-block prints at that time. The predominant color of black (or blue-black) used in producing the figure drawings and contour lines was of fine carbon particles, or India ink. Perhaps the finest

Figure 1.8 The tattooed visage of the chief of Tahuata, Marquesas Archipelago, observed by Captain James Cook and drawn in 1775 by an expedition artist.

ink had a sacred plant connection, such as the ink from Nara that was made from the soot collected from stone temple lanterns burning sesame oil. The Japanese tattoo masters used four principal colors in their designs: red, indigo, green, and yellow. A transparent red was extracted from the flowers of safflower, but reds were also obtained from mineral pigments. Blue was primarily indigo, obtained by boiling old rags originally dyed with this plant material. Violet could be produced by mixing these two pigments. Green was obtained from mineral pigments. Yellow, used less frequently, was obtained from sulfur and arsenic, but some yellows were also obtained from gardenia and gamboge. These pigments were mixed together to produce still other shades.

The modern practice of tattooing in the United States and Europe owes much to our early contact with the Japanese (fig. 1.10). However, the traditional dyes have been replaced by modern synthetic ones, the products of the revolution in organic chemistry, very lightfast and presumably much less toxic than some of the traditional minerals containing mercury, arsenic, and chromium.

Figure 1.9 Tattooed bandit killing a wayfarer, depicted in a wood-block print by the Japanese artist Yoshitoshi. From *28 Famous Murders with Verse* (1866), © John Stevenson.

Figure 1.10 A contemporary tattoo, with largely synthetic pigments (the actual identities are trade secrets) and a plant motif (water lily), on my daughter Katy's leg.

The Business of Plant Colors

Plant colors were exploited to add richness to the lives of the inhabitants of each of the early civilizations, which resulted in trade and commerce. Since many plant colors are very ephemeral, considerable research was necessary to find hidden sources of color or to make the ephemeral colors more permanent. These colors were valuable in coloring our bodies, as in tattooing, or in more ephemeral ways, as in cosmetics. As the civilizations matured and spread their cultural influences, the techniques and commodities for plant coloring became economically important. Plant materials, used for color, had been traded commercially among these civilizations since antiquity.

Egyptians during the time of King Tut and much earlier used oils, perfumes, and plant and mineral pigments to decorate their hair and areas around their eyes. Most of these items were produced locally and were undoubtedly bought and sold in local markets. Some were brought to Egypt by caravans or supplied as gifts or tribute by kingdoms under the pharaoh's influence. King Tut's wife would have had her own beauty kit, with perfumed oils for her hair and body, and would have carefully applied eye shadow and liner as well as lip color. Similar makeup was used in Europe in the nineteenth century and spread quickly to the United States after the end of the Civil War and the Victorian era. Lipstick had its origins in the ancient civilizations of the Near East. Lip color was used in Greek and Roman society. In Japan the geisha applied white (rice powder) makeup to the face and then diminished the size of lips by painting with a bright red pigment derived from safflower, the same dye used for printmaking and tattooing. Indigo was used as a blue eye shadow by women in Indian society. Thus our contemporary use of cosmetics such as eyeliner and lipstick is the continuation of an ancient practice.

The Challenge of Coloring Fabrics

Color was applied to leather and fabrics for garments that protected their wearers from the harsh climates where they lived. Fabrics were turned into clothing that marked the social divisions within the towns surrounded by settled agriculture, and soon distinguished among classes within the early civilizations. Coloring the fabrics and leathers was not a simple matter. Many of the colors used in body decoration were pigments, such as orange and red ocher, that did not adhere well. Bright animal colors were often produced by structures and not pigments; these could only be applied to fabrics as parts of feathers and so forth. The pigments that produce the brilliant and varied colors of flowers and fruits do not adhere to fabrics and wash off quickly; they lack the important quality of fastness. In addition, these pigments, once isolated from the plant, tend to fade quickly in direct sunlight. Thus they make poor dyes. Cultures experimented with the use of plants to dye fabrics and leather. The light colors, browns and tans, were relatively easy to obtain from a large variety of plants. Brighter and faster colors were harder to come by, particularly bright blues and reds.

Pigments were used in the decoration of everyday objects as well as for painting (as in the cave paintings in Lascaux, France, some 17,000 years ago). Because of their stability, almost all pigments were finely ground minerals, such as ocher. Since they were not found near the caves, their use is evidence of long-distance transport and probably trading. Such pigments are still used in the production of paints. In some cases, plant-derived pigments were used, by absorbing the pigment extracts in fine powders to produce solid colors called lakes. The lakes could then be ground and used just as the mineral pigments were. In Japanese printmaking, vegetable dyes such as indigo and safflower were carefully mixed with rice flour, and printed on paper.

Weaving appeared at about the same time as agriculture, around 10,000 years ago in different parts of the world. Woven materials depended upon the materials at hand. Most of the early fabrics were woven from fibers obtained from the stems of plants, after a long period of soaking to cause the nonfibrous tissue to ret. In the Near East, the first vegetable fibers used for fabrics and rope were such retted or bast fibers, particularly flax and hemp. Hemp was also available in Asia, along with other fibers such as jute and ramie. Asians had also domesticated another fiber plant that produced the finest fabrics of all: cotton. Cotton fibers had been discovered in India (now Pakistan) by the Indus Valley civilization, dating back at least 4,500 years. Silk, a fiber obtained from the cocoons of the silk moth, was also developed in Asia, and moved west much later. Wool, principally from sheep, was widely used in weaving.

Making dyes fast meant discovering how to bind the dye molecules to individual fibers. Dyes tend to bind more easily to animal-based fibers, such as wool, and much less to vegetable fibers, such as flax and cotton. Chemical pretreatment of the fabric, later to be called mordanting, was developed in all of the weaving and dyeing traditions. Various mordants were used, including fermented urine, vegetable ashes, and mineral salts. The most successful mordant was alum, the sulfate salt of aluminum and potassium. These treatments made a variety of dyes adhere to the fiber and made fabrics hold their color much longer. Some mordants also changed the hue of the dye.

Such techniques were developed in Egypt, and resulted in the production of the wealth of fabrics in King Tut's tomb. The fabrics (linens) were all woven from flax, using a variety of techniques, and were dyed different colors. As mentioned previously, fabric was dyed blue with an extract of the pods of *Acacia*, a thorny tree of dry climates. Fabric was dyed red with safflower and yellow with madder.

The search for bright colors to dye fabrics in early civilizations in the Near East and Asia led to the discovery of new dye plants. Subsequently, they made the fabrics of these regions valuable in trade with the West, and the dyes became important trading items in their own right. Ultimately, explorers of the New World found some substitutes for the Asian dyes, and these new items became important items in trade as well.

Some valuable dyes were found locally. A source of purple dye was found in the nearby Mediterranean, and became available to civilization along its shores, ultimately the Greeks and Romans. The source was the marine gastropod *Murex* and its relatives. Abundant sources of these organisms were limited and soon exhausted; the price of this dye then rose astronomically. Such "royal" purple was only used by the upper strata of society.

Madder grew wild in the region (fig. 1.11). Its roots produce both yellow and red dye. When ground into a powder, mixed into a paste, and mordanted with alum, madder produces a bright and very fast red color. It was used for festive clothing, and eventually became widely available. Woad was also grown in Europe. However, its production as a crop depleted soil nutrients, its preparation as a dye was complicated, and the resulting blue color was not very bright.

Despite the availability of some local plants, materials were imported from afar (initially from Asia) for the dyeing of fabrics. The most important of these dyes were indigo (for dark blues), cochineal (for bright red), and brazilwood (for dark reds). Indigo was obtained from the leaves of the indigo plant. Its leaves had been harvested in India for dyeing since antiquity. The actual dye chemical, indigotin, is closely related in structure to the purple of *Murex* and the lighter blue of woad. Many centers for weaving colored fabrics and printing fabrics developed in Asia. These traditions (figs. 1.12, 1.13) continue in the region, as in parts of India and Indonesia

Figure 1.11 Madder (*Rubia tinctorum*), one of the most ancient dye plants, used in west Asia since antiquity.

Figure 1.12 Ikat weaving by the Iban of Sarawak, Malaysia. Cotton yarn is tied and dyed with traditional plant sources prior to setting up the loom.

(Bali). Much later, indigo cultivation extended into the Middle East, but it continued to be an important item of trade from Asia to the West. The foliage of indigo does not directly yield a blue dye. The precursor molecule is extracted by mashing the leaves in water. Beating the slurry (by hand in earlier times) aerated the mixture and caused a blue precipitate to form. This precipitate was collected, dried, and compressed into pellets or flat sheets. The dried precipitate was traded, shipped, and ultimately used to dye fabrics in the West.

Cochineal is a dye derived from a group of sap-sucking scale insects, parasitic on various plants. Such scale insects feeding on the common Eurasian kermes oak were discovered to produce a satisfactory red dye. These insects were plentiful in West Asia, for example in Armenia (an important early source of the dye). The insects were collected from plants at the right stage and compressed into a raw dye material known as kermes. Cochineal dye produced a range of reds and was relatively fast. Kermes was shipped into Europe from throughout Asia, and was later also produced in the south

Figure 1.13 Top, traditional weaver in Ubud, Bali (Indonesia), in 1975. *Bottom,* a detail of a fabric used as a traditional headdress, traditional dyes on silk from West Sumatra (Indonesia).

of Europe. In the Spanish conquest of Mexico, a New World form of the cochineal insect was found feeding on cultivated prickly pear cacti in peasant gardens; the dye was also important in Central America among the Aztecs. Soon the Spaniards were shipping cochineal from the New World to Europe, upsetting the traditional trading patterns of cochineal from Asia to Europe.

Dyes obtained from the dark heartwood of certain trees had been used in India since very early times, and were shipped to the West. The two most-valued species were the red sanders tree and sappan wood, members of the legume family. Both of these produced a water-soluble red dye that provided the brilliant colors of Indian textiles, prized throughout the classical world. In sappan wood, the pods from the tree were also an important source of the dye. These were also called "Brezil" wood. In the discovery and conquest of the New World, similar dye sources were found in the heartwood of certain forest trees, particularly in the legume family and especially in the genus *Haematoxylum* (or logwood; fig. 1.14). Interest in exploiting this dye wood was so great that the new colony of Brazil obtained its name from it. Dye wood chips from the New World soon made their

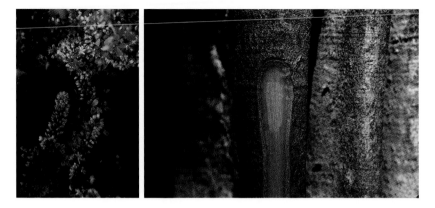

Figure 1.14 Brazilwood (*Haematoxylum campechianum*). The dye in this plant is obtained from the heartwood and was exported from Latin America to Europe after European colonization. *Left*, flowers and foliage (with pealike flowers of the Fabaceae). *Right*, accumulation of the pigment in the wood of a branch.

way to European factories, and the pattern of global trade was adjusted once more.

The Rise in the Trade of Dyes and Textiles

Dyes and colored fabrics were prized commodities, luxury goods. They were transported in caravans, and then on ships, from very early times. The limited market was the religious, political, and commercial elite in Europe. Add the value of other products—medicinal plants, perfumes, spices, and precious jewels and metals—and the notion of global trade is given a new historical perspective. The volume of trade increased over time, particularly from the twelfth century on, supplying the growing wealth in the population centers of Europe. The trade was not dominated by essential goods, but by exotic and luxury items. This trade also helped establish the prosperity of Islamic civilization. Various cities became important, and famous, for the quality of tanned hides and dyed fabrics. Thus Fez, in Morocco, has been famous for its leather and dyed leather and fabrics for centuries (fig. 1.15). The trade also promoted the growth of city-states, such as Genoa and Venice, in Europe. A sense of the diversity of trade is seen in a list written by Francesco di Balduccio Pegolotti in Florence, 1310–40. Pegolotti listed 288 "spices"—exotic materials traded in Florence and of oriental or African origin. Spice had a different meaning than it does today, for this list included the variety of commodities that would fetch high profits in trade, luxury goods of all sorts: fragrances, fine textiles, dyes, mordants, medicines,

Figure 1.15 Traditional tanning and dyeing in Fez, Morocco, 1977. *Left*, vats employed in tanning and dyeing leather; *right*, dyers handling yarn.

and food flavorings. Pegolotti's list included nineteen dyes (four indigos, six brazilwoods, five madders), eight sources of cotton, and twelve sources of alum. Our word *spice* comes from the French *épice*, which originally meant goods and survives as the French name for a grocery store: *épicerie*. The list also included a variety of sources of mineral pigments, used in cosmetics and in painting.

The complex processes of dyeing fabrics to produce different colors and intensities were well established in the Venetian Empire, and Venice was an important place for dyeing fabric. These processes were described in a practical how-to book on dyeing, the *Plictho*, which was first published by Gioanventura Rosetti in 1548 (fig. 1.16). The preface poetry of Rosetti's book reveals the ties among the journeymen dyers and hints at the subtleties of the dyer's art.

> The green, the blues, and scarlets
> And those that carry the emblem of fortune.
> Of velvets, damasks and all styles
> Furnished by Art
> This Plictho of dyeing if to you gives
> Right, and truth then the world will see,
> That this art deserves a crown;
> And without working with it one will not me believe,
> Serve yourself with heart, with this good Work
> And let reason in the soul take seat.

Figure 1.16 The dyers of sixteenth-century Venice. Illustration in *The Plictho*, by Gioanventura Rosetti. Courtesy of MIT Press.

The powerful appetite for the exotic and the economic push to capital-ize on it drove geographic exploration of the East by European seafarers, then colonization and enslavement. This obsession to find and appropriate new products continued during the nineteenth and twentieth centuries until after the end of the Second World War, and still influences much economic policy today. Tourists continue to be lured to the exotic sources of these goods, in the tropics. Lévi-Strauss described the seduction in *Tristes tropiques*:

> In the old days, people used to risk their lives in India or in the Americas in order to bring back products which now seem to have been of economically little worth such as [brazilwood and pepper, which] added a new range of sense experiences to a civilization. . . . [From] these same lands our own Marco Polos now bring back the moral spices of which our society feels an increasing need.

As the volume of trade and demand for products increased, competi-tion for quality in colorfastness and design increased apace. Techniques in dyeing and mordanting improved, and the scale of production (yes, in-dustrialization) grew. Exploration of routes to the East (and inadvertently to the West) was motivated by economic interests: to find the source of these products and monopolize their trade. Routes were discovered and battles were fought over the control of the seas and access to the areas of production in India and farther east. The Portuguese arrived first and were pushed out by the Dutch, and then the British moved in as well. Trading depots, or "factories," turned into colonies. The colonies then provided the raw materials feeding the Industrial Revolution in Europe, including fiber and dyes for the great textile mills—dark and satanic. The colonies were additionally viewed as locations for the production of raw materials first produced elsewhere, and they were also developed as markets. Thus the British established the cultivation of indigo in South Carolina in the eighteenth century, to provide dye materials for the British mills, along with the cultivation of cotton throughout the South. Cotton, along with sugarcane, became the principal economic justification for the trade in slaves to the New World. At the same time, colonists were forbidden to weave their own fabric and were forced to buy from British merchants. What seemed a harmless act, the spinning of cotton yarn by Mahatma Gandhi during the struggle for Indian independence, symbolized the fight against the economic imperialism of the British.

The height of this development is illustrated by the Royal Cloth Manufactory in France, founded by a Dutchman, Josse Van Robais, in 1665. This concern operated in the town of Abbeville until 1804. Initially the work was partitioned among several large buildings constructed in different parts of the town, but a large central facility was constructed by 1713. Some 3,000 workers toiled there, primarily recruited from near the town. Security in the facility was tight, and work was highly organized. Fifty-two sequential processes were involved in the production of the finest cloth. The factory kept its own ships, which sailed up the Somme to the town, delivering the raw materials. Despite the scale and efficiency, there was competition from other manufacturers. Salaries varied considerably among different classes of workers, and worker dissatisfaction led to a strike in 1716, quelled by a military detachment. The markets, then as now, surged and sagged.

Dyes and the Establishment of Chemical Technology

The incentives to produce more attractive fabrics to clothe a more prosperous populace stimulated the development of better mordants and more fast dyes, more complicated multicolored patterns, and scientific curiosity that resulted in the growth of chemical research and the establishment of organic chemistry.

By the mid–nineteenth century the discipline of chemistry was well advanced. The periodic table had been proposed, and a variety of chemical products were being manufactured, some for the dyeing of fabrics. Furthermore, a tradition of training chemists for research and development was established, inspired by the school of Justus von Liebig in Germany. Two of Liebig's students, August Wilhelm von Hofmann and Friedrich August Kekulé von Stradonitz, would figure large in the unfolding story of the artificial production of dyes.

Kekulé established the theory that made the later discoveries, and the foundation of organic chemistry, possible. Kekulé and Archibald Scott Couper independently established that a carbon atom can bond with four other atoms. Later, Kekulé established the structure of ringed carbon compounds of high stability (fig. 1.17). He determined the structure of the organic compound benzene, a closed ring with alternating double bonds between the adjacent carbon atoms. Such an arrangement is an "aromatic" ring. I show the most important of these structures in this chapter, to put them in a historical context, and will cover the chemistry of plant pigments in detail in chapter 3.

Hofmann had investigated the residues of coal tar (the substance left after coal burns) and had discovered benzene and other compounds in the tar. He had detected the presence of aniline in coal tar, the same substance that could be distilled from indigo dye. He also established the Royal College of Chemistry in London, which, from 1845 until his return to Germany two decades later, trained British youth in chemical principles. His students learned the theory and its applications in industry, of which the most important were the dye works.

William Perkin, an unusual young man, was admitted to the Royal College of Chemistry at the age of fifteen, in 1853. His teacher, a former student of Hofmann's, had noted his talent for art and interest in photography and all sorts of inventions. He arranged for Perkin's entry into the prestigious college. After rapid progress, he was given a research project, to prepare an amino derivative of anthracene, a coal-tar hydrocarbon. In continuing his experiments in his home laboratory, his intent was to produce a substitute for quinine, a naturally produced medicinal alkaloid (another impetus to the development of organic chemistry was the quest of synthesizing the

Figure 1.17 The benzene ring, which provided the understanding of carbon bonding that led to the discovery of the structure of dyes and their synthesis.

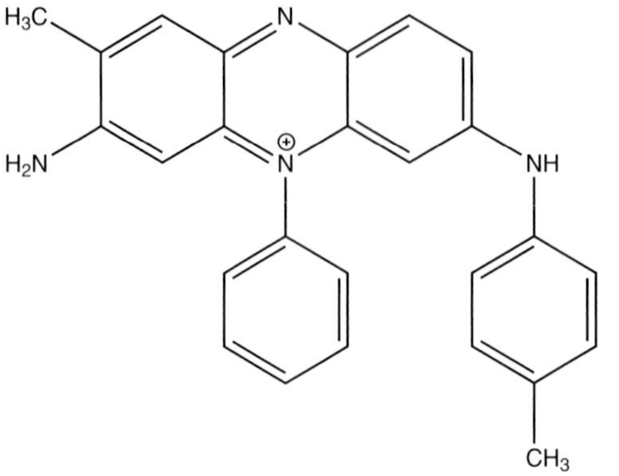

Figure 1.18 Mauveine, the first synthetic dye, discovered by William Perkin in 1856.

compounds in medicinally active plants). In his experiments on various coal-tar products, Perkin noticed a blackish residue that when dissolved in alcohol produced an intensely purple color. He found that this substance could dye silk to a beautiful purple and that the fabric resisted fading. Perkin then joined forces with his brother and father to commercialize the new dye. He resigned from the college in 1856 to concentrate his energy on the new invention, to the disappointment (and continuing support) of his mentor. The invention was patented, the process for the dye's production was commercialized and marketed as the new color mauve, and the artificial dye was a spectacular success (fig. 1.18). His achievement was very quickly picked up by chemists in France and Germany, who were working toward a similar end, and other aniline dyes were created.

The second aniline dye was discovered in this very competitive environment: Albert Schlumberger in 1859 discovered an efficient process for producing aniline red (called azalein). Then a third dye, aniline blue, was discovered in England in 1860. Hofmann then published the actual chemical structure of these new dyes. In 1865 he left the Royal College to take up a professorship at the University of Berlin. Organic chemistry, the chemistry of carbon compounds such as aniline, was becoming an important new discipline, partly because of the commercial realization of its new inventions.

The next chapter of this story, and the one that had the most profound impact on the commercial trade in dyes, was the discovery of the principal dye of madder, alizarin. The work progressed very rapidly and logically in the laboratory of Adolf von Baeyer at the Gewerbe Institut in Berlin, assisted by his research associate, Carl Graebe. In 1868, Graebe presented the results of research demonstrating the structure of alizarin. Then Graebe and Carl Liebermann, the son of the owner of a calico-printing factory, worked intensively on the synthesis of alizarin, and on producing variations in the anthracene structure with the end of understanding which variations in chemical structure produced colors. They announced the results of the synthesis of alizarin less than a year later. Work to commercialize this discovery, aided by Perkin and Sons, quickly followed. Baeyer discovered the structure and means of synthesis of the last important natural dye, indigo, in 1880.

These and related discoveries profoundly affected the trade in natural dyes and the production of textiles. New manufacturing companies formed and prospered by making these dyes, principally in Germany and Switzerland: Badische Anilin- und Soda-Fabrik (BASF — "we make the things you don't see"), the Deutsche Chemische Gesellschaft (aided by Hofmann), Bayer and Company (yes, Bayer Aspirin), Farbewerke Hoechst, Gesellschaft für Chemische Industrie Basel (Ciba, soon to join with Geigy and now Novartis), British Dyestuffs Corporation (to become Imperial Chemical Industries, or ICI), and Aktien-Gesellschaft für Anilin-Fabrikation (AGFA). All of these are multinational corporations today. Many of the scientists involved in these discoveries helped the companies directly (and even helped establish some), and they profited from their inventions in the way that biotechnologists have during the past two decades. The United States did not have enough expertise in organic chemistry to compete in the race for artificial dyes, and we ended up buying 90 percent of our dyes overseas. After World War I, the Germans were stripped of their patents, and U.S. companies such as American Cyanamid and DuPont moved into dye manufacturing.

The result of the production of these new dyes, brilliant in color and permanent in fastness, was the destruction of many centers of natural dye production, and the profound alteration of the trade routes that had helped reinforce the colonial economies. Madder cultivation, much of it centered in southern France and the Mediterranean, collapsed. The production of indigo, particularly in India, continued somewhat because of its high quality and economy relative to the synthetic version. It is presently the only natural plant dye of any economic importance.

A final consequence of this period of discovery, technological innovation, and rapid commercialization was the establishment of a new discipline that put the study of the production of color in nature on a scientific footing. It not only altered the economy of dye production and the coloring of fabrics with an infinite variety of new colors, but it changed the way we understood the production of colors by plants.

Although diminished by the development of synthetic dyes, colors from plants still retain some economic importance. Of course, many of the traditional uses of plants in producing colors, such as dyeing fabrics, have been replaced by synthetics. Now we have fabrics of natural fibers dyed synthetically, fabrics of synthetic fibers and dyes, and even artificial plants made from those fabrics and from plastics directly. Perhaps the synthetic substitutes in our homes, businesses, and institutions make the natural production of color even more desirable. Although what is "natural" has been replaced by what is synthetic or "artificial," there are enough of us that value these natural fabrics and dyes to keep these traditions alive.

I am not an anthropologist, nor a historian, and I wrote this book primarily from my perspective as a botanist. However, preparing this chapter made me even more aware of the importance of plants in human culture, not only historically but also in contemporary society. This experience certainly increased my motivation to write about the science of plant colors, with an additional perspective of the importance of these colors in our past and in our contemporary life. The history of use of plant colors goes back to our origins as humans, and it occupied an important place in economics and global trade in world history.

Chapter Two Light, Vision, and Color

....

Color rings the doorbell of the human mind and emotion and then leaves.

FABER BIRREN

The whole of life lies in the verb *seeing*.

PIERRE TEILHARD DE CHARDIN

....

Perceiving and enjoying the colors in nature, such as the eye of a peacock feather (fig. 2.1) or the color of a flower, is a profoundly subtle act. The colors are derived from light and its interaction with nature. We, along with virtually all other animals and plants, are creatures of light. Plants capture light as their source of energy, and use light as an environmental signal to direct their growth and physiological responses. Many animals use light as their most important sensory signal. Therefore it is important that we start with some principles that will inform our understanding of how plants interact with light, and how animals perceive that interaction as form and color. For humans, color takes on more importance than mere perception, and our relationship with plants is strongly influenced by our color perception. In order to understand the origin of color, I provide a basic overview of some very general concepts, in a historical context; many may be familiar to some of you, but it is important that we start on the same footing.

Figure 2.1 Peacock feathers interact with light structurally, producing intense metallic colors. Feathers and butterfly wings were an inspiration to Renaissance scientists studying the nature of light. We will learn in chapter 10 that such colors are produced by plants.

Light as Particles

My early experience of light as a child was as a beam coming to me from a source, like a ray of morning sunshine filtered through a window, or a flashlight penetrating the forest darkness on my way back to the cabin at my summer camp after the evening campfire (fig. 2.2). Most ancient philosophers, the first scientists, viewed light as such a ray, or beam, consisting of tiny particles (or corpuscles). By carefully observing rays of light, they deduced some basic rules about light's behavior. Euclid observed that the angle of incidence of light reflected from a flat surface was the same as the light striking it—the law of reflectance (fig. 2.3, top). Much later this law was modified to accommodate the effect of differences in the density of transparent materials on the degree of reflectance, and on that critical angle where the surface reflects all of the light impinging on it (fig. 2.3, bottom).

Later, scholars observed that the path of light from the air into a trans-

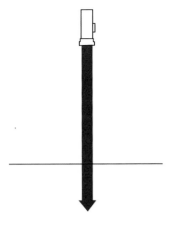

Figure 2.2 A beam of light as emitted by a flash-light.

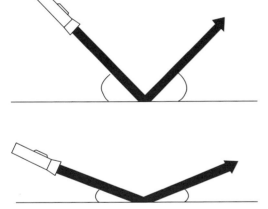

Figure 2.3 The effects of a reflecting surface on the path of a flashlight beam. *Top*, the beam is reflected at the same angle as it strikes the reflective surface. *Bottom*, depending upon the density of the reflective surface, at a critical angle all light is reflected by the surface.

parent medium, such as water or glass, was not a straight line if the angle of incidence was not perpendicular to the surface. Ptolemy carefully observed the angles of incidence and the angles of refraction for the contact between different media (fig. 2.4). Much later, the Arab philosopher Alhazen studied these angles of refraction in a more systematic manner, showing the power of magnification by a curved glass lens as well as by a curved mirror. Light refracted through (or by) such structures produced an image in focus at some distance beyond the lens or mirror (the focal length). Both Alhazen and Roger Bacon, the English philosopher and scientist, wrote with enough detail about the curvature of a lens refracting light and re-

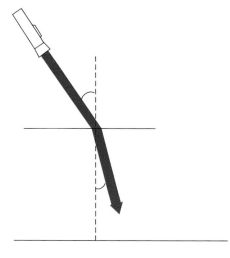

Figure 2.4 When a beam of light passes from air through a denser medium, the beam bends (refracts). The ratio of the two angles is the index of refraction of the medium compared to that of air.

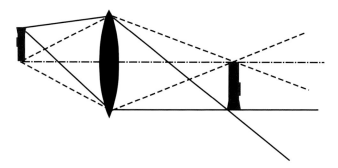

Figure 2.5 When a beam of light passes from air through a denser medium with curved surface (a lens), it refracts to form an image at a distance beyond the lens.

producing an image that others were able to produce lenses with practical benefits (fig. 2.5).

The optical instruments that became popular during the Renaissance were the products of the theory of refraction married to the knowledge of materials and technique. Such knowledge was shared among Europeans and contributed to the inventions of the microscope and telescope. We are reminded of the contributions of van Leeuwenhoek and Hooke with the microscope, and Galileo with the telescope. In developing these instruments, the inventors became aware that imperfections in the lenses could cause distortions in the resolution of the virtual image (spherical aberration), particularly at the edges of the image. Furthermore, the slightly different focal planes of the colors that made up the images led to poor color resolution and edges of pure color around the objects. Experimentation led to the development of improved compound lenses that corrected for these defects.

It wasn't until the seventeenth century that the correct laws of refraction were discovered by the Dutch scientist Willebrord Snell. His law made it possible to calculate the bending effect of a material on a ray of light, giving its index of refraction (the ratio of the incident angle to the angle within the medium), all in relationship to the path of light through air.

Many deduced that light coming at us from a source, such as the sun, must consist of particles traveling at a finite speed. The French mathematician Pierre de Fermat postulated that the laws of reflection and refraction could be explained by the path of light in which the time of travel was minimized. However, detecting the velocity of light was a tall order for scientists at this time (Galileo made an unsuccessful attempt), and would await further scientific development.

Newton on Light

Isaac Newton held to the view that light consisted of corpuscles. His experiments on light led to his famous treatise *Optiks*. Newton studied the splitting of a ray of sunlight into separate rays of different colors, showing that a prism split "white" light into colors that were displaced at different angles to each other, blue much more than red (fig. 2.6). He also showed

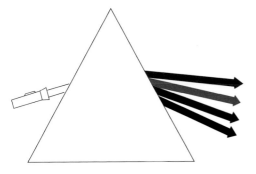

Figure 2.6 Newton observed that a prism separates white light into a spectrum of colored light, with the blues bent more than the reds.

that these separate colors could be fused back together to form white light. Newton was thus able to explain how different objects had different colors—some colors were reflected by the object, and some absorbed.

Despite his masterful treatise on light, some of his observations were inconsistent with the particulate nature of light. Newton became aware that a parallel beam of light produced a shadow that was not completely sharp (as predicted by particles), but had a slightly fuzzy edge. He also observed the production of colors from thin plates such as the wings of insects and the feathers of birds (like the peacock, fig. 2.1), not explainable by the particulate nature of light.

Light as a Wave

For me, the concrete realization of light as a wave came in 1960. This was not long after the Soviet Union launched the first Sputnik satellite in 1957. The Cold War fear that this achievement engendered led to a new physics curriculum, first used by Mr. McIntee in the physics course at my small hometown high school. Our teacher, much more knowledgeable in chemistry, put even more of his considerable energy into our physics course. My best memory is the tank he built for the study of waves, similar to the textbook's approach to the explanation of the behavior of light. In the tank we could observe waves interacting with each other, augmenting and cancelling their heights.

At about the same time as Newton, Christiaan Huygens proposed that light was propagated as a wave, not as individual particles in a ray (fig. 2.7). One prediction of the wave theory of light was that its velocity should be reduced when traveling through a medium of greater refractive index. Although the velocity of light was estimated from planetary observations by

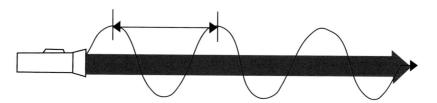

Figure 2.7 A diagram of light propagated through space as a wave. The frequency of vibrations (at the speed of light) is related to the distance between the wave peaks, or the wavelength.

Ole Rømer in 1678, it was not until the mid–nineteenth century that the velocity of light was measured experimentally, and shown to be reduced in water, thus confirming the wave theory.

Understanding light as a wave made it possible to understand why the shadow edge from a beam of light is slightly blurred. It made it possible to understand that the colors of a beam split by a prism, or by a diffraction grating, are the result of waves of shorter wavelengths being bent more than others. It made it possible to understand other phenomena, noticed but not explained by Newton and others. Newton's rings arise when light shines on a lens sitting on a flat glass mirror. Such conditions produce concentric alternating light and dark circles. The wave theory explained this as the alternating cancellation (peak and trough) and augmentation (two peaks or two troughs) of parallel waves, just as I saw with the waves interacting in the tank in my physics class, and that we see when the waves produced by two rocks thrown into a pond interact with each other. Early in the nineteenth century, Thomas Young performed the experiment, similar to the wave tank we used in high school, of passing a beam of light through two adjacent narrow slits, revealing the alternating light and dark bands we now call Young's fringes. Newton had remarked on the metallic colors of insects and birds, and Hooke had speculated on their cause, in his work on microscopy. Young found that the wave theory accounted for the production of such colors, as in a peacock feather (fig. 2.1). In thin-film interference, a layer whose refractive index is different than that of the medium below interacts with waves passing through and reflected internally in the medium (fig. 2.8, top). Light at a certain wavelength is interfered with destructively, such that its passage through the medium is augmented (fig. 2.8, bottom). Light at a shorter wavelength (depending upon the thickness and refractive index of the medium) is augmented at its reflectance from the surface, producing an intense color at that particular wavelength. Examples of such interference colors include the scales of butterfly wings, the barbules of bird wings, and even the oily film on the surface of a rain puddle. Some plants also produce color structurally (chapter 10). The angle at which the beam of light strikes the surface also affects the wavelength and color of constructive interference. A grating of fine lines on a flat surface diffracts the light of different wavelengths in a similar way, producing the spectrum of colors, just as in their separation by a prism (fig. 2.9). Gratings produce color in a few invertebrate animals, mainly marine.

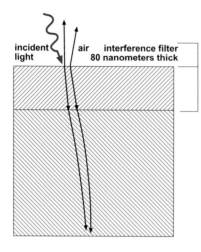

incident
light

air   interference filter
       80 nanometers thick

Figure 2.8 Thin layers (or very small structures) interfere with the wavelengths of light passing through (*top*), causing some wavelengths to be augmented through constructive interference and other longer wavelengths to cancel each other through destructive interference (*bottom*). This phenomenon is known as interference and produces intense colors, as does the thin layer of oil on a mud puddle.

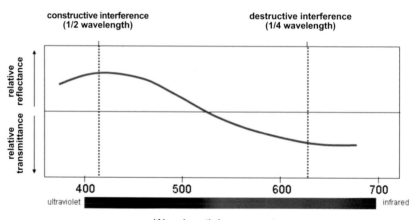

constructive interference
(1/2 wavelength)

destructive interference
(1/4 wavelength)

relative reflectance

relative transmittance

400   500   600   700

ultraviolet   infrared

Wavelength in nanometers

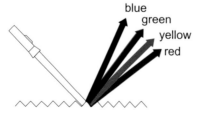

blue
green
yellow
red

Figure 2.9 A grating (closely etched lines on a surface) diffracts a beam of white light (a mixture of different wavelengths) to produce beams of pure color. As with a prism, the blue (shorter) wavelengths bend more then the red (longer) wavelengths.

Light as Wave *and* Particle

These two fundamentally different views on the nature of light uneasily settle on us today in the modern understanding of light being a small portion of a spectrum of radiation with electronic and magnetic components, electromagnetic radiation. At the heart of this understanding is the fundamental paradox of light (as a part of a broader spectrum that includes radio waves and X-rays) being simultaneously a discrete particle, or quantum, and a continuous wave. Viewing light as a wave explained the classical problems of optics, such as diffraction, Newton's rings, and iridescent colors. A wave moving in a parallel direction, as in a beam, could also account for the classical laws of optics, such as reflectance and refraction. However, there were some residual problems.

The problems come with the production and destruction of light. The basis of the material world is atoms and molecules, that is, particles (made of other particles). These atoms and molecules produce light when energized, as when heated, and that light is not produced in the even fashion predicted by wave theory—a particle producing a wave, or a wave turning into a particle. Also, light is destroyed, or transformed, when it strikes molecules.

Three phenomena hinted at the need to find a relationship between the wave and particulate nature of light. In the nineteenth century, scientists became aware that objects emitted light as they were heated. Furthermore, large heated objects produced more light at higher temperatures, with a corresponding decrease in the peak in wavelength of that light. Thus the embers of a campfire produce an orange glow, compared to the white light of the flashlight and even whiter light of the sun. Furthermore, the shapes of these emission curves were never symmetrical, but were more abrupt at shorter wavelengths and tapered off at higher wavelengths.

The second phenomenon was the emission of specific bands of radiation of very narrow wavelengths produced by heating pure elements. The discovery that each element produced a unique emission spectrum led to the establishment of spectroscopy as a science.

These separate phenomena were explained by the German physicist Max Planck at the beginning of the twentieth century. Planck showed the inverse relationship between wavelength and the energy content of a quantum of light, or photon. The shorter the wavelength of electromagnetic radiation is, the greater is the energy content of photons at that wavelength. Planck's discovery revolutionized our understanding of physics and light, and is a fundamental principle today.

The third phenomenon, observed late in the nineteenth century, was that intense light (the shorter in wavelength the better) could produce electrical currents. This phenomenon stimulated Albert Einstein to solve the puzzle of the photoelectric effect, and helped him produce his theory of relativity a few years later.

So it will be useful for us to consider light as a small portion of the spectrum of electromagnetic radiation, simultaneously having properties of a wave and properties of a particle (or photon). When we deal with light in general, our wave understanding will be foremost in our minds. However, when concerned with the effect of light on individual molecules, such as plant pigments, our concept of light as a particle will be more illuminating. In the spectrum of electromagnetic radiation, the visible portion is a narrow region. We measure wavelengths of visible radiation in billionths of a meter, or nanometers. Despite its narrow range, light (the electromagnetic radiation visible to us) makes up about 50 percent of the electromagnetic energy coming to us from the sun (fig. 2.10).

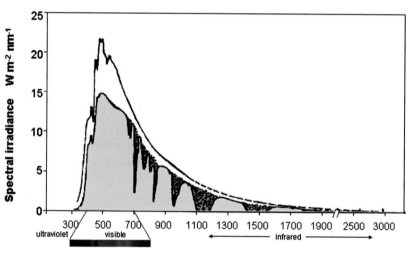

Wavelength in nanometers

Figure 2.10 The electromagnetic spectrum of solar radiation. The *top line* is the radiation arriving at the outer atmosphere, and the *bottom line* is that at sea level. The difference between the two lines is due to the scattering of all wavelengths by small particles and the absorbance of specific wavelengths by gases. Those that absorb in the infrared region are known as greenhouse gases, such as carbon dioxide and methane. The wavelengths are measured in billionths of a meter, and the units of energy are the same as we use for our electric bill, watts, arriving in a square meter in a nanometer increment.

In its wave form, we usually measure the strength of this radiation by its energy content, as in watts per square meter (like our electricity bill in kilowatt-hours). On the surface of the earth, we are exposed to sunlight that has passed through the atmosphere, subjected to the filtering action of the ozone in the stratosphere. Therefore, most of the short ultraviolet (UV) wavelengths are absorbed. The UV that does pass through the atmosphere requires that we be diligent in protecting ourselves from sunlight. The atmosphere, with its mixture of different gases and water vapor, also absorbs bands of solar radiation at longer wavelengths. Part of the radiation is in the visible portion of the spectrum, absorbed particularly by water vapor, but most of it is in the infrared region, where it is absorbed by different gases. In the visible region of the spectrum, the shortest wavelengths are violet and blue, followed in the longer wavelengths by green, yellow, orange, and finally red. We are visually most sensitive to the green portion of the visible spectrum. Plants are fairly sensitive at all visible wavelengths, but particularly in the red portion of the spectrum.

In contrast, we often measure radiation spectra in number of photons arriving in a certain area in a certain amount of time, such as a square meter per second, as for the spectrum of radiation in a Malaysian rain forest (see fig. 12.6). These measurements are more useful than watts per square meter for understanding how radiation interacts with individual molecules in a plant, or in our eyes.

Vision

Both plants and animals use light to orient themselves in a spatial environment. Single-cell algae with eyespots swim toward or away from light. Various plant parts respond to the direction of light, growing toward or away from it. Of course, the spatial acuity of a plant is limited compared to animals, which have evolved a variety of visual organs.

Surrounding much of an eyespot with a dark pigment improves the alga's directional responses. A hollow circular depression lined with light-sensing cells provides even better spatial resolution, depending upon the number of cells and the width of the aperture (fig. 2.11). Eye cups are found in a variety of animals, including flatworms, annelid worms, mollusks, and even some animals with backbones. Such a structure is likely to be the evolutionary precursor to the eye. Of course, other animals improved their spatial resolution with compound eyes, as among the insects. A narrowing of the opening of the eye cup produces a pinhole that can produce an image focused

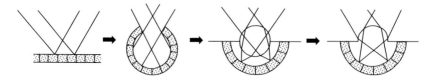

Figure 2.11 The evolution of eyes in animals. This diagram shows the steps that could have occurred to produce a complex visual organ, starting with a light-sensitive surface on the left, putting the sensory cells in a hollow pit to improve directional acuity, and eventually producing a lens projecting an image onto a surface of sensory cells, as in the eyes of vertebrate animals.

on the sensory cells, much like the early pinhole cameras manufactured by Kodak. Two animals that have such a visual apparatus are the chambered nautilus (*Nautilus pompilius*) and the abalone (*Haliotis* spp.).

It is easy to imagine (but there is no direct evidence) that the eye cup could have led to a structure with a lens at the aperture, and then to the evolution of the eye. Strikingly sophisticated eyes are found in an amazing variety of organisms, including spiders, cuttlefish (such as the octopus), and all vertebrates. I will focus on the human eye, but it is also important to consider some variations that affect the vision of other animals that interact with plants, relying on plants for nutrition (and plants relying on the animals for services they provide). Such animals include birds (which feed on nectar and fruits), insects (which feed on all plant parts), and nonhuman primates (which feed on foliage and fruits). These animals carry pollen, fertilize flowers, and disperse seeds for colonizing new sites.

The outer eyes of humans and other vertebrate animals consist of the transparent cornea, protecting the iris inside (fig. 2.12). The curvature of the cornea helps direct the light toward the interior of the eye. The iris contracts and expands, changing the aperture (pupil), and thus the width, of the beam of light that penetrates into the interior of the eye. The lens, highly transparent and flexible, consists of cells filled with a transparent lens protein, crystallin. The beam of light is refracted by the lens, whose plane of focus is altered by the contraction and relaxation of muscles. The refractive index of the lens is not uniform, and its variation helps reduce the spherical aberration otherwise produced by its curvature. Light passes through the transparent medium (the vitreous humor) filling the interior of the eye. Most of that interior is lined by a special sensory tissue, the retina. The retina is supplied with blood vessels and a network of nerve cells that coalesce to form the optic nerve connecting the eye to the brain. Beneath

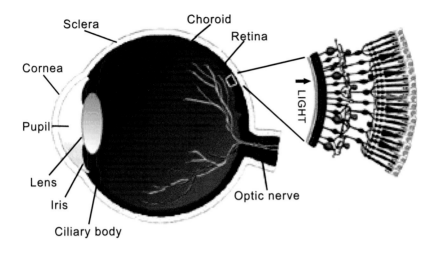

Figure 2.12 A diagram of the human eye. Expanded view on the right shows the distribution of cells of the retina. © Helga Kolb.

the nerve cells and vessels lie the sensory cells. These are of two types, rods and cones. Rods are narrower in diameter and far more numerous. Rods are much more sensitive to light and are dominant in most areas of the retina, particularly toward the periphery. Rods are responsible for vision in low-intensity light (and the pupil is enlarged). The peak sensitivity of the rods is 507 nanometers, and their response spans the entire range of our sensitivity to light. The cones, as we will see in a bit, are responsible for color vision.

Such a description demeans the function of such a magnificent organ, of 6.4 million cones (some 200,000 per square millimeter in the fovea) and 12 million rods, and of some 1 million nerve-cell extensions entering the optic nerve. Although we have had some notion of the function of the eye for centuries, we are still learning new things about it—such as that some nerve cells respond to light and help organize the visual responses within the eye itself. The range of spectral sensitivity of the human eye is impressive, from 370 to 730 nanometers, and most sensitive in low light in blue-green. The eye is so sensitive that just a few photons will trigger a response. The visual acuity of the human eye is impressive, but fuzzy compared to that of certain animals, such as raptors—including hawks and eagles. Technically, this visual acuity is defined by the spatial resolution in degrees, one minute of one degree. For us this means that we can, with normal uncorrected vision, resolve a fly at 13.8 meters. Our eyes are not superlative at any given task, such as acuity or sensitivity to dim conditions, but are remarkable in their ability to accommodate different conditions.

Color Vision

Our ability to detect different wavelengths, and consequently to perceive color, depends on special sensory cells, the cones. Inspired by Newton's discovery of different colors of light within a single white beam, at the beginning of the nineteenth century the British scientist Thomas Young theorized that color vision would be possible if the eye possessed three different color-sensing cells, predicting that they would be red, green, and violet. Only in the mid–twentieth century were the absorbance properties of the three cells actually measured: blue (or S) cells with a peak at 475 nanometers, green (or M) cells at 500 nanometers, and yellow (or L) cells at 525 nanometers.

The distribution of rods and cones in the retina, with the latter concentrated in a small dense central area, the fovea, provide flexibility of sight under different conditions. At low light intensity, as at dusk and at night, the iris opens to allow more light to enter the eye. It is focused on a broad area of the retina and is detected primarily by the extremely sensitive rods. This provides vision of high sensitivity, but of lower resolution and with little color information. During the bright conditions of daytime, the iris constricts, and a narrower beam of light focuses primarily on the fovea. The density of cells in this area provides an image with excellent spatial resolution. The S, M, and L cone cells add the dimension of color to the image.

Comparative research on the eyes of different animals has revealed quite different color sensitivities and different absorbance properties of the cone cells. Some organisms have but one type of cell, others two, some as many as five. Thus most animals perceive color quite differently than we do (fig. 2.13). Although we share the vision of three color-sensing cells with the apes, most of the other primates have only two cell types, and cannot discriminate red colors. Many birds have evolved a class of cones with peak sensitivity at 370 nanometers, giving them sensitivity well into the ultraviolet. Some bees and dragonflies have cones with a peak sensitivity of 340 nanometers. A bat was recently discovered to possess only a single cell type, with range from our visible light into the ultraviolet, and no color discrimination. No animals are able to detect wavelengths very far into the infrared. As we analyze the spectral sensitivity of animals (particularly insects) with greater care, the visual responses appear more complex. Given

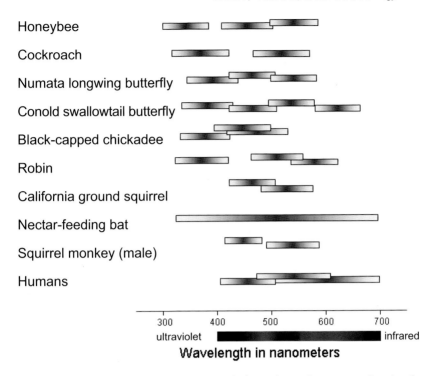

Honeybee

Cockroach

Numata longwing butterfly

Conold swallowtail butterfly

Black-capped chickadee

Robin

California ground squirrel

Nectar-feeding bat

Squirrel monkey (male)

Humans

300   400   500   600   700

ultraviolet ████████████████ infrared

Wavelength in nanometers

Figure 2.13 Color vision in different animals. The *bars* indicate the ranges and peaks of sensitivities of different types of cone cells in the retina, providing different ranges of spectral sensitivity and different degrees of color discrimination. Color discrimination is possible only when there are cone cells with different spectral sensitivities, even if they overlap.

the great sweep of the electromagnetic spectrum, the narrowness of our visual and color sensitivity is probably due to the energy content of photons at different wavelengths. Those in the ultraviolet wavelengths with higher energy are more likely to damage visual pigments, and those in the infrared wavelength lack sufficient energy to change the shape of the visual pigment to start the visual stimulus.

The biochemical basis for color sensitivity is fairly simple. Both rods and cones have light-sensitive pigments, rhodopsins. These consist of a pigment, retinol, that is embedded within a long protein, folded back on itself. Retinol is not made by animals, but is derived from yellow carotenoid pigments produced by plants and algae (described in chapter 3); Thus we rely on plants for our vision. When photons strike the retinol molecule, its shape changes, and that change sets in motion the cascade of events resulting in visual detection. The differences in color sensitivity, between rods and cones, and among the S, M, and L cones, are due to slightly different amino acid sequences in the proteins, or opsins. These are ancient molecules, whose sensory functions go back some 600 million years. Colored oil droplets in the cone cells of some animals, such as turtles, further affect sensitivity to color.

The Perception of Color

The retinal layer contains cones and rod cells, and nerve cells that connect to the cones and rods. Although these networks coordinate some of the information from rods and cones, most processing of this information to create a colored image occurs in the brain, in the visual cortex. Color perception is much more than the overlapping sensitivities of the three cones in our retinas, and there is still controversy about how perception works.

Color perception is correlated with the three types of color-sensing cells. Thomas Young suggested that color is perceived though the additive effects of these cells. He observed that three primary colors (the sensory cells) could be combined to make all the colors through these additive effects. His ideas were further developed by Hermann von Helmholtz to form the three-color theory of color vision, with the primary colors of blue, yellow, and red. Von Helmholtz commented that the three cell types did not have to be sensitive to only one color, but only more to one color than to others. Mixing two colors (with wavelengths affecting two cone types) produces secondary colors, such as green and red producing yellow. Mixing all three colors (affecting the three cone types) produces white light. Color television works in a similar way, by mixing small points of primary colors on a screen, and reproducing all colors through the various mixtures.

In middle-school art class, students learn that mixing pigments produces quite different results than blending rays of color (fig. 2.14). Blending the three subtractive primary colors of cyan, magenta, and yellow produces a dark brown-black color. Mixing magenta and cyan produces blue; mixing cyan and yellow produces green. These subtractive effects of pigments produce color by reflecting only a portion of the spectrum and absorbing the rest; yellow absorbs blue and reflects red and green. When rays of color are mixed together, as in light passing through color filters, the effects of the mixing are additive. Thus mixing the primary additive colors of blue, green, and red produce white (colorless) light, and mixing green and red produces brown. Virtually all of the color effects of pigments in plants are

Figure 2.14 Primary colors mix
to produce secondary colors.

produced subtractively. So we need to consider the subtractive effects of
mixing colors when we consider later the complex colors of different plant
organs, such as leaves and flowers.

The Young-Helmholtz theory of color was criticized by the German
psychologist Ewald Hering, who established an alternative theory of color,
one that still carries some weight today. Hering took as a starting point that
there are four primary colors, blue, green, yellow, and red. He based this on
surveys of color perception in different groups of people. He also noted that
human vision seems to function in pairs of colors, red-green and yellow-
blue. These colors complement and oppose each other. Colors appear in
such pairs when one color is observed for a long time and the other occurs
as an afterimage. For instance, looking at a red panel for many seconds,
we notice a green object of the same shape after we stop looking. We have
much to learn about color perception; experts still disagree on what color
systems are the most accurate and explanatory.

We can perceive many, many different colors. Presumably, animals with
color-sensitive cone cells can also perceive many colors. Beyond the basic
colors, we have developed a rich vocabulary to describe the shades of color.
There are some 800 color terms described in the *Oxford English Dictionary*. I
must admit that I quite often have to look in a dictionary to learn the description of a color word with precision (I stumble with words like *mauve*).
These words describe only a small percentage of the million shades of color
we can differentiate.

Our names for colors were influenced by our past experiences in nature, and particularly with plants. This is another way in which plants are profoundly embedded in our culture, either forming the names of colors or the adjectives that give more specificity to basic color names. These words are full of poetry, recalling memories of earlier times for each of us: rose, lavender, heliotrope, heather, iris, fuchsia, saffron, sage, indigo, madder, alizarin, dahlia, marigold, absinthe, carrot, grass, mustard, pansy, peony, poppy, primrose, olive, periwinkle, orange, mahogany, wisteria, goldenrod, peach, carnation, plum, asparagus, lime, pomegranate, melon.

It is not enough to name these individual colors with the words that have evolved in language. Such words may have different meanings for different people. The situation is analogous with the common names and the single scientific name of a plant. A plant that has a broad distribution takes on the various names of the cultures (and languages) that encounter it. The annatto encountered by the Secoya (*bayo bōsa*) has a different name than that used by the Shuar (*ipíak*). The annatto is a small tree growing naturally throughout the Amazonian region and into Central America. We may not recognize the Spanish word *annatto* and use *lipstick plant* instead. Despite these differences, the botanist recognizes one name for this plant, *Bixa orellana* L. (in the family Bixaceae), so that an amateur naturalist or scientist of any nationality will know its identity. This plant was known by the Swedish botanist Linnaeus (thus the abbreviation L. after the binomial name) and was included in his great work, the *Species Plantarum*. So it is with color. Systems of color description have been developed for centuries; even the German writer Goethe made a major contribution.

Our color sensitivity has three dimensions. We are aware of the *hue* of the color, or its predominant wavelength. Hue allows us to classify color into its primary names, such as red, blue, and so forth. We are also aware of the degree of *saturation* or richness (or chroma) of the color. Color may be diluted by its mixture with gray, which reduces the intensity of color perception. Finally, colors may vary in *brightness* (or value). Various systems have been developed to demonstrate the interactive effects of these three characteristics on color production. The most widely used today is the system of the Commission Internationale de l'Éclairage (CIE). In this international system each gradation of color can be assigned a quantitative value (fig. 2.15). Each color is located in a three-dimensional space (but diagrammed in two dimensions) where the range of values are its hue, saturation, and brightness. You might see the most practical application of this system when you choose a specific paint color for the interior of a

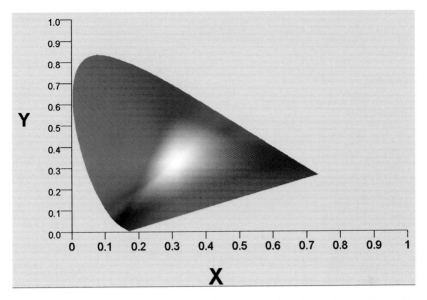

Figure 2.15 The CIE chromaticity diagram demonstrates the effects of saturation and brightness on hue, and suggests the enormous number of combinations that can result in unique colors. Courtesy of Hong Zhang.

home or apartment. The artist Albert Munsell arranged colors on charts in relationship to each other. The Munsell book of colors is used by scientists to characterize colors observed in nature, such as those of flowers, and to present color cards to individuals to conduct research on the perception and naming of colors.

Our names of colors are a rather rough description of our experience of the world of color. We generally think of certain color names as representing primary colors and having universal use. For us these colors would be blue, green, yellow, and red. In many cultures (and languages), however, there may be fewer color names, so that one name might include two of our colors. Many of the colors named in different cultures allow their members to discriminate among natural and useful objects in their environments, and different environments present quite different colors and color combinations. The Secoya (chapter 1) recognize green, yellow, and red, but not blue. The Hanunóo of the Philippines, studied by Harold Conklin in the 1950s, recognized only four colors: black (*mabiru*), white (*malagti*), red (*marara*), and green (*malatuy*). Red and green represented other physical properties as much as color, red being dry and old and green being moist and fresh. Harley Harris Bartlett, an ethnobotanist, studied the language of

Malays living in Sumatra in the 1920s. He found that that they recognized the same colors as we do. When he analyzed in more detail the words they employed, however, he found that they derived two color terms from their contact with Europeans. When I picked up Malay in the 1970s to help with my work in Malaysia, I learned that blue was *biru*, and only much later appreciated that this word was recently adopted from the English *blue*. In addition, Malays adopted the word *tjokòlat* from the European *chocolate* for the color brown. Previously that color was lumped with gray. Although some had concluded that these people did not have the physiological capacity to actually see the colors missing from their languages, it is clear today that they have the same color receptors as we do.

Using standardized color charts (such as the Munsell color charts or the CIE system) has made possible standardized surveys of color naming in widely different cultures, a research program first established by Brent Berlin and Paul Kay at Stanford University in the 1960s. Their initial results were supplemented by comparable field results from other anthropologists, to form the World Color Survey. They found a relatively small number (eleven) of basic color terms used to partition the color spectrum. These are called "fundamental neural response categories" (FNRs). Others have applied these methods to add languages to the World Color Survey, well over 100 languages by now. These results document the strikingly different uses of color terms in different cultures, and demonstrate links between language groups, suggesting stages in the cultural or linguistic evolution of color description.

Clearly such fundamental neural response categories skim the surface of the depths of color perception. Cultures have quite different uses of color. A Polynesian island group, the Bellonese, had little use of color in their society and language. When Danish anthropologists arrived to conduct their survey, an informant told them, "We don't talk much about color here." Yet a related Polynesian people, the Maori of New Zealand, have considerable color sensitivity and a rich color vocabulary. The perception of color remains mysterious; it is as much a philosophical as a psychological phenomenon. Clearly, basic neurological processes affect this perception. Synesthesia, where color sense is mixed with some other sense, such as letter recognition, smell, or taste, is particularly common among artists and may be explained by some mingling of sensory pathways in the brain. The strong emotional responses to red may be explained by the color's association with blood and arousal and perhaps has some deep physiological and neurological basis. I was not surprised to read that red sports uniforms give a significant edge to competitors, or to see bright red $50,000 sports cars selling like hotcakes in a nearby Miami dealership. Perhaps other colors (such as green, discussed in the final chapter) have other emotional connotations. Maurice Merleau-Ponty wrote profoundly about color perception, a process that allows instantaneous judgments of color values while taking into consideration the sensory input, the light environment, and the immense memory of other experiences: "We see too that colors (each surrounded by an affective atmosphere which psychologists are able to study and define) are themselves different modalities of our coexistence with the world."

We have learned much about how we see and discriminate the objects in our surroundings by their colors. Such knowledge helps explain our visual sensitivity, but does not diminish my continual amazement at the subtlety and beauty of our perception of forms and colors, particularly of plants. These general concepts will help us to understand the science of color in plants, starting with the molecules that plants synthesize to produce color.

Chapter Three Nature's Palette

···

I perhaps owe having become a painter to flowers.

CLAUDE MONET

The gods are growing old;
The stars are singing
Golden hair to gray
Green leaf to yellow leaf,—or chlorophyl
To xanthophyl, to be more scientific.

EDWIN ARLINGTON ROBINSON

···

A collection of late summer wildflowers, from roadsides and stream margins in central Massachusetts,
gives some idea of the extraordinary diversity of plant colors (fig. 3.1). These colors are produced by a great variety
of pigments, different kinds of molecules that absorb and
transmit light to produce such color. Although plants are
much simpler organisms than humans, no brains and ner-
vous system (and thus no worries), they have the capacity to produce whole families of molecules that we can't
make, some of which are required as vitamins and amino
acids in our diet. Many of these molecules are quite poisonous, and some are also beneficial to our health. Before
I begin to describe the variety of pigment molecules, the
palette that plants use in producing color, it is important
to explain why plants have such a capacity.

As sedentary creatures of light, plants have evolved
their amazing biochemistry to defend against, and obtain
the services of, mobile animals. Animals can move around
to escape being eaten and to obtain their own food. Plants

Figure 3.1 An assembly of wildflowers, collected along roadsides and stream edges in Petersham, Massachusetts, in August 2004, illustrates the amazing range of colors pro-duced by plants, due to the pigments that they produce. See notes for a guide to the identity of different flowers.

are bound in place as a consequence of the enormous surface areas required to fill their "dietary" needs. The foliage surface supplies the light energy that they use to fix carbon dioxide and make sugars. The root surfaces, particularly the root hairs, provide a comparable surface for absorbing mineral elements and water. The dietary needs of plants are quite simple compared to ours. They need no vitamins, only a few elements. They ob-tain carbon (C) and oxygen (O) from carbon dioxide in the atmosphere when the tiny pores in their leaves, the stomata, are open. They obtain hydrogen (H) from the water that they split apart in photosynthesis. In lesser amounts, they assimilate nitrogen (N) from the dissolved ions of ammonium or nitrate in the soil water, or from bacteria in their roots. They also obtain smaller amounts of phosphorus (P) and potassium (K) as ions in the soil-water solution, and they assimilate sulfur (S) as sulfate from the soil water as well. The rest of their nutritional requirements are metallic elements dissolved at low concentrations in the soil water (iron, magnesium, calcium, manganese, zinc, and molybdenum), as well as a bit of chlorine. That's it; no worry about nutritional supplements, amino acid balance, and so forth.

Being in one place provides opportunities for plants, but it also presents challenges. Their sophisticated biochemical capabilities have provided them with services and defenses normally associated with mobility. Being creatures of light has also made them sensitive to this cue for the control of their development.

Both plants and animals have much in common. Both groups of organisms are made of cells with similar compartmented organization. Both have true nuclei and organelles that provide specific metabolic services. Both have mitochondria that process sugars into forms of energy that can be used throughout the cell. Plant cells also develop plastids, whose functions are principally associated with light. After cell division and during specialization, proplastids become the mature and functional plastids: chloroplasts, chromoplasts, and amyloplasts. Chloroplasts absorb light energy for photosynthesis. Amyloplasts store starch in cells, and chromoplasts accumulate plant pigments, producing color in various plant organs.

In addition to the normal metabolic pathways, many of which are very similar to those of animals, plants can produce extraordinary molecules, including pigments. We consider the normal metabolism as *primary*, and the extra capacities as *secondary*. The secondary pathways of plant metabolism produce a variety of molecules that are toxic to the animals seeking them as food, including tens of thousands of nitrogen-containing alkaloids and nonprotein amino acids. In addition, plants produce a great variety of compounds that absorb light and produce colors.

Although plant pigments are most familiar to us as the molecules that produce the visual colors, the greens, reds, yellows, blues, and so forth, plants also produce pigments that have little visual effect, at least to us, but that are important for the control of their development and responses to natural light environments. Some pigments may also provide visual signals for other animals, many with broader radiation sensitivities than ours (fig. 2.13). Finally, many pigment functions in the plant are unrelated to their color production. Some may protect against herbivory or disease, or as signals in interactions with beneficial microbes.

The Chemistry of Plant Pigments

The word *pigment* calls to mind the sorts of products we mix to produce commercial paints, or mix on a palette in creating an oil or acrylic painting. Such pigments, particulate and opaque, are contrasted to other molecules in solution, or dyes. Pigments reflect the wavelengths of light that produce a color, and absorb the other wavelengths. The pigments we mix for oil painting are primarily mineral compounds with such characteristics. Dyes, on the other hand, are in solution (not particulate) and produce color by allowing certain wavelengths to pass through the solution. Thus in watercolor painting, we introduce pigments into the water solution and paint them on paper. The vibrant colors in these paintings are produced when light passes through the dried film of dyes and reflects off the white cellulose fibers of the paper. In both, colors are produced in a subtractive fashion (see chapter 2). The wavelengths that are not absorbed pass through the cells and are scattered and reflected by the plant tissues. The colors that we will discuss here, produced by plants, are equivalent to industrial dyes. They produce color by allowing certain wavelengths to pass through them, and scattering of those wavelengths produces the enormous variety of plant colors.

The Atoms of Plant Pigments

All plant pigments are compounds whose molecules contain some proportions of the elements of carbon, hydrogen, and oxygen. Additional atoms may form parts of certain pigments: nitrogen and sulfur. Magnesium (Mg) represents a variety of metallic ions that interact with some pigment molecules and affect their absorbance of different wavelengths of radiation. For instance, magnesium ions (Mg^{2+}) attach to chlorophyll to make this pigment molecule functional—and green.

The atoms of each of these elements contain some combination of particles, protons and neutrons, in their nuclei. The charge of each nucleus is balanced by the electrons in shells around the nucleus. Simple rules govern the number of electrons in each shell. The outer electrons are important in explaining the tendencies of atoms to bond with the atoms of other elements, and in explaining the absorbance of electromagnetic radiation by atoms and molecules. What is most important in explaining the chemistry of organic molecules, including plant pigments, is that the outer electron shell of carbon is stable when it shares four electrons with other atoms. Thus carbon has a valence of four. Such a valence makes possible the diversity of carbon molecules as the basis for life, including the complexity of plant pigment molecules.

Organic Molecules

Plant pigments, like all organic molecules, are held together by covalent bonds between carbon and the other atoms. We are accustomed to seeing each of these molecules described as a chemical formula. Each formula lists the actual number of atoms in a single molecule. Two molecular examples are the simple structures of benzene and glucose. Benzene is a hydrocarbon, C_6H_6, with six carbon and six hydrogen atoms, its structure first described by Kekulé (fig. 1.17). In the simple diagram of this structure, the hydrogen atoms are assumed to be present unless replaced by another atom. The branch origins are carbon atoms, and the lines represent a shared pair of electrons.

Another example is the simple sugar, glucose (fig. 3.2), present in every cell. Glucose contains carbon, hydrogen, and oxygen in the following ratio: $C_6H_{12}O_6$. The British chemist Walter Norman Haworth discovered the actual three-dimensional structure of glucose. Haworth borrowed from the earlier achievements of Kekulé with benzene. We will use molecular diagrams, like those of glucose and benzene, to depict the various pigments produced by plants. (Many of these appear in appendix A.) My biochemist friend and colleague Kelsey Downum encourages nonchemical audiences to view these molecules as a form of chicken wire, particularly those with multiple rings of carbon, as in pigments. The vinblastine alkaloids of the roots of the Madagascar periwinkle are a complicated example. They probably defend the plant against herbivores in nature, and they are also an effective chemotherapeutic agent against forms of childhood leukemia. Vinblastine, with a molecular weight of 811 and a chemical formula of $C_{46}H_{58}N_4O_9$, contains eight rings, five of them partially consisting of nitrogen. It is one of the most effective cancer treatments: a complicated structure with an important human benefit. Rather than worrying about your chemical understanding, you could focus on these structures simply as elegant abstract designs and impressive intellectual achievements. We take the amazing achievements in chemistry in the eighteenth and nineteenth centuries for granted (I recommend reading Oliver Sacks's remarkable autobiography, *Uncle Tungsten*, for a greater appreciation), remembering that the motivation to discover these molecular structures was sparked by the commercial value of natural plant dyes.

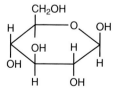

Figure 3.2 Haworth's structure for glucose.

Electrons and Light

The ability of different atoms and compounds to absorb and produce light depends on the characteristics of electrons in the outer shell. Simple atoms or very small molecules absorb very narrow wavelengths of radiation, and they emit electromagnetic radiation of narrow wavelengths when heated. The narrowness of these bands is a consequence of the particulate character of photons. When a photon of the appropriate wavelength—and energy—strikes an electron, the electron jumps to a higher orbital level. The energy of that photon, and its luminance, is absorbed by the electron. Its return to the original level may produce light at a longer wavelength, or fluorescence. Most likely, the production of electromagnetic energy from the return of the electron to its original orbital is at very long wavelengths, or in the infrared region as heat.

Differences in the spectrum of solar radiation above the atmosphere and at sea level (see fig. 2.10) are caused by absorbance of gas molecules, including water, oxygen, and carbon dioxide. Most organic molecules, like glucose, absorb little of the visible wavelengths of electromagnetic radiation.

Most electrons in orbit around individual atoms are not of the right energy level to capture photons in the visible spectrum. Photons at shorter wavelengths, and higher energies, may be absorbed at much shorter wavelengths by many organic molecules, but the energy is so intense that it may ionize or irreversibly change the organic molecule. At longer wavelengths, such as above the visible region and in the infrared, the energy content of the photons is insufficient to raise electrons to higher energy levels, although it may affect the rotational (twisting) movement of the electrons and individual atoms. This may cause absorbance in the infrared wavelengths.

How Pigments Absorb Light

Plant pigments are rather elaborate carbon-based organic molecules, and most contain ringed structures. By definition, these molecules contain bonds that absorb radiation in the visible wavelengths. For instance, a single carbon-to-carbon double bond, $C=C$, absorbs strongly in the ultraviolet wavelengths at about 180 nanometers (nm). However, the three carbon double bonds in the closed ring of benzene absorb strongly at the longer wavelength of 255 nm. Other bonds, such as nitrogen doubly bonded to

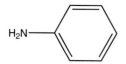

Figure 3.3 Aminobenzene. The amino group shifts absorbance of electromagnetic radiation by this molecule to longer wavelengths than benzene alone.

oxygen, $-N=O$, absorb in the visible portions of the spectrum, at around 665 nm. Still other structures added to the basic molecule often shift the wavelength of absorbance toward the red. A single ammonium group added to the benzene ring shifts the absorbance to 280 nm (fig. 3.3). Thus absorbance of *light* by plant pigments is due to the repetition of basic structural units that move the peak of absorbance toward the visible, modified by chemical structures attached to the basic molecule that may further affect absorbance in the visible wavelengths.

The complexity of the interactions of electrons in these large molecules dampens the specificity of each electron's interaction with photons, so that the capture of light by the pigment molecule is broadened considerably. It may not be possible to detect the narrow bandwidths of absorbance by specific groups attached to the pigment molecule, and the peak of absorbance may not be very revealing of the identity, or the molecular structure, of the pigment.

Our visual pigment, retinol (fig. A.14) absorbs a rather smooth and broad peak in the visible wavelengths at 520 nm that reveals little of the underlying structures that contribute to its overall absorbance.

Absorbance Defined

I have been using the term *absorb* in a commonsense way, to indicate if the pigment in solution, or in the plant tissue, does not allow light to pass through it at a certain wavelength. We can define this term with a little more precision to allow us to compare the *capacity* for different pigment molecules to transmit light more, or less, effectively. This concept was discovered and made practical during the eighteenth and nineteenth centuries. The great French scientist Pierre Bouguer (who was one of the first geophysicists) discovered that a beam of light is attenuated in an *exponential* fashion as it passes through a uniform medium. Bouguer also studied the sensitivity of human vision and the attenuation of light with distance, and created the unit of the candle as a measurement unit. We still use the term *foot-candle* as a measurement of human visual sensitivity. The German-French mathematician Johann Lambert developed a more useful form of this relationship, which was further improved by the German scientist Albert Beer in 1852 as an analytical tool in chemistry. So today we know of this relationship as the Lambert-Beer law (or even the Bouguer-Lambert-Beer law), but usually just as Beer's law. Imagine that you have placed a pigment in solution, and that you have measured the relative intensity of light passing through the first centimeter of solution compared to that outside as 0.5. That same beam passing through a second centimeter of solution should have the same effect, and so on. However, each incremental 0.5 of attenuation reduces the already attenuated beam, so that $0.5 \rightarrow 0.25 \rightarrow 0.125 \rightarrow 0.0625 \rightarrow 0.03125 \rightarrow 0.015625 \rightarrow 0.073125$, and so forth. By plotting the path distance versus the attenuation (I_i/I_o), the decreasing exponential relationship can be clearly seen. A positive percentage interest on a bank account would show the opposite, an increase. Beer refined this relationship as A (absorbance) $= \varepsilon \cdot c \cdot l$, where ε is the extinction coefficient of the pigment/molecule at a wavelength; l is the distance through the medium, conveniently kept to 1 centimeter; and c is the molecule's concentration. The value of this equation is that it can be used to calculate the concentration of a molecule in solution or to compare the abilities of different pigments to absorb light. Although certain pigments, such as chlorophyll, may have more absorbance per quantity of molecules, they may also be larger (have larger molecular weights). On a per mass basis, the major pigments we discuss in this book—the chlorophylls, carotenoids, and anthocyanins—have roughly the same extinction coefficients at their greatest wavelengths of absorbance.

The Types of Plant Pigments

Generally, plant pigments are the products of several major metabolic pathways: (1) the porphyrin pathway, (2) the isoprene pathway, (3) the phenylpropanoid/flavonoid pathway, and (4) the betalain pathway. With the exception of 1, these pathways are poorly elaborated in humans, but produce an amazing variety of secondary compounds in plants. These metabolic pathways are assisted by their location within a small volume of a cell (and quite often in specialized cells and tissues within a plant).

The pathways described below are the synthesis of a century of research by specialists throughout the world. Evidence supporting the existence of such pathways comes from a variety of sources. First, the pathway may occur within a compartment, often the chloroplast organelle (shown in fig. 4.2). Second, the intermediate steps of the pathway may be detectable. Third, the major catalytic agents of the pathway (the enzymes) can be isolated and studied. They may be in a compartment in the cell, or bound together in a large multienzyme complex. Fourth, the pathway is usually regulated by key reactions, and production may be slowed down by an accumulation of the final product. Finally, introducing radioactive carbon to the structure of intermediate molecules in the living cells or tissues allows tracing of the label into the final products. This method was developed by the chemist Melvin Calvin after the Second World War. He traced the pathway for fixing carbon in photosynthesis and later was rewarded with a Nobel Prize in Chemistry for his discovery. We will take up each of these pathways very briefly and describe the principal pigments produced in each. A few rare pigments of unusual biosynthetic origin will also be introduced.

Porphyrin Pigments

Plants are green because they produce chlorophylls, pigments with a structure called the porphyrin ring. Other porphyrin pigments are produced in plants, and these have functions that are not associated with the wavelengths they absorb. Our most visible porphyrin molecule is hemoglobin, which produces the red color of blood. Chlorophyll molecules are obviously pigments; we can see them easily because they absorb all of the visible wavelengths except green, and those wavelengths are reflected out of leaves and other plant organs. Their wavelengths of absorbance are essential to the capture of light energy in photosynthesis, the process that converts light energy to chemical energy, ultimately in the form of sugars.

All porphyrin pigments, including ones made by us (hemoglobin), are produced by the same pathway of synthesis. This pathway begins with the conversion of glutamic acid (present in all cells and known to us as the flavor-enhancer monosodium glutamate) in three steps to aminolevulinic acid (fig. 3.4). Two of these molecules fuse together to form the pyrrole ring of porphobilinogen, four of which fuse together to form the basic protoporphyrin ring. In animals, the addition of iron (Fe^{2+}) produces the heme group that makes up hemoglobin. In plants and the cyanobacteria, magnesium (Mg^{2+}) is added, and five additional steps complete the formation of chlorophyll *a*, including its long phytol tail (fig. 3.5, top). All higher plants produce two chlorophyll molecules; chlorophyll *b* (fig. A.2) is derived from *a* by a single chemical step.

These two molecules, the most abundant pigment molecules on earth, have slightly different absorbance spectra. Both absorb in the blue and red wavelengths, but in chlorophyll *b* the blue peak absorbs at a longer wavelength and the red peak absorbs at a shorter wavelength than chlorophyll *a*

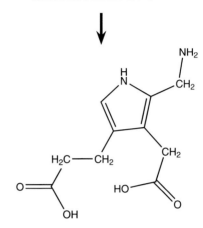

glutamic acid   **aminolevulinic acid**

Figure 3.4 In the synthesis of a porphyrin ring, glutamic acid is converted to aminolevulinic acid, and two of these molecules condense to form the single ring molecule porphobilinogen. Four of the latter are then assembled into a larger porphyrin ring.

porphobilinogen

(fig. 3.5, bottom). Chlorophyll *a* is an ancient molecule, having originated as the pigment of light absorbance in the first oxygen-producing photosynthetic organisms well over 3 billion years ago. In time, other photosynthetic organisms evolved additional chlorophyll pigments, but principally produced chlorophyll *a*. Early in evolution, non-oxygen-generating photosynthetic bacteria appeared. Their absorbing pigment, bacteriochlorophyll, absorbs at much longer wavelengths than chlorophyll *a*, and the colonies of these bacteria look purple instead of green.

Chlorophylls are produced within the chloroplasts of all plants (except some parasitic ones) and algae. In plants, the final production of functional chlorophyll is triggered by brief exposure to light. The pigments, with their long phytol tail, are strongly water repellent, and they are associated with proteins on membranes within the chloroplasts.

The basic structure of chlorophyll *a* was discovered in 1913 by the German organic chemist Richard Willstätter, helped by a succession of international collaborators. He was awarded the Nobel Prize in Chemistry just

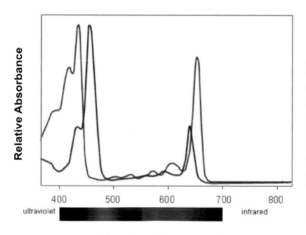

Figure 3.5 Top, struc-
ture of chlorophyll
a. Bottom, absorbance
by chlorophyll a and
b. Both molecules
absorb in the blue and
red wavelengths and
are transparent in the
green wavelengths.
The blue line is chloro-
phyll a, and the green
line is chlorophyll b.

two years later, in 1915. (Given the physical and medical biases of these awards, it is particularly striking when one is given to a scientist working on plants, and I suppose I demonstrate my plant bias by mentioning them.) Willstätter took special enjoyment in working with fresh plants collected from nature. He wrote the following about his research on chlorophyll.

> On my way to a lecture one morning I gave some quick instructions for an experiment with grass and alcohol. Mr. Meig asked, "Shall I order the grass from E. Merck-Darmstadt?" I led my associate to one of the windows of our workroom, which opened onto the green lawn of the old Botanical Garden, and pointed down there.

In addition to chlorophylls, plants make other porphyrin pigments that function in the transfer of electrons in both photosynthesis and respiration: the cytochromes. These are hemes that sequester iron. Finally, plants of the legume family (Fabaceae) make a pigment molecule, leghemoglobin, which looks superficially like the hemoglobin of animals. Leghemoglobin regulates the oxygen environment around the nitrogen-fixing bacterium (*Rhizobium*) that inhabits the nodules of the roots of many legumes. These molecules do not affect the appearance of plants, and their functions are independent of any color production.

Some algae and cyanobacteria produce pigments associated with chlorophyll *a*. Some, such as chlorophyll *c* and *d*, are variants of chlorophyll *a*, and others are derived from the same pathway as the chlorophylls, but the four rings are not closed. This results in a linear arrangement of the pyrrole rings, and they are called tetrapyrroles. The red tetrapyrrole phycoerythrin makes red algae look distinct from the other algal groups (figs. 3.6, A.3).

Figure 3.6 Unusual accessory pigments color the brown and red algae. Phycoerythrin is the pigment in the red alga *Tricleocarpa cylindrica* of the Caribbean region (*left*), and fucoxanthin in brown algae such as this kelp, *Nereocystis* sp., from the California coast (*right*).

All plants produce another tetrapyrrole, associated with a special protein, which forms a family of important sensory pigments: the phytochromes (fig. A.4). These absorb in the blue and red regions of the spectrum, and a slight chemical modification produces a large shift in the absorbance in the red end. These pigments help plants sense the light environment in a variety of ways, particularly to detect whether they are being shaded by other plants. Phytochromes do not affect the appearance of plants, because they are present in such low concentrations. I'll have more to say about these pigments in the final chapter.

The Isoprene Pathway

The pigments of chromoplasts (see fig. 7.17) are produced by a distinct plant biosynthetic pathway known as the isoprene pathway. The basic subunit that starts and adds to this pathway is isopentenyl diphosphate (IPP, fig. 3.7). The predominant pigments of this pathway are the yellow-orange carotenoids. In this pigment synthesis, the IPP subunits are joined together, each adding five carbon atoms. These additions produce the intermediate molecule phytoene, with forty carbons (fig. A.6). When the ends of the long chains are closed to form rings, β-carotene is synthesized (fig. 3.8). This pigment is present in chloroplasts and assists in photosynthesis by absorbing light and transferring it to chlorophyll. It also helps protect the photosynthetic mechanism from high-intensity light. β-Carotene converts to other pigments, violaxanthin, zeaxanthin, and others, that absorb excess sunlight even more effectively. All of these pigments produce yellow-orange colors in plant tissues. They absorb the green to blue wavelengths and transmit the longer wavelengths. When hydrogen atoms are removed from the phytoene skeleton, producing carbon-to-carbon double bonds, the result is the carotenoid pigment lycopene, which produces the red color of tomato fruits (figs. 3.9, A.7). Carotenoid pigments are not soluble in water and are bound to the membranes of plastids (chloroplasts in photosynthetic cells and chromoplasts in the cells of colorful organs like fruits and flowers), where they are also synthesized. Carotenoids are abundant in green leaves (associated with photosynthesis), and are produced in certain root vegetables, particularly

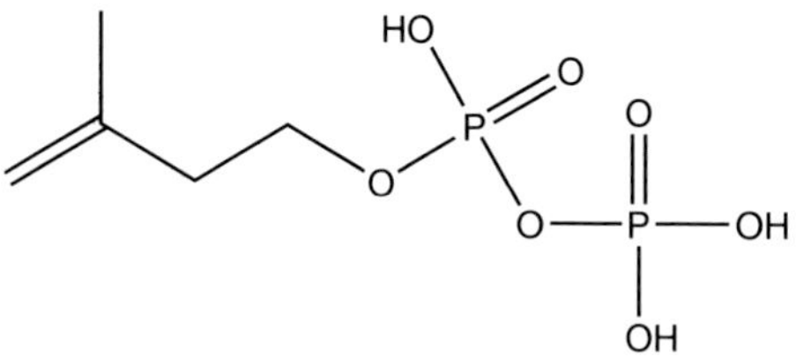

Figure 3.7 Isopentenyl diphosphate (IPP), the building block of the isoprene pathway, including the carotenoid pigments.

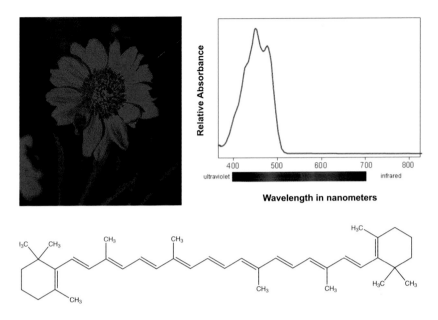

Figure 3.8 *Top left,* the principal pigment coloring the ray flowers of the balsam root (*Balsamorhiza sagittata*) is β-carotene (*center*). *Top right,* absorbance spectrum of β-carotene.

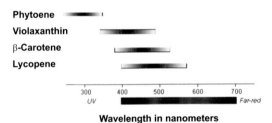

Figure 3.9   The ranges and peaks of absorbance of these four carotenoid pigments show how modifying the basic pigment skeleton changes the absorbed wavelength.

carrots. The principal color of the brown algae (the Phaeophyta) is due to the presence of the carotenoid fucoxanthin (figs. 3.6, A.9).

Carotenoids are essential in our diets, because we use them to produce the visual pigments in the rod and cone cells in the retina. Carotenoids are also moved to the macula of the eye, at the center of the retina (fig. 2.12), and their loss is a cause of macular degeneration, a common condition that can lead to blindness in the elderly.

The yellow and orange colors of many flowers are caused by the accumulation of carotenoid pigments in the chromoplasts of petal cells. The wavelengths of absorbance of these pigments are influenced by their length and the numbers of double carbon bonds, so they can produce a fairly broad range of colors (fig. 3.9). The increased absorbance in visible

Figure 3.10 Astaxanthin, a modified carotenoid pigment, was complexed to a protein in this lobster grown by Michael Tlusty of the New England Aquarium in Boston. This particularly intense color is produced by a special diet. Photograph courtesy of Michael Tlusty.

wavelengths provides a good display for humans and other animals. The structural modifications limit their absorbance to the blue and green wavelengths, producing yellow to red colors. However, some crustaceans obtain carotenes in their diets, modify them, and attach them to special proteins (fig. 3.10). These changes produce unusual blue pigments, the astaxanthins, which appear during reproduction in some crabs and lobsters. Unusual carotenoids, associated with special colors in plants, will be mentioned later in the book.

The Phenylpropanoid/Flavonoid Pigments

We previously saw that the double bonds of the benzene ring, distributed around that ring, absorb radiation in the ultraviolet region very effectively, and modifications to the ring can move that absorbance toward longer wavelengths. The early land plants, derived from green algal ancestors, evolved biochemical innovations that made them successful on land. The physical challenges of the land environment selected changes in chemical pathways to produce an enormous variety of ringed compounds. These modifications produced the cements (lignins) that make plant cell walls stronger on a weight basis than steel. They also produced a variety of pigments that intercept the harsh ultraviolet radiation on land. During plant evolution, still other complex ringed compounds were produced as defenses against being eaten by animals or attacked by fungal diseases. The molecule-building block for this large family of molecules, many of them pigments, is the simple molecule phenylpropane (fig. 3.11). This molecule is similar to phenol, used as a disinfectant. The molecules produced via this pathway are frequently called phenolics, and the pathway the phenolic

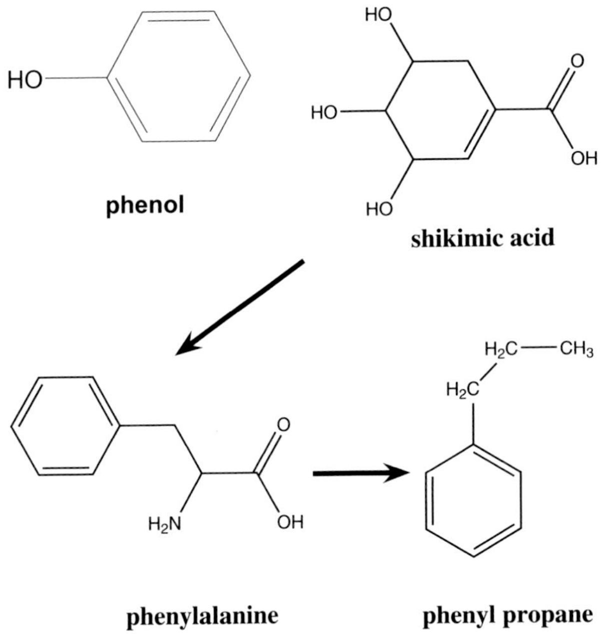

Figure 3.11 Intermediates of the phenylpropanoid, or shikimic acid, pathway.

pathway. The pathway begins with the formation of shikimic acid (which gives the pathway one of its names). Five steps lead to the production of the essential amino acid phenylalanine. We incorporate this amino acid into our proteins, and we obtain it from plants, or from animals that eat plants. Phenylalanine is converted by an important controlling enzyme, phenylalanine ammonia-lyase (PAL), to phenylpropane through the loss of its amino group. Phenylpropane is the starting point for the synthesis of a rather bewildering variety of plant pigments. The major types, with examples from familiar plants, are described in the following sections.

QUINONES · Quinones are produced in considerable variety throughout the plant kingdom, but only a small minority absorb radiation in the visible wavelengths and produce color. Various side groups are added to the basic structure, a single quinone ring. Certain of these molecules, such as plastoquinone in chloroplasts (fig. A.16) and ubiquinone in mitochondria, help transport electrons in photosynthesis and respiration, similar to some of the tetrapyrrole pigments mentioned previously. Vitamin K (phylloquinone, fig. A.17), whose absence from our diet causes symptoms of poor blood clotting and bone loss in the elderly, is a naphthoquinone with a long hydrocarbon side chain.

Some plant quinones are vegetable dyes, economically important until the end of the nineteenth century. The dye of the madder plant (alizarin, fig. A.15) is an anthroquinone. Lawsone (fig. A.19) is a naphthoquinone (a quinone with an aromatic ring added, fig. A.18), produced in all parts of the henna plant, and is a dye used in fabric coloring, hair tinting, and skin decoration (fig. 1.4). Juglone is a similar compound produced in walnuts (in the heartwood and nut shells), which produces a dark brown fabric dye and a fine artist's ink (fig. A.20).

THE FLAVONOIDS · A single pathway is responsible for the production of an enormous variety of plant pigments, most of the pinks, reds, and purples we observe in the plant kingdom. This pathway begins with the production of coumaric acid from phenylpropane, and the addition of three 3-carbon units (malonate), to form naringenin chalcone (fig. 3.12). The enzyme that catalyzes this conversion, chalcone synthase, is an important controlling point in the synthesis of flavonoid pigments, along with PAL.

A few metabolic steps from naringenin chalcone yield chalcones and aurones. Although these plant pigments absorb strongly in the ultraviolet regions of the spectrum, part of this energy fluoresces in the yellow-orange wavelengths. These molecules are produced in the petals of certain flowers, such as the aurone aureusidin in snapdragon flowers (fig. 3.13, left; fig.

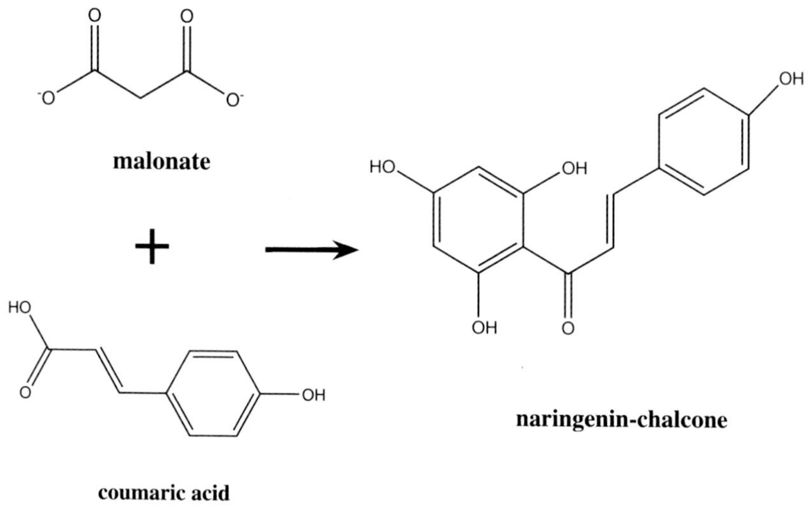

malonate

+

coumaric acid

naringenin-chalcone

Figure 3.12 The flavonoid pathway begins with joining coumaric acid and malonate to form naringenin chalcone.

Figure 3.13 The principal yellow-orange color of many snapdragons (*left, Antirrhinum* 'Luminaire Harvard Red' [photograph courtesy of Ball Horticultural Company]) is due to aureusidin, an aurone pigment. *Right,* the brilliant orange color in the flame of the forest tree (*Butea monosperma*) is due to the production of butein, a chalcone pigment.

A.32). The spectacular orange of flowers of the flame of the forest tree, native to the seasonal tropical forests of Asia, is due to the production of the chalcone pigment butein (fig. 3.13, right; fig. A.31). Both chalcones and aurones produce yellow to orange colors in flowers. Many of these plant pigments vary in function and color. Many are principally important for the absorbance of ultraviolet wavelengths and not visible to us.

ANTHOCYANINS · In the flavonoid biosynthetic pathway, the basic chalcone molecule is modified in a series of steps, each step producing a type of flavonoid pigment, from flavanones, to dihydroflavonol, to flavone-3,4-diol, to 3-hydroxyanthocyanidin, and finally to the anthocyanins (fig. 3.14). The flavanone molecules absorb strongly in the ultraviolet spectrum and provide no visible color. However, plants may produce them to protect against damage by ultraviolet light. Further modifications in this pathway produce more absorbance in the visible wavelengths, and therefore the ability to produce color. The flavones produce ivory or cream colors, and the flavonols produce yellows. The anthocyanins, exceedingly important in plants, produce orange-red to violet colors (fig. 3.15).

We take the understanding of such a complicated plant metabolic path-

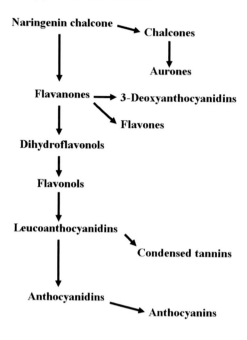

Figure 3.14. (top) Intermediates in the flavonoid pathway, starting with naringenin chalcone and leading to the synthesis of anthocyanin. Molecule structures appear in appendix A.

Figure 3.15. (bottom) A comparison of absorbances of phenolic and flavonoid molecules, showing the shift toward absorbance at longer wavelengths in the flavonols and anthocyanins.

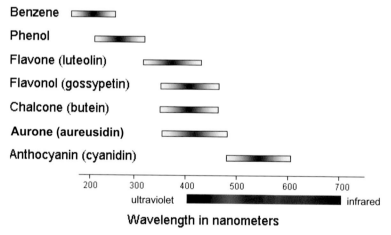

way for granted. The structure of anthocyanidin was discovered by Willstätter about ninety years ago, another impressive achievement. Willstätter grew the plants for his research in his own garden.

There, and in the spacious grounds of the Kaiser Wilhelm Institute, in the summer of 1914 grew the plantings of dark wine-red double cornflowers, of deep crimson and scarlet dahlias, large-flowered asters, red salvias, red-leaved red beets, deep purple pansies, petunias 'Karlsruhe Rathaus', and a little later of bronze-colored chrysanthemums.

The first structure was worked out with a red variety of cornflower (then named *Cyanopsis*), although Willstätter completed most of the research with rose petals. He named these molecules after the plants from which they were isolated; the first named molecule was cyanidin (fig. 3.16). This discovery and the rediscovery of Mendel's laws of inheritance (Mendel studied the inheritance of anthocyanins in pink and white pea flowers) led to rapid progress in understanding the pathway of synthesis of this molecule. The basic details were elucidated by many scientists a half century ago, using radioactive carbon labels to determine the intermediates in the synthesis of anthocyanins. Today, with advances in molecular genetics, we know the genes, and their control, for the enzymes that catalyze the steps in the synthesis of anthocyanin. Similar intellectual histories could be constructed for each of the pigments described in this chapter, which is beyond the scope of this book. This is just to emphasize that these disembodied facts are the product of human passion, insight, and persistence.

The steps in the synthesis of flavonoid pigments, leading to the terminal products, the anthocyanins, have involved modifications in structure to shift pigment absorbance toward longer wavelengths, similar to shifts in the carotenoids (fig. 3.15). Most flavonoid molecules have attached sugars that increase their solubility. Flavonoid synthesis occurs in the cytoplasm of the cell. Once sugars are attached, these molecules are transported into the vacuoles of cells. Most sugars are added at a single position in the flavonoid skeleton, and sometimes a short sequence of sugars may be added. This has little effect on absorbance, but a slight shift in wavelength occurs if sugars are attached at other positions.

Anthocyanins are the most important color-producing pigments in plants. These molecules are responsible for most of the colors produced in flowers and fruits, and in the reds of developing and senescing leaves. There are several major types of anthocyanin molecules (figs. A.39–A.44), which will be discussed in more detail in chapter 7. So far, some thirty different anthocyanidin molecules have been discovered. The number of anthocyanins (pigment attached to sugar) is very large, because of the types of sugars and their positions of attachment. All of them share the basic strong absorbance peak in the middle of our visual spectrum, with minor (but important) differences among different types (see fig. 7.12). Flavonoids are found throughout the plant kingdom, from bryophytes to flowering plants. Anthocyanins, although

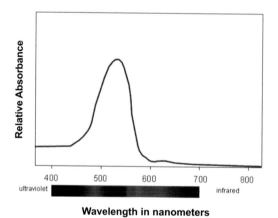

Figure 3.16 Top, a wild rose, *Rosa gymnocarpa*, source of the anthocyanin cyanidin-3-glucoside (*middle*). *Bottom*, absorbance spectrum of this anthocyanin.

not as broadly distributed as the flavonoids as a whole, are found in some bryophytes, some ferns and gymnosperms, and most of the flowering plants.

Betalains

Although widespread in flowering plants, anthocyanins are conspicuously absent from families in a single order of plants, the Caryophyllales — sometimes termed the Centrospermae. This group is distinguished from the other flowering plants by betalain production and several other characters. Betalain-containing plants include the cacti, *Bougainvillea* and relatives, and *Portulaca*. These plants can produce the major groups of flavonoid pigments, but not the anthocyanins. They produce yellow to red pigments that are fundamentally different from anthocyanins in structure and mode of synthesis (fig. 3.17). Betalains are produced by joining two nitrogen-containing

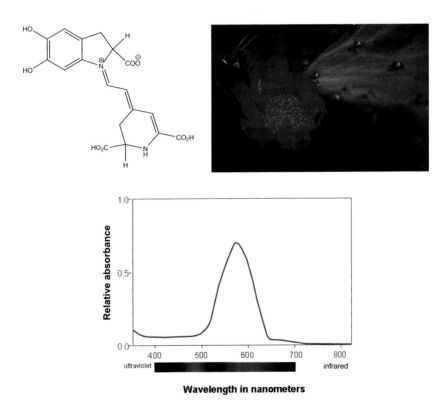

Figure 3.17 Betalain pigments. *Top left*, betanidin, the red pigment found in the roots of the red beet plant. *Top right*, the prickly pear cactus (*Opuntia* sp.) produces both types of pigments: a betaxanthin in the flowers and a betacyanin in the fruits and bases of spines. *Bottom*, absorbance spectrum of betanin.

molecules, derived from the amino acid tyrosine, with an indole structure. The red pigments are called the betacyanins (e.g., fig. A.47), and the yellow pigments betaxanthins (e.g., fig. A.48). Although these pigments have color effects similar to carotenoids and anthocyanins, they are fundamentally different structurally, and metabolically more expensive to produce because of their nitrogen content. These molecules have intense absorbance peaks in the visible wavelengths, and little absorbance in the ultraviolet. Like anthocyanins, they are attached to sugars, which increase their solubility in water, and they accumulate in the vacuoles of cells. Their presence within a single order of the flowering plants is a great mystery, as is the evolutionary assembly of their pathway of synthesis. Interestingly, enzymes that attach sugars to the pigments appear closely related to enzymes that perform the same work for flavonoids. Betaxanthins are strongly fluorescent, and may add color to plants that produce them, such as four-o'clock flowers.

Other Pigments

Although most pigments we observe in plants, hundreds of different molecules, are produced by the four pathways described above, other pathways produce more rarely observed pigments. Many of these pigments are economically and culturally important enough to deserve mention here. These pigments can generally be grouped by the basic molecular structures from which they are elaborated. Although we are limiting this discussion to plants, an amazing diversity of these minor pigments is produced by fungi, and many lichens (plant-algal symbioses) produce the raw materials for "vegetable" dyes.

INDOLES · Indigo dye is a pigment derived from the indole nucleus (fig. A.46). This dye is found primarily in the legume genus, *Indigofera*, but similar pigments have been obtained from other plants. This dye, and its production, was discussed in chapter 1.

QUINONE METHIDES · Quinone methides are similar to the quinones. Some of these pigments help produce the dark red heartwoods of the rosewoods (*Terminalia* spp.). Another commonly observed pigment of this group is coleone E (fig. A.49), which contributes to the variegated red colors in leaves of coleus (*Plectranthus* spp.). Two such molecules are condensed to produce the elaborate pigment of the dragon's-blood plant (*Dracaena draco*) from the Canary Islands. The bright red sap of this plant contains the pigment dracorubin (fig. A.50). Santalin (fig. A.51), the pigment that colors the heartwood of red sandalwood (*Pterocarpus santalinus*, of the Fabaceae family), was used as a dye in India and later exported to Europe.

PHENALONES · Mostly produced by fungi, the phenalones are also synthesized in the roots of members of the Haemodoraceae and a few other families among the monocots. Some of these, particularly red-purple molecules from the roots of Carolina redroot (lachnanthofluorone), have been employed as vegetable dyes. Many pigments in this group produce bright yellow colors. Curcumin is the persistent yellow pigment in the rhizomes of the turmeric plant (fig. A.54).

PYRONES · The pyrones contain the simple pyrone ring, and generally produce orange and red colors. The pigment mangostin (fig. A.55) contributes to the yellow latexes of the family Clusiaceae. This tropical family is familiar as the pitch apple native to south Florida, and the bitter yellow latex of the purple mangosteen fruits—my favorite as the most delicious of all tropical fruits.

THIARUBRINES · The thiarubines are odd pigments, perhaps the only ones containing sulfur. They are found only in members of the plant family Asteraceae. These are polyacetylenes, named for the three triple carbon-carbon bonds in its structure. Thiarubrine A (fig. A.53) has been identified in roots and other plant organs of various species, including the common ragweed, that late-summer scourge of asthma sufferers. These molecules are biologically very active and may constitute a plant defense against bacterial and fungal diseases. Their color may be a mere by-product of this function.

Browns and Blacks

The pigments described previously absorb at specific wavelengths in the visible spectrum, and thus produce characteristic hues: blues, greens, yellows, and reds. Plants can combine these separate pigments to absorb more evenly across the visible spectrum, producing browns grading toward blacks, depending upon their balance and concentration. Plants also produce molecules that give brown to black colors; most are complex molecules derived from the fusion of simpler molecules. The rapid production of brown (and eventually black) in plants is seen in many injured tissues; a cut apple turns brown. Sap from some plants, particularly in the sumac family (Anacardiaceae), can quickly turn black. These brown and black colors are produced from the oxidation and condensation of more simple molecules, primarily products of the phenylpropanoid pathway, and they absorb all across the visual spectrum.

Tannins, present in many plant tissues, oxidize when the tissues are exposed to air upon wounding (fig. A.25). More simple phenolic compounds are joined into larger molecular aggregates, soluble tannins. Condensed tannins result from the joining of more complex flavonoid molecules, such as leucoanthocyanins. These molecules attach to proteins, causing them to bind together (fig. A.26). This may explain their biological function as defenses against being eaten, because these precipitated proteins cannot be digested by animals consuming such plants. They were discovered very early as "tanning" agents in the production of leathers (fig. 1.15).

Tannins are mostly brown colors, but what about black? In many animals, including humans, brown and black pigmentation are produced by complex aromatic molecules, melanins. Melanins are synthesized via the joining of indole groups (for similar structures see fig. A.45). Sulfur-containing forms of melanin produce the tints of red in our hair. The details of the biochemical synthesis of melanin in animals are not completely understood, but we do know that melanin is produced in special organelles, melanosomes, and in special cells, melanocytes. One of my colleagues, Lidia Kos, studies the migration of melanocytes in the mouse embryo. The complexity of such molecules provides them with rings and groups that absorb at many different wavelengths across the spectrum, all in one large molecule, producing browns and blacks.

What about plants? Certain structures in plants appear black. Some fleshy fruits turn from red to blue and grade toward purple-black from the high concentration of anthocyanins. High concentrations of condensed tannins may also produce a dark color. Other structures, such as seeds, are a pure and permanent black. Some limited analysis of the chemistry of these pigments indicates that the molecules are devoid of nitrogen (in contrast to the pigments of animals). These complex molecules (fig. A.29) are tannins and may be derived from the condensation of catechols, and their structures are only partly understood.

We have briefly surveyed an amazing variety of molecules of pigment molecules that produce colors in plants. We will take up their roles in color production in various plant organs, along with their functions, in later chapters.

Chapter Four The Canvas

....

Empty Canvas. In appearance: truly empty, keeping silent,
indifferent. Almost doltish. In reality, full of expectations.
A little frightened because it can be violated. But docile.
Wonderful is the empty canvas—more beautiful than some
paintings.

WASSILY KANDINSKY

I always thought of the white paintings as being, not passive,
but very—well, hypersensitive,—so that one could look at
them and almost see how many people were in the room by
the shadows cast, or what time of the day it was.

ROBERT RAUSCHENBERG

....

The tissues in which pigments are produced affect the
colors we observe in plants. The analogy with a blank
canvas or other medium upon which artists apply their
palette is appropriate for plants. Robert Rauschenberg's
white paintings, completed by 1961, were on display in
the summer of 1961 at Black Mountain College, in North
Carolina. John Cage saw them, and they revolutionized
his direction of musical composition. He saw the white
paintings as "airports for lights, shadows, and particles."
Perhaps he had read Kandinsky's comments on the blank
canvas, written thirty-four years earlier. The canvas, or
any medium of artistic composition, is important in the
creative work; it is not totally blank. Pigments, by themselves, are inadequate to explain the vividness of colors,
whether in a plant or a work of art. Artistic pigments are
adapted to the medium of painting, translucent or opaque

in oils or acrylics, highly translucent in watercolors. In oils, the thickness of the paint applied and the density of the pigments require a surface on which the medium (the oil paints) can dry. The completed painting may consist of layers of such paints. In watercolors, the paper may be quite important in the overall effect aimed for by the artist, and the whiteness of the paper provides a background that diffusely scatters the light back toward the viewer and through the dry and translucent layer of pigments. Thus the wavelengths of absorbance and transmittance produce the colors we perceive in viewing the painting; the canvas—or paper—is important in helping us perceive those colors.

Plants are much more like a watercolor than an oil or acrylic painting. The plant organs, particularly leaves and flower petals, are structurally complex features that efficiently absorb or reflect different wavelengths of light, depending on the requirements of the plant. Thus leaves and flowers are optical organs (fig. 4.1). Given that plants are creatures of light, depending upon it for their source of energy as well as visual cues in controlling their physiology and development, it should not be very surprising that their organs have sophisticated optical properties. Also, as immobile organisms often exposed to harsh sunlight day after day, many plants have evolved elegant structures to reduce the absorption of light.

The principal interactions of electromagnetic radiation with matter (chapter 2) are all important in plants: reflection, refraction, absorbance, transmission, and even interference. These effects occur in the different

Figure 4.1 The optical properties of the leaves of the satin leaf tree (*Chrysophyllum oliviforme*) are strongly influenced by hairs that develop on the undersurface. Tannins in the hairs produce the bronze color so characteristic of this Caribbean tree. *Left*, details of the undersurface; *right*, foliage and flowers.

layers of a leaf or petal, or any other plant part, starting at the surface and continuing through the internal layers.

Starting with the Cell

Plant cells are different from those of animals in several ways, mostly as a consequence of their sedentary lifestyle (fig. 4.2). First, plant cells are bound by a wall composed of cellulose. This simple polymer of glucose molecules is the most abundant molecule on earth. Cellulose consists of multipolymer "ropes" whose components are secreted by the cytoplasm and assemble to form the cell wall. Cellulose has some very interesting optical properties. It has a refractive index somewhat greater than that of water, depending upon the angle at which light passes through the polymer and on the degree of absorption of water. Fully moistened cellulose may have a refractive index as great as 1.40. The wall makes the cell rigid and helps protect the plant from the loss of water due to drought. Thus individual plant cells (in contrast to many animal cells—particularly during development) don't move around. They maintain intimate contact among neighboring cells by the pores (plasmodesmata) that connect the cytoplasm of plant cells. The cytoplasm of the cell, inside the wall and membrane, is a complex environment. It contains the nucleus, where genetic information is stored and which is transcribed into messages and translated into products that run the cell and organism (more on this in chapter 5). The cytoplasm also

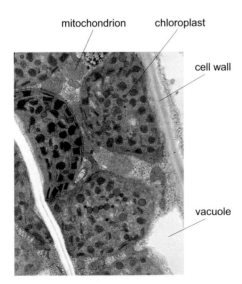

Figure 4.2 Photosynthetic cell from the leaf of an umbrella tree (*Schefflera actinophylla*). Various parts of the cell are labeled, but certain features mentioned in the text are missing from this photograph taken with a transmission electron microscope.

contains several organelles. Most prominent in plant cells are the plastids, particularly the chloroplasts, sites of photosynthesis. Plant cells also contain mitochondria, found in virtually all cells, animal and plant. Finally, plant cells contain a skeleton (cytoskeleton), consisting of linked proteins that provide some rigidity and a means for the movement of organelles within the cell.

Cells primarily consist of water. Water has a refractive index of around 1.30, much greater than air. We see the effect of this index when viewing the bending of a shaft of light in water. Adding molecules in the solution of the cytoplasm increases the refractive index to more than 1.34. Finally, the membranes within the cell, such as in the chloroplasts, have refractive indexes similar to that of cellulose, around 1.40. The slight differences in refractive indexes of the different cell components provide conditions of weak reflectance at boundaries between them, as between the cytoplasm and a chloroplast, or even between structures within the chloroplasts. However, these differences are small compared with that between the refractive index of air (1.00) and the cell wall (around 1.40). The greater density of cell contents, and the higher refractive indexes of cell components, means that light is strongly reflected from this surface when it contacts air. Even within the cell, differences in refractive indexes cause some scattering of light.

Surface Effects

When light strikes the surface of a plant, some of it is reflected because of the higher refractive index of the plant tissues. This effect increases at oblique angles. However, most plant organs are not smooth, like glass or the surface of a still pond. So the nice clean angles of reflectance or diffraction are not so easily observed (fig. 4.3). Irregularities on the surface, due to differences between adjacent cells or the curvature of the outer cell wall, violate the classical smooth reflectance we observe on flat surfaces. The small projections on the plant surface scatter light in various directions, not just at the predicted angle, producing what we call diffuse reflectance. Diffuse reflectance is reduced when the cell structure on the outside of an organ, such as a leaf, is unusually smooth. Such smoothness may be increased by the cuticle, a waxy layer whose secretion by the surface cells reduces evaporation of water. In such a case, it is more likely that light reflects at oblique angles purely from the surface. Such reflectance gives the organ a shiny, colorless appearance.

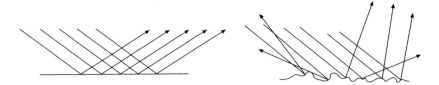

Figure 4.3 Surface effects on reflectance. A smooth surface (*left*) produces reflectance with light waves reflected at the same angle, in parallel. An irregular surface (*right*) produces diffuse reflectance with light waves reflected at different angles.

Figure 4.4 Effects of surface waxes on scattering light. The undersurface of leaves of willows, such as the southern willow (*Salix carolin-iana, left*), produces wax particles, seen in a scanning electron microscope (*right*).

A plant surface magnified under a compound microscope may reveal some multicolored, rainbowlike colors from reflected light. It is likely that some regular diffraction lines form on the surface, but the color effects are cancelled by the irregularities in the larger surface.

Waxes, Hairs, and Scales

Many plants reduce the capture of radiation with surface structures. Particles of wax may accumulate on the surface, reflecting and scattering light. In willows, such as the southern willow common on tree islands in the Everglades, the undersides of the leaf are white due to deposits of wax particles that scatter light at all wavelengths (fig. 4.4). Plants also produce hairs, simple or branched, which provide a complex network of surfaces that reflect and scatter light with some efficiency. Hairs also provide an insulating layer of still air that may isolate the leaf or stem from higher or lower temperatures in the surrounding atmosphere. In the satin leaf tree

Figure 4.5 Effects of scales on scattering light, as seen on the undersurface of leaves of the Jamaican caper (*Capparis cynophallophora*). *Top*, differences between the upper and lower surfaces; *bottom*, a detail of the undersurface.

(fig. 4.1), the undersurface hairs are colored light brown by deposits of tannins. The resulting reflectance is a bronze color, and the flat and parallel arrangement of the hairs gives the leaves their satin appearance. Scales, flat multicellular structures attached to the plant organ, may be particularly effective at reflecting light, producing an organ that looks silvery or bronze when reflecting light (fig. 4.5). In chapter 6, I will provide other examples of such structures on leaves, including scales, and will discuss their usefulness in different environments.

Cells as Lenses

The curvature of the surface, or epidermal, cells of plant organs may produce conditions for the coordinated refraction of light into the plant. Depending on the shape of the cell surface, the way in which light is absorbed and refracted varies (fig. 4.6). Cells whose outer wall rises into a pointed tip may be particularly effective in absorbing light in a diffuse environment. Such

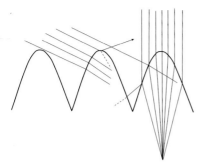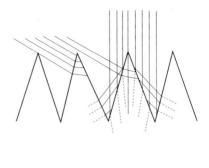

Figure 4.6 Surface projections affect light interception. Rounded surfaces may refract light into a focused area (left). Pointed projections may decrease surface reflectance at oblique angles of radiation and increase absorption of diffuse light (right).

a shape reduces the likelihood of surface reflection at oblique angles. Side illumination is more likely absorbed by the steeply pitched walls, and direct illumination glances off the surface more directly into an adjacent cell.

A more rounded external wall can function like a miniature lens, re-fracting light into a place in the interior of the organ — into a spot of higher intensity. Such lenslike cells would be less effective in capturing diffuse radiation than the cone-shaped external cells above.

Much was learned about the optical properties of plant surfaces by the physiological anatomists of the late nineteenth century. These botanists, many German and trained in the morphology and physiology of plants, ventured to distant lands to study the function of plants in relationship to their natural environments. When they found relationships between certain plant structures and specific environments, as in the understory shade of a tropical rain forest, they attempted to explain the functional significance of these structures. A good example of such early work is that of George Ernst Stahl (fig. 4.7), who joined many European scientists at the Foreigner's Laboratory of the Dutch botanical gardens in Buitenzorg (now Bogor) in Java. In some ways, this early work is now being duplicated, but with more sophisticated and quantitative techniques.

Leaves as Lenses

In special cases in succulent and thick leaves, the surface curvature of the entire organ refracts light into the leaf, focusing it on specially oriented chloroplasts. That is the case for a small *Peperomia* plant, found growing on tree branches in the forests of French Guiana (fig. 4.8). The leaves seem like

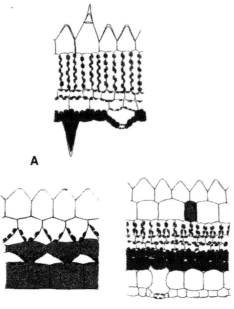

A

B.   **C**

Figure 4.7 Illustration by the physiological anatomist George Ernst Stahl, working on the unusual anatomy of tropical rain-forest under-story plants in Java at the end of the nineteenth century. Stahl speculated on the focusing effects of the surface projections of leaves, and these diagrams include the distribution of anthocyanins.

Figure 4.8 Top, *Peperomia rotundifolia*, a small epiphyte from tropical rain forests in South America, with round jewellike leaves. *Bottom, Lithops otzeniana* from South Africa, a living stone.

luminescent gems of jade or emerald. The smooth curvature, coupled to the large, clear cells beneath (which may also store water), allows the refraction of light onto the photosynthetic cells inside and beneath. In the living stone plants of South African deserts (*Lithops* and other genera), the curved leaf tip collects light and refracts it to where chloroplasts are displayed, well to the interior and under the soil surface.

Internal Effects

The importance of the internal structure of plant organs can be appreciated by comparing them with the simple structure of an alga, such as the flimsy and flat thallus of the sea lettuce that I easily find among the roots of red mangrove trees lining Biscayne Bay, in Miami. The body of the alga is a few cells thick, with no air spaces, and with chloroplasts abundant in all cells. Sea lettuce is light green, not easily seen in a surface view, but more easily observed by allowing light to pass through it (fig. 4.9). Light is scattered to a small extent, because of slight differences in the indexes of refraction among the cell components. In such structures it is best to measure the transmittance and reflectance of light by collecting it at the different scattering angles. For this we use an integrating sphere attached to a spectroradiometer (an instrument that measures the spectral distribution of any source of radiation). This instrument can be used to analyze the intensity of transmittance through the structure and the reflectance of light from the surface. I use the word *absorption*, and not *absorbance*, to distinguish between the behavior of pigments in solution (following Beer's law, chapter 3) and the scatter of light in a complex tissue with some pigments on membranes (absorption, or absorptance). For the following examples I will use the range of electromagnetic radiation to which we are sensitive (light) and to which plants respond in photosynthesis, 400–700 nanometers. Absorption of light by the sea lettuce is that which is not reflected or transmitted (fig. 4.10). Grinding sea lettuce in acetone solvent (appendix B) puts the pigments, chloroplasts and xanthophylls, into solution. Then it is possible to compare the optics of the sea lettuce to the transmittance of the pigment solution.

What we see in this analysis is that the absorption of light by the structure is somewhat similar to the transmission of light through an extract of the pigments (fig. 4.10). The simple structure of the sea lettuce thallus has little effect on the transmittance and reflectance of electromagnetic radiation. The sea lettuce thallus, two to three cells thick with no

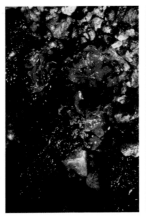

Figure 4.9 Sea lettuce (*Ulva*). *Left*, the whole "plant," about 6 cm across. *Right*, a microscope photograph of the surface of the thallus, showing the closely packed cells.

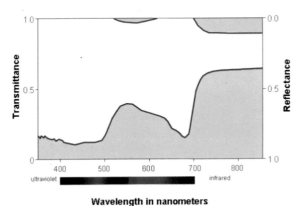

Wavelength in nanometers

Wavelength in nanometers

Figure 4.10 The optics of sea lettuce. *Top*, the gray areas represent the transmittance (*lower area*) and reflectance (*upper area*) of the green portion of a leaf; absorption is the white area. *Bottom*, the *blue-green line* represents the absorption (absorbtance) of radiation by the alga, and the *lime-green line* represents the transmittance of the pigment extract from the tissue. These two lines resemble each other.

air spaces between these cells, reflects little light (less than 1 percent) and allows transmittance of about a quarter (25.1 percent, or 0.25) of the light through the thallus. Since the light absorbed by the thallus is that which is *not* reflected or transmitted, sea lettuce absorbs about three-quarters (74.8 percent) of the light. *Ulva* transmits poorly in the blue and red wavelengths (wavelengths well-absorbed by chlorophylls *a* and *b*) and transmits well in the green wavelengths and beyond the visible portion of the spectrum.

The sea lettuce makes a good comparison with the optics of plant tissue, as a leaf. The plant leaf we will use has the advantage of being variegated, containing normal green areas, and white areas where no chlorophyll is present. Dumb cane grows in the understories of rain forests in the New World tropics—and in professional offices and hotel lobbies throughout the world (fig. 4.11). This plant shows the effect of chlorophyll on the leaf optical properties. Examining the anatomy of this leaf, we can see that the chlorophyll is present in chloroplasts in the middle tissue of the leaf, the mesophyll, and is absent from the outer layers, the epidermis. The mesophyll is differentiated into two layers. In the upper layers, the cells are elongated and contain many chloroplasts. The lower layers contain cells with fewer chloroplasts and a number of irregular air spaces between the cells. These air spaces have a major effect on the optical properties of the leaf (fig. 4.12). Green areas of the leaf reflect much more visible light than the sea lettuce (7.9 percent) and transmit much less (8.1 percent). The result is that the leaf absorbs 84.0 percent of the light striking it. Much of this absorption is due to the presence of chlorophyll in chloroplasts of the mesophyll. The leaf absorbs more in the blue and red wavelengths, but the difference with the green wavelengths is more modest compared to the sea lettuce. We can see the effect of chlorophyll on absorption by comparing the optics of the green areas with the white areas. In white areas both reflectance (41.0 percent) and transmittance (27.3 percent) are higher than in green areas, and absorption is reduced dramatically—to 31.7 percent. The increase in absorption at 660 nanometers suggests that a tiny amount of chlorophyll was present in the white areas.

Finally, the white area was measured in the same way as the other area, except that the air spaces of this leaf were infiltrated with water by immersing them in a partial vacuum. The air spaces, all with a refractive index of 1.00, were replaced with water, with a refractive index of 1.30, closer to that of the cells. In the infiltrated leaves, average reflectance was reduced (to 23.8 percent), and average transmittance was increased (53.0). Losing the air spaces reduced the internal reflectance of light within the structure,

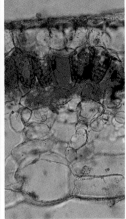

Figure 4.11 The dumb cane (*Dieffenbachia picta*), so named because accidentally eating the plant causes the throat to swell and temporarily stop speech. *Left*, the variegated plant; *right*, a hand section of a leaf, revealing the layers, distribution of chloroplasts, and presence of intercellular spaces.

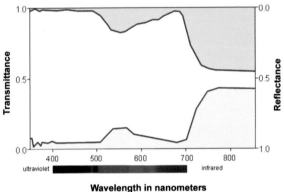

Wavelength in nanometers

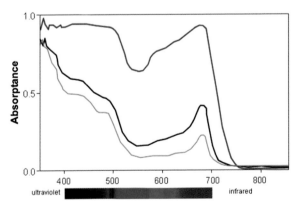

Wavelength in nanometers

Figure 4.12 The optics of the leaf of the dumb cane. *Top*, the gray areas represent the transmittance (*upper area*) and reflectance (*lower area*) of the green portion of a leaf; absorption is the *white area*. *Bottom*, leaf absorption; the *green line* is the normal green leaf, the *black line* is from a white area of the leaf, and the *gray line* is from a white area of the leaf in which the intercellular spaces were filled with water.

causing a decrease in reflectance and an increase in transmittance, so that the average absorption by the cleared white leaves was only 23.2 percent.

These differences in light absorption, between chlorophyll in solution compared to a structure with pigments in plastids, and between a structure with air spaces and one cleared of air spaces, are due to two important optical properties of plant tissues: scattering or path-lengthening effects, and packaging or sieving effects.

Scattering Effects

Air spaces within a plant organ reflect and scatter light at different angles, depending upon the shapes of spaces and loosely packed cells in the interior. When pigments are located within this region of scatter, the likelihood of light absorption is increased, even at poorly absorbed wavelengths, because the complex patterns of scattering increase the path length of a beam of light (fig. 4.13). For chlorophyll, the dumb cane leaves absorb a significant amount of radiation in the green wavelengths, although less than at the blue and red ends of the light spectrum. In solution, chlorophyll absorbs very little of the green wavelengths, and the likelihood of absorption in these wavelengths is enhanced by the greater probability of striking a chlorophyll molecule in a longer optical path. In sea lettuce, where the path-lengthening

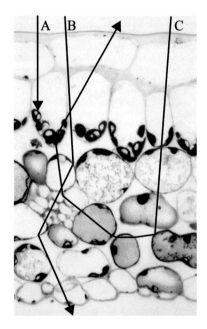

Figure 4.13 Packaging and scattering effects demonstrated in a transverse section of the leaf of a tropical rain-forest understory herb, *Triolena hirsuta*. In light ray A, the ray enters the leaf and is absorbed by a chloroplast. In light ray B, the ray enters the leaf, is slightly scattered, and passes through the leaf without encountering a chloroplast. This path is made possible by the packing of chlorophyll in chloroplasts. In light ray C, the ray enters the leaf, is scattered, bounces back, is partly absorbed by a chloroplast, and exits the upper surface of the leaf. This ray produces the green color we expect to see in normal leaves.

effects are minimal in the absence of air spaces, much less light in the green wavelengths is captured by chlorophyll—and there is more intense color produced in the green wavelengths.

The effect of path-lengthening on absorption by pigments requires that the pigments be in the midst of the region of air spaces in the plant organ. If the air spaces are near the outside, then much of the radiation scatters directly back to the exterior without alteration by absorption. In such cases, the reflected light looks silver despite the presence of pigments beneath the spaces. Such spaces also decrease the amount of light transmitted through the organ.

Cell shape helps determine the extent and shape of air spaces. Columnar cells reduce the amount of scatter due to air spaces. Light passes efficiently through the vacuoles and along the sides of these cells to regions beyond. Such cells are normally limited to superficial layers in an organ, as in the palisade mesophyll cells of a leaf. Layers beneath are likely to be more irregular in shape and more likely to scatter some light back through the columnar cells.

The effects of such cell spaces on the optical properties of a plant organ are dramatically demonstrated by comparing leaf samples from the dumb cane plant with internal spaces filled with water or air (fig. 4.14). The net result, photographed on a dark background, is that the leaf turns a dark, dull green. Perhaps such a change in color would seem less dramatic to us if we compared it to a bank of white snow glistening in the winter sunshine. The weather warms, and the snow turns to water, freezes again at night, and turns into clear ice. The pure white of snow is similarly caused by the

Figure 4.14 The effect of air spaces on color production in the dumb cane leaf. *Left*, a normal leaf. *Right*, a leaf treated under vacuum, so that water fills all air spaces within the leaf.

scattering of light among the complex crystals, much of it returning toward the viewer's eye.

Packaging Effects

In the art of watercolor, water-soluble pigments are brushed onto the paper and coat the cellulose paper fibers evenly. When sunlight illuminates the painting, the scattered light passes through the pigmented coatings and is reflected from the paper fibers to produce brilliant colors. In plant cells, however, pigments are not so evenly distributed. Pigments are always sequestered in a portion of the cells. The flavonoid pigments accumulate in cell vacuoles, allowing some light to pass through the external cytoplasm and cell walls. Some flavonoid pigments may accumulate in the cell walls of certain plants as the bright red sphagnorubins of sphagnum moss and some liverworts (see fig. 11.2). These produce quite brilliant colors.

Chlorophyll and carotenoid pigments are not soluble in water and are bound to the membranes within plastids. They are present in limited and discrete packages within cells. Chlorophylls are packaged in chloroplasts, located in the cytoplasm at the periphery of photosynthetic cells. Their location is easily seen in the transverse section of the dumb cane leaf in figure 4.11. When chromoplasts form, to produce color in petals of certain flowers, those organelles are also filled with the appropriate carotenoids. Chromoplasts are located around the periphery of the cells that produce them, surrounding the empty and large central vacuole (see fig. 7.17). Thus the packaging effects of the organelles are increased by their location in a small part of the plant cells producing them.

Anthocyanin pigments may be packaged in the vacuole to produce similar effects. They may form vacuolar inclusions. Some early studies of these inclusions led to their being called anthocyanoplasts. These are not membrane-bound organelles, but rather associations of pigment molecules with special proteins. No matter how they are formed, the optical effects of such inclusions may be similar to organelles. In flowers of lisianthus, anthocyanins aggregate as small particles in the vacuole, looking a little bit like chromoplasts but in the wrong place in the cell. In these plants, special proteins are made that bind to the anthocyanins (fig. 4.15).

The optical implications of packaging effects are important. When pigments are in small particles, it is more likely that light will pass through a structure, even when scattered, without encountering a pigment molecule. Path-lengthening effects may actually decrease the capture of photons at wavelengths most effectively absorbed by a particular pigment. Thus pack-

Figure 4.15 Packaging effects on anthocyanins in lisianthus (*Eustoma grandiflorum*). *Left*, typical flowers of this ornamental plant. Photograph courtesy of Ball Horticultural Company. *Right*, detail of petal anatomy, showing the anthocyanins as bodies in the vacuoles.

aging (sometimes also called sieving) helps explain the reduced efficiency of chlorophyll absorbance by leaves in the blue and red portions of the spectrum. The concept of sieving effects was explored in suspensions of algae in water. An effective model of light absorption, necessary to explain the photosynthetic productivity of such communities, required an understanding of how photons would be captured in a suspension of pigment granules.

Another implication of the packaging of pigments into structures within cells is that their positions at different times may affect the amount of color that is produced (fig. 4.16). Consider large round cells in which pigment granules (inclusions or organelles) are located in the cytoplasm surrounding the large central vacuole. If the organelles are placed along the sides of the cells, in relation to the surface of the structure and the direction of illumination, the probability of their absorbing light is decreased, and the light is more effectively transmitted through the tissue. Such tissue would produce little color. However, if the organelles are placed along the outer and inner margins of the cell, more parallel to the surface of the organ, the likelihood of their capturing light is much greater. Such an arrangement would more effectively intercept light and could produce color both

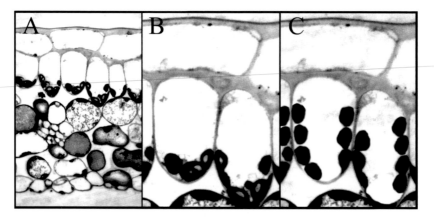

Figure 4.16 Influence of package (chloroplast) movement on the absorption of light and the color of a plant organ, using a leaf of *Triolena hirsuta* (fig. 4.13). A is the entire leaf section. B shows the normal daytime position of the chloroplast in shade, with a higher likelihood of intercepting radiation. C shows the position of chloroplasts along the side walls of a palisade cell, after exposure to light. This position reduces the likelihood of light interception and makes the leaves look lighter.

through transmission and reflection of the light. Thus the movement of organelles within cells is an effective mechanism for changing their optical properties.

Modeling Plant Optics

Knowing how the plant canvas works, how plant organs interfere with—reflect, transmit, or absorb—light, has some important consequences.

The economic value of some plant organs, flowers and fruits in particular, is enhanced by the intensity of their colors. Since color production is not simply the result of the production and concentration of pigments, knowing how structures interact with light, how they display the pigments in an optimal fashion, should help us improve the display of colors by plants.

Plants are the most visible organisms on our planet, and they provide the structure of terrestrial ecosystems. The plants in these ecosystems vary in optical properties in subtle ways. Knowing how environmental factors may affect optical properties, how lack of water could decrease reflectance, how an increase of nitrogen could increase green, how a disease or nutrient deficiency could decrease green, makes it possible to observe plants remotely to learn about their condition. Thus it should be possible to fly over a crop in an airplane or to photograph a forest from a satellite to obtain information about the health of crops and ecosystems. We still have much to learn about the optics of plant structures—and certainly a long way to go to interpret their spectral signatures.

The physiological anatomists had some appreciation for the refraction of light within plant organs. The classical—and still authoritative—study of plant tissue optics was first published by Richard Willstätter and his long-time associate Arthur Stoll. They published in 1917 a model of the optics of a leaf, along with the results of the analysis of the chemistry of chlorophyll (fig. 4.17). This model was the first to consider the complex patterns of internal reflectance within the leaf because of the odd cell angles and the accompanying air spaces. Before and since this publication, others have attempted to understand the leaf as a homogeneous medium, much as a solution of dye would absorb light in the exponential fashion explained by Beer's law. Still others have built on this old model, trying to make it more realistic and in the process making it more complex and difficult to program.

Because of the complexities of plant organs, there are no completely satisfactory models of their leaf properties to date. It is not difficult to under-

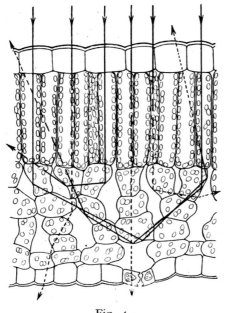

Fig. 4.
Gang der Lichtstrahlen im Blattgewebe.

Figure 4.17 The first model of leaf optics was developed by the chemist Richard Willstätter and his associate Arthur Stoll at the beginning of the twentieth century. This diagram accompanied the description in a comprehensive monograph published in 1918.

stand why. Once light enters the leaf, it encounters internal cell structures and intercellular spaces at crazy angles. These surfaces cause the partial reflectance of light, lengthening its path through the leaf. Furthermore, the reflectance from these surfaces is best described as diffuse, and not specular, adding to the complexity of the patterns. Such scattering of light, and the possibility of refraction's locally focusing light within the plant organ, may produce higher light intensity at certain points within the structure than that on the surface. These patterns of light absorption may be particularly important for certain plant organs. We will take up the optics of individual organs, along with their pigments, in individual chapters later in the book.

Chapter Five Patterns

....

We cannot remember too often that when we observe nature, and specially the ordering of nature, it is always ourselves alone we are observing.

G. C. LICHTENBERG

Flowers construct the most charming geometries: circles like the sun, ovals, cones, curlicues and a variety of triangular eccentricities, which when viewed with the eye of a magnifying glass seem a Lilliputian frieze of psychedelic silhouettes.

DUANE MICHALS

....

Plants produce very precise and elegant patterns, something like an artist placing pigments on a canvas. These include the repetitive elements of leaves, flowers, and flower parts, and they also include the patterns of color within those organs. These patterns of color are caused by the presence or absence of pigments, and other changes in optical properties (fig. 5.1). The patterns are also the result of developmental processes peculiar to plants, and the driving force is natural selection. Some patterns, such as the production of chlorophyll in the middle of the leaf and not on the surface, are not visible. Others, such as the intermittent chlorophyll distribution in leaf tissue of the dumb cane, discussed in chapter 4, make complex color patterns highly visible to us. We like these patterns, and have become agents in selecting them for our gardens.

These patterns are under genetic control, at least to some extent, and may involve different mechanisms of controlling pigments and structure. Since many of them

Figure 5.1 The complex patterns in the leaves of *Episcia cupreata* are set early in leaf development. Separation between the outer (epidermal) and adjacent layer produces the silver colors, and variations in the production of both chlorophyll and anthocyanins enrich the pattern.

are a result of the basic laws of inheritance, and of development unique to plants, it is helpful to review these subjects to gain a better understanding of color patterns.

Inheritance

The information that controls the division of cells and the formation of tissues, organs, and entire organisms is maintained in the nuclei of individual cells. It consists of polymers of DNA (deoxyribonucleic acid), composed of four different nucleotides (adenine, thymidine, cytosine, guanidine) whose specific sequence is the basis for this information. In the nucleus it is normally found as a double helix, along with a second strand of antisense nucleotides. (As I trot out the basic DNA dogma, a bit more than fifty years after its discovery by James Watson and Francis Crick, I am reminded that my high school biology text made no mention of DNA.) In the expression of genetic information, lengths of DNA are unraveled, which allows the copying (transcription) of this information to a complementary length of messenger RNA (mRNA; uracil substitutes for guanidine in these molecules). Each sequence of nucleotides is information for the production of a unit of inheritance, or a gene. The actual product is almost always a protein, an enzyme or a molecule that alters the activation of other genes. Genes reside within the long polymers at specific locations (or loci) on chromosomes. In the normal cells of a plant (and in us as well), such chromosomes occur in pairs, as do the genes. Our cells have forty-six chromosomes, twenty-two pairs and an additional pair that determine sex. The chromosome numbers of different plants vary greatly.

In individuals, slightly altered genes may occur at a single place on a chromosome. These may express different characteristics for a trait, such as flower color or leaf shape; these are called alleles. The laws of the expression of these genes, and alleles, were established by Gregor Mendel in the nineteenth century, ignored and forgotten, and then rediscovered and appreciated early in the twentieth century. Mendel found that some alleles are dominant in their expression, and others are recessive. He found that the allele for pink pea flowers (P) was dominant over an allele for white flowers (p). White flowers occurred only when both chromosomes contained the p allele (homozygous recessive). However, pink flowers occurred when both chromosomes had the P allele (homozygous dominant), or with a mixture of P and p (heterozygous).

On a molecular level, that particular sequence of DNA is transcribed as mRNA, and then translated into a protein gene product (perhaps an enzyme in the anthocyanin pathway) that begins to operate in the cytoplasm (fig. 5.2). The sites of the protein synthesis (translation) are ribosomes, on membranes in the cytoplasm, the endoplasmic reticulum.

Later on, many exceptions were found to Mendel's rules of inheritance for combining genes. First, Mendel was lucky in that all of his characters (flower and seed color, etc.) were controlled by genes on different chromosomes. If they had been on a single chromosome, they would have mainly been inherited together. Second, not all alleles function in a dominance-recessive relationship. In some cases a dosage of alleles is important. So the homozygous gene for anthocyanin production in red flowers (RR) might be weakened in a heterozygous combination (Rr) to produce pink flowers, with rr producing white flowers. Finally, the neighborhood of a locus on a chromosome may affect the expression of alleles at that position, switching them off or reducing their activity.

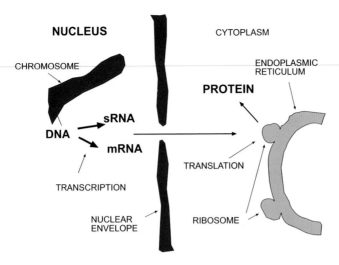

NUCLEUS   CYTOPLASM

CHROMOSOME   ENDOPLASMIC RETICULUM

PROTEIN

sRNA

DNA

mRNA

TRANSLATION

TRANSCRIPTION

NUCLEAR ENVELOPE   RIBOSOME

Figure 5.2 The central dogma, the expression of genes from DNA polymers in the nucleus, transcribed as RNA, and translated as proteins in the cytoplasm.

Slight changes in a normal gene alter its function in a disruptive way, usually lethally. Such changes, caused by a nucleotide substitution, loss of a nucleotide, or a break in the polymer due to physical damage of the chromosome, are called mutations. A small percentage of such mutations may be valuable under certain circumstances, and accumulate in the process of natural selection.

The coordinated activity of many genes makes enzymes within a metabolic pathway that leads to the final product, such as an anthocyanin. This activity involves the expression of genes that promote making enzymes, and involves genes that suppress the production of enzymes when the pigment's concentration is high. Although the vast majority of genes controlling plant color are located in the nucleus, a small minority are associated with organelles, especially the plastids, chloroplasts and chromoplasts. Plastids contain a small amount of DNA, enough for a few genes, although most genes making plastid parts are in the nucleus. In those cases it is possible for pigmentation in tissues to be disrupted by a mutation in the plastid.

This is a very brief summary of an enormous topic, taught in every middle school and high school biology course. It is also the minimum necessary for discussions on the control of patterns in this chapter, and in later chapters.

Plant Development

In sexual reproduction, flowering plants form a single fertilized egg, or zygote, within the ovule (fig. 5.3). There may be one to thousands of ovules within the ovary. This zygote undergoes a single transverse cell division, the distal cell becomes the embryo of the seed, the seed develops from the ovule, and the rest of the fruit is derived from the ovary wall. During development, the embryo is sustained by the surrounding endosperm tissue. At maturity the embryo consists of an embryonic root, an embryonic shoot, and one or two embryonic leaves, or cotyledons. These regions of cell division are called meristems. The root meristem produces the root system of the plant. The shoot meristem produces the branch system, leaves, and ultimately flowers and fruits. The cotyledons function only for a short time, and fall away.

The fertilized egg contains all of the genetic information necessary for the entire organism to develop. Since all other cells result by mitosis from this one cell, during which the genetic material is copied, every cell contains the same genetic information and has the capacity to develop into an independent organism; every cell is *totipotent*. Totipotency involves the timely expression of genetic information present in the nucleus, to coordinate the development of the organism, through the division, elongation, and specialization of cells. The selective expression of certain genes makes different cells, tissues, and organs.

The copying of DNA before cell division, and its transcription and translation, are highly conservative and accurate processes. During cell

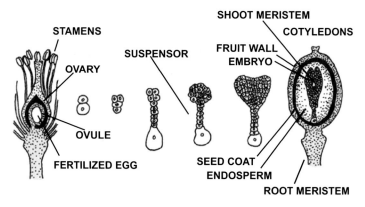

Figure 5.3 The development of an embryo from a fertilized egg.

division, mistakes are rarely made, and they may be retained as mutants in individual cells of the growing tips of the plant, the meristems. Genes are also present in organelles, and their distribution among cells is a consequence of their multiplication and, subsequently, of cell division. Thus mutations may also accumulate in organelles. If present in the meristems, a mutation (such as a chloroplast that can't make chlorophyll) will spread with cell division. If a mutation resides in organelles in the sexual cells, it will spread mainly from the maternal side, since the egg has most of the cytoplasm where organelles can accumulate.

The process of embryo development in plants is fundamentally different than that in animals. When plant cells divide, they retain their relationships to adjacent cells. Their walls make them relatively inflexible, and connections between the cells allow interactions between the cytoplasm of these adjacent cells. In contrast, in animal embryo development, cells are flexible and move. Cells move to another part of the embryo to specialize as the tissue of an organ. Furthermore, animal development is largely determinate. The organism develops into a definite size and shape, only slightly modified by the environment in which it grows. Plant growth is largely indeterminate. The shoot meristem develops into the aerial portion of the plant, which grows indefinitely. The root meristem similarly develops into a root system. The key to understanding this indeterminate development is to look into the structure of the embryonic tip of the young shoot.

The apical meristem, or bud, consists of layers and regions of dividing cells (fig. 5.4). The outer layer tends to divide anticlinally, or parallel to the surface. The inner layers tend to divide periclinally, or perpendicular to the length of the shoot. Occasionally, cells in each region will divide in the opposite plane. Divisions near the tip produce outgrowths of tissue, the leaf primordia, which develop into the future leaves of the shoot. Different layers of meristematic tissue, from superficial to deeper, are involved in the production of leaves and flower parts. The outer layer divides mainly horizontally and produces the outer layer (epidermis) of these organs. The inner layers divide both horizontally and vertically and produce the internal tissues of the organ, including the conductive tissue (xylem and phloem). Leaves and petals thus contain cells derived from both the outer and inner layers of the meristem.

Lateral buds form at the base of every leaf. Derived from the shoot tip, they look like the meristem at the tip of the shoot. They can develop into leaves, flowers, or lateral branches. As the seedling grows, it initially produces an erect shoot with leaves along its length. Most plants develop lateral

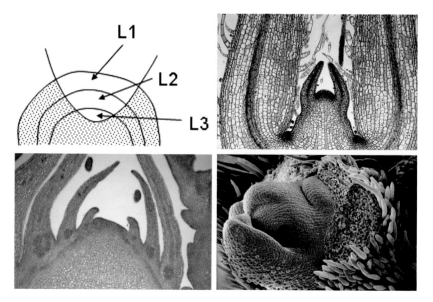

Figure 5.4 Shoot meristems. *Top left*, the three embryonic layers. *Top right*, coleus (*Plectranthus*), a radial section clearly showing leaf primordia, lateral meristems, and the meristem. *Bottom left*, a radial section of castor bean (*Ricinus communis*), showing the shoot meristem and leaf primordia. *Bottom right*, a scanning-electron-microscope photograph of the shoot meristem of green champa (*Artabotrys hexapetalus*), a fragrant vine of India. Courtesy of Usher Pozsluszny.

branches, and a spatial architecture, when the lateral buds become active. Eventually, buds on the lateral branches produce secondary branches, and so on. The pattern of branching is under genetic control, and is influenced by growth-regulating molecules. All meristems remain active until they are physically damaged or until they convert to reproductive meristems, producing flowers and fruits. Once the fruits develop, the meristematic activity is lost. If all of the buds become flowers and then fruits, the plant dies. A few large palms may grow 30 meters high in forty years, flower and fruit, and then die. The key to understanding the indeterminate growth of a plant is to see the plant more as a population of individual meristems, each with some autonomy. A large tree contains many thousands of such meristems in its aerial branches, and a similar number in its roots. Shoots, with the meristematic region at the tips, continue to produce leaves, grow lengthwise, and thicken, during the life of the plant. The meristematic tissue is lost only if physically damaged, or if the bud is converted from vegetative growth to reproduction.

Leaves in a vegetative bud, and flower parts in a floral bud, form in a similar fashion. In the leaf production, a small sector of tissue pushes out just beneath the apical dome. Such outwelling occurs from the periclinal divisions of cell layers beneath the surface. As the projection continues to expand into a peglike structure, its upper and lower sides are established. It continues to lengthen, and the blade of the leaf forms from the division and expansion of cells along the side. In contrast to the bud, the leaf is a determinate structure; it completes its growth and eventually dies and falls off the plant.

The leaf blades of the two fundamental divisions of the modern flowering plants, the monocots and eudicots, generally develop differently. In eudicots the leaves establish and develop as above, with cell division along the sides of the growing leaf. In monocots, such as the grasses, a meristematic region normally forms near the base of the developing leaf. This continues to produce files of cells that expand and mature toward the tip. Such growth is the basis for the suburban ritual of lawn mowing. Excellent examples from temperate gardens are hostas, of numerous varieties, several with white stripes all along the outer margins of the leaves. In tropical areas, something similar is seen in the variegated varieties of the ti plant, with stripes of color or white along the length of the sword-shaped leaves. Those white, or red, regions are derived from cells in the basal meristem. They continue to produce the same cells that accumulate for the length of the leaf.

Patterns and Development

Given the genetic control of our development, it is easy to envision the coordinated promotion of genes in pathways to synthesize pigments that give plants color. Producing green leaves from the synthesis of chlorophyll in chloroplasts, or making flower petals pink from accumulating anthocyanins in cell vacuoles, seems an easy biological problem to solve. All cells in the plant possess the genes for enzymes to synthesize chlorophylls, and cells in most plants also possess genes for the enzymes to synthesize anthocyanins. Those genes are controlled by promoter or inhibitor genes, so that pigments are produced in certain organs. Thus chlorophyll is produced in chloroplasts of the leaf, but not in the petals of the flower of the same plant. Special carotenoids may be produced in chromoplasts of the petals to create an orange-red color, or anthocyanins may accumulate in the vacuoles of petal cells to produce a pink or red color. These genes are normally inactive in leaf tissues.

Plant tissues form through the influence of growth-regulating molecules, along with gradients of sugars and other common molecules, beneath
the meristematic region of the shoot. The veins of leaves and petals are
established in this way. These specialized tissues may then influence the
production of pigments in their cells, compared to cells of different tissues
surrounding the veins (fig. 5.5). Color, or its absence, is often associated
with veins in plants. Such color may also be affected by the timing of the
organ's formation, so that color may vary from the earliest portions of the
organ to form, to the last. Thus color patterns of many flowers, as in hybrid
orchids, and in some variegated-leaf plants, as in the small understory fittonia, are closely associated with veins (fig. 5.6).

In dumb cane, the plant whose optical properties were explored in chapter 4, varieties have different patterns of variegation. The variety 'Wilson's
Delight' has a striking contrast between the white midrib and major veins
and the dark green leaf blade (similar in appearance to figure 4.11, but with
a larger area of white). The white veins are controlled by a single dominant

Figure 5.5 Patterns of leaf variegation in monocots. *Left*, pink, white, and green stripes
in a ti plant variety (*Cordyline fruticosa*). *Right*, white stripes on margins of a hosta (*Hosta*
sp.) derive from the outer leaf layer early in development. Cells derived from a basal
leaf meristem produce the long stripes.

Figure 5.6 Many patterns in leaves and flowers are associated with the underlying distribution of veins, which, in turn, are established in gradients of growth regulators and nutrients. In silver nerved plant (*Fittonia verschaffeltii, top*), the bright pinks along veins stand out strongly against the dark green of the remainder of the leaf. In the orchid hybrid *Ascocenda* (*bottom*), darker purples on the petals denote the underlying veins.

nuclear gene. When plants of 'Wilson's Delight' are self-reproduced, they produce plants with a ratio of white to green midribs of three to one. This pattern is controlled by a single nuclear gene (with dominant and recessive alleles), obeying the rules of Mendelian inheritance.

Most often, however, the pigments or other visual effects (like air spaces just beneath the surface) are produced in very small areas of a leaf, petal, or fruit. Such patterns, associated with the development of organs in the plant, are much more difficult to comprehend. Since plant colors are often produced in such patterns, we will spend some time discussing the mechanisms for their formation. These fine-grained patterns occurring within an organ are mostly related to the mechanism of cell division and formation of the organ. The patterns of color production in all parts of the plant depend upon the process of development in the shoot meristems of the plant. Any mechanism that makes certain cells of the meristem different, particularly in the production of pigments, can produce color patterns in the mature organ because of the fates of products of cell division. Such color patterns may be caused by infection, the accumulation of somatic mutants in plants (both in the nucleus and in organelles), the expression of nuclear genes, and the presence of very unstable and mobile genetic elements that affect the activity of normal genes.

Infections

Occasionally infections from disease may produce persistent patterns in plant tissues. Some may border on attractiveness, although most make plants look blotchy in ugly ways. Most notable among pathogens that produce patterns are the viruses, particularly those that produce mosaics or that cause flowers to "break," or make white streaks in otherwise colorful petals. Viruses are infectious particles, containing a core of DNA (or sometimes RNA that is reverse-transcribed into DNA) inside a capsule of proteins. Viruses may infect plants through insect vectors, particularly those that suck their sap. Some viruses cause wounds (like ring-spot viruses) near the sites of infection. Others may spread throughout the plant. Once inside the plant, particles may move from cell to cell to reach different regions, or may only attack certain plant organs. Many viruses may be transmitted through pollen grains or other means, to infect the embryos and be transmitted through the seeds. Since the germinating seeds appear to produce the same patterns as their parents, such patterns appear to be inherited, but they are not.

Many of the viruses that produce patterns in leaves and petals infect cells of the meristems. Those infected cells divide to form sectors of tissues of different colors, usually yellow or white from the destruction of chloroplasts. The patterns of white on pigmented organs help to reveal when the virus infected the meristem. Larger patterns suggest a much earlier infection. Viruses that cause such patterns are the mosaic viruses, widespread in many crop plants. These viruses are similar in the symptoms they cause. They infect some of our most common vegetables, such as cucumber mosaic virus, tobacco mosaic virus (which infects tomatoes and other vegetables), and maize streak virus. One of the most popular root vegetables in Miami (and throughout the Caribbean and Latin America) is malanga (or dasheen). This member of the arum (or aroid) family is grown by farmers south of the city, and crops are frequently infected with dasheen mosaic virus. It produces pronounced variegation in the leaves of the plant. The pattern results from the inability of infected cells to produce functional chloroplasts, and the areas of white result from the division of those infected cells.

Some viruses produce very striking and attractive patterns in petals. Such mosaic viruses affect the reproductive meristems by attacking individual cells, which continue to divide and spread the virus to the daughter cells. Most fa-

Figure 5.7 Broken, or virus-infected, flowers of the common tulip (*Tulipa* hybrid). Some of the flowers are normal, with solid colors. Those infected by the breaking virus have streaks of different colors running the lengths of the petals. Photograph courtesy of Allan Meerow.

mous among these is the virus that causes the petal-color streaks in tulips, the tulip-breaking virus (fig. 5.7). This virus can be spread from infected to normal plants by small sucking insects, particularly aphids. Infected cells are deficient in their ability to produce anthocyanins, and they spread this to daughter cells through division. The infection gives the petals the finely streaked appearance. When tulips were the rage in Holland in the seventeenth century, broken tulips were particularly valuable, and were selected for many bouquets in the still-life paintings of the Dutch masters. Such plants still attract our curiosity but must be kept away from the intensely colorful cultivars, to prevent the purity of those colors from being "broken."

Chimeras

With so many thousands of regions of dividing cells throughout the aerial branches and root system of a plant (such as a large tree), and thus millions of dividing cells, the probability of mutations arising in these tissues and passing the errant genes to daughter cells is not negligible. Most of these mutations are lethal to the cell, which perishes before dividing. Some mutations could be deleterious if present in the entire plant, but may persist

among normal cells in the dividing shoot apex. Horticulturists have identified many shoots with advantageous mutations, and have propagated them
to produce new plant varieties.

Such mutations that alter pigmentation may produce striking patterns
in plants. Most notable are mutations that prevent the proper development
of chloroplasts and the production of chlorophyll (fig. 5.8). These produce
areas of tissue that are a brilliant white on a dark green background. Such
mutations may also alter the production of anthocyanins to form variegated
patterns of reds and pinks (fig 5.9).

The location of the cell with the mutation in the meristem determines
the type of pattern. The mutation may occur throughout an entire layer of a
meristem, or in a small sector of a layer. Since these layers may divide parallel
or perpendicular to the surface (or possibly both), the layer and divisions
will determine the extent of the blotches of altered color. Three basic types
of chimeras (tissues with different genotypes) may be produced, based on
cell layers in the meristem (fig. 5.4). If the mutation occurs solely in the outer
layer of the meristem, subsequent divisions will produce a surface entirely
of mutated cells, with interior cells that look normal. If only certain cells are
established in an area of this outside layer, the result is a pattern of color. If
mutated cells occupy a sector of the organ, large blotches of different color
appear. The timing of formation and presence of different cells in the layers
and sectors of the meristem produce different color patterns (fig. 5.10).

Such chimeras may occur in plants in the wild, or on a single branch of an
otherwise normal shrub or tree. Variegated leaves of hosta are the result of
chimeras. The white margins of the leaves are derived entirely from the superficial layer, and the rest of the leaf from the superficial and deeper layers.

Figure 5.8 Many variegated ornamental plants are selected for presence and absence of
chloroplasts in leaves. Mutants can be under genetic control of the nucleus and due
to mutations in the chloroplasts. *Left*, the variegated *Pieris japonica* shrub is popular in
gardens in the northwestern United States. *Right*, the variegated variety of the golden
dewdrop (*Duranta erecta*) is commonly grown in the tropics and south Florida.

Figure 5.9 Very repetitive patterns are seen in certain plants. *Left*, the larger lower petal of the poinciana tree (*Delonix regia*) is always mostly white in the center, with flecks of white and yellow. *Right*, leaves of different coleus varieties (here *Plectranthus* 'Kong Red Coleus') have unique, accurately repeated patterns. Photograph courtesy of Ball Horticultural Company.

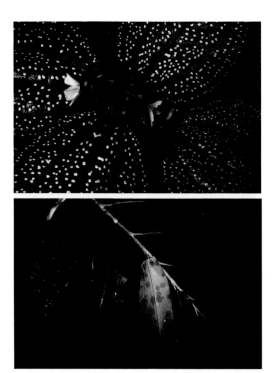

Figure 5.10 These small points of color probably occur later in the expansion of leaf primordia, and expand very little because most of the cell division has already occurred. *Top*, in *Sonerila margaritacea*, a rain-forest understory plant from Southeast Asia, white areas are derived from cells that lack chloroplasts. *Bottom*, in *Aframomum longipetiolatum*, a ginger from West Africa, the reds start out as very small points and then spread out in all directions. The red spots eventually fade in older leaves.

Control by Plastids

Many color patterns in plants are due to varying capacity to produce colors, such as the greens of chlorophylls in chloroplasts or the yellows and oranges of carotenoid pigments in chromoplasts. Much of that capacity is controlled by genes within the cell nucleus, but a small part is present within the organelle in its small loop of DNA. Plant cells have relatively small populations of plastids, few to a dozen or so. If one of the plastids is mutated, so that it cannot produce pigments, more mutated plastids may accumulate in the cell from the division of the plastids. As cells divide, some daughter cells accumulate more, or all, of the mutated plastids. Such cells lose their ability to produce color, and produce colorless sectors in organs that developed from that single meristem. Such mutations can pass from generation to generation, because mutated cells could become a sexual gamete in plants. However, since the male gamete (the sperm nucleus) usually carries very little cytoplasm during fertilization in contrast to the much larger egg cell, such plastid mutations are inherited maternally.

Many color patterns in leaves are caused by cells without functioning chloroplasts, thus creating white blotches in the leaf blade. As the normal and mutant chloroplasts are sorted out with successive cell divisions, the color patterns start out in fine textures and intermediate colors and end in large blotches of pure white in a blade of green. Such patterns are very common in ornamental plants; just about every popular ornamental has some white variegated varieties.

Patterns under Nuclear Control

Rather than being caused by somatic mutations within the apex, some effects on cells in the shoot meristems are controlled directly by the nucleus. This can be determined by crossing experiments, where the contribution of the color pattern comes from either the male or female parent. Nuclear genes can produce patterns in several ways. Disruptions in cell connections produce air spaces and areas of metallic or silvery color. Anthocyanin accumulation in cell vacuoles produces areas of intense red on a pale background, or pale areas on a red background. Finally, nuclear control of pigment production in plastids can result in absence of color in certain cells, and in areas of the plant organ. Sometimes control of anthocyanin synthesis, plastid function, and the formation of air spaces can together produce even more complex color patterns.

In the dumb cane (see chapter 4), wild plants from the Neotropics have variegated patterns of white, yellow, and green on the leaf blades. The white color is caused by a lack of chlorophyll in plastids. The capacity to produce such variegation is not maternally controlled, but is promoted by a single dominant nuclear allele. This is seen in simple breeding experiments of different varieties. If the cultivar 'Perfection' is self-pollinated, the seedlings produce variegation and solid green appearance in a ratio of three to one. This suggests that a variegated and green allele are present in this cultivar. There are some subtle differences among these cultivars in the size and number of white blotches, suggesting that still other genes may modify the expression of variegation. Much mystery surrounds the actual mechanisms of expression of the nuclear genes controlling color in small areas of a plant organ. Although the organization of the shoot tip helps explain how patterns can form as the organs develop, the mystery is how the nuclear gene can target individual cells in the shoot tip.

In the understory of tropical rain forests of Southeast Asia, aglaonema grows as an ecological substitute for the New World dumb canes. These plants are also variegated and have been studied genetically to improve their pattern formation. In aglaonema, two separate loci, both with dominant and recessive alleles, control two different color mechanisms, and together produce some very subtle patterns of variegation (fig. 5.11).

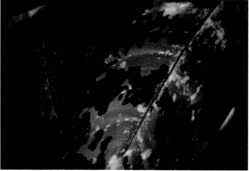

Figure 5.11 Patterns of variegation in aglaonema (*Aglaonema commutatum*), a rain-forest understory herb of the aroid family native to Southeast Asia. *Top*, the pattern of variegation, silver on green, is due to a slight separation of cells between the upper and middle cell layers, and is under the control of a single locus. *Bottom*, the complex pattern of variation in this variety is due to a partial loss of chloroplasts in two different layers of the leaf, both under separate genetic control, so that the differences may be superimposed.

Mobile Genetic Elements

Normally mutations that produce color variations in plants occur at a very low frequency. Breeders increase the rate of their appearance by promoting mutations through irradiation or by treating with chemicals that induce mutations. Many of the color patterns of horticultural varieties were formed by such means, and selected by breeders. In some cases, such mutations occur at rather high frequency. Their analysis has led to the discovery of natural means of enhancing the formation of mutants through the action of mobile genetic elements.

The discovery of mobile genetic elements was one of the great scientific detective stories of the twentieth century. It resulted in the award of the Nobel Prize in Physiology or Medicine (of all things) for Barbara McClintock's research on inheritance in corn (*Zea mays*). This is the only Nobel Prize awarded for fundamental biological research on plants.

McClintock studied inheritance patterns in the kernels of maize cobs. Anthocyanins accumulate in the outer layers of the kernels in some plants. Kernels are actually individual fruits, or grains, on the long cob. Intensity and hue of color are affected by the amounts of anthocyanin in layers of the kernels. The patterns of color, particularly the blues and reds, were of interest to Native Americans in the Southwest. Colored maize kernels were used in ceremonies of worship, such as the corn dance of the Hopi. The varieties kept by Native Americans were the stocks for breeding experiments to learn about the inheritance of color patterns. Different-colored grains are produced in crosses of races with dominant and recessive genes for kernel color. These are striking in appearance (fig. 5.12) and are inherited in simple Mendelian ratios.

McClintock made such crosses, and observed that some of them caused variegated color patterns *within* individual kernels, as well as among the kernels on a corn cob. She focused her attention on the nature of the genetic controls on these patterns within kernels, and this led to the discovery of mobile genetic elements, transposable genetic elements—or transposons for short. She discovered genes that could move from location to location on a chromosome, or jump to another chromosome (jumping genes, as some called them), and alter the expression of the gene adjacent to where it was inserted. Her discoveries were met with dismay by traditional geneticists, and she worked patiently and brilliantly in obscurity. Her discoveries were aided by her intuitive knowledge of the maize plants she studied. She published her work in respected journals, but not many read the articles.

Figure 5.12 Inheritance patterns in maize grains. *Top*, differences in the production of anthocyanins in the grains is under genetic control, and we are seeing the patterns of inheritance among different individuals. *Bottom*, patterns of color seen within grains are controlled by a transposable element. Photograph courtesy of Virginia Walbot.

Much later, mobile elements were discovered in bacteria, and then in the fruit fly, *Drosophila*, and in other animals. The techniques of molecular genetics made it possible to analyze these elements as short, unique sequences of DNA with the capacity to insert themselves into the DNA on chromosomes, and the mechanism of the action of transposons was revealed. So, many years after making her original discoveries, McClintock received her Nobel Prize. Virginia Walbot later elucidated the mechanism of color production of the *Mu1* transposable element *Seeker* in maize, as illustrated in figure 5.12 (bottom).

Transposable elements can be important in producing color patterns in plants because they can induce mutations in cells of the apical meristems. Subsequent cell divisions can then produce patterns of variegation similar to those produced by other means. Several such systems have been discovered and analyzed in plants. In addition to maize, we have learned quite a bit about the control of flower color and pattern in two plants (fig. 5.13), the petunia and the snapdragon.

In snapdragons, an allele called *pallida* encodes an intermediate required for anthocyanin synthesis, and it makes the flowers red. Adjacent to the

Figure 5.13 Patterns of color formation in the petals of the petunia (*Petunia hybrida*, *left*) and the snapdragon (*Antirrhinum majus, right*), model organisms for the study of pigmentation and pattern formation in plants. Both photographs courtesy of Ball Horticultural Company.

pallida allele (upstream, as we say) lies another gene that promotes the synthesis of this product. The more actively the promoter gene is transcribed, the redder the corolla becomes. When a transposable element, called *Tam3*, is inserted upstream from the promoter and *pallida* allele, no anthocyanins are produced in that particular cell (fig. 5.14). As a transposon, *Tam3* is very unstable and can easily be excised from chromosomes of individual cells of the reproductive meristem. When removed from cells in the very young petals, those cells divide and expand and then produce a red spot on the petal. The instability of this transposon can thus produce a series of different color patterns in snapdragons. The transposon is suppressed at high temperature. Thus the cells where it appears can be induced to produce anthocyanins when the temperature is increased during development. Timing the suppression of the mobile element at different stages of petal formation produces different patterns. Earlier suppression would produce fewer and larger splotches as the affected cells divide, and later suppression would produce more and smaller dots of red on the pale pink petal. The transposon thus makes it possible to look at the patterns and directions of cell division in the snapdragon petal, and perhaps in other plants and in leaves as well. This research was led by Ann-Gaëlle Rolland-Lagan in the laboratory of Enrico Coen, one of the world leaders in research on the genetic control of patterns in plants.

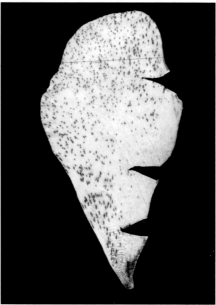

Figure 5.14 Patterns produced by a transposable element in the snapdragon. *Left,* flowers of the *pallida* mutant, with the transposon producing streaks of color. *Right,* flecks of red (anthocyanin production) appear later in a petal that had been warmed to suppress the transposable element earlier in development. Photographs courtesy of A.-G. Rolland-Lagan, J. A. Bangham, and E. Coen.

Why Patterns?

We certainly seem to be obsessed with patterns. Any unusual pattern observed in plants in natural conditions is noted. Such plants, particularly if they differ among more uniform neighbors of the same species, are collected and brought into cultivation. These plants are the raw material of breeding programs to create other horticultural curiosities. Such newly patterned plants are constantly appearing in the marketplace, to compete with their more plainly decorated predecessors. We buy them to decorate our homes and offices.

Certainly the novelty of these plants is attractive to us, adding to the enormous variety of plants brought into cultivation from distant parts of the globe. Where the patterns involve color, as in flecks of gold, pink, or red on normally somber green foliage, our fascination perhaps increases. Such plants may allow us to put more color into our gardens more permanently, without having to bother with the ephemeral displays of flowers. Perhaps we empathize with the order of certain patterns in plants, hinting at some animal-like intelligence. Perhaps it even resonates with the order of neural networks in our brains, like some psychedelic vision brought on by consuming a plant elixir.

In the following chapters we will see these patterns in the various organs of plants; many of them are striking and bizarre. Our task will be to understand the natural history of color and patterns, how they might have evolved in plants, what selection pressures led to their establishment, and the possible physiological advantages conferred on the plants—their basic raison d'être.

Chapter Six Leaves

···

I believe a leaf of grass is no less than the journey work of
the stars.

WALT WHITMAN

The turning spindles of the cells
weave a slow forest over space,
the dance of love, creation,
out of time moves not a leaf,
and out of summer, not a shade.

KATHLEEN RAINE

···

We care enough about greenery to grow plants like
the umbrella tree (fig. 6.1) in our yards in Miami,
far from its natural home in the tropical rain forests of
northern Australia. It finds the local climate agreeable and
does well here. In the farmlands south of Miami, dozens of
nurseries grow these plants, pack them into semitrailers,
and ship them to the discount stores of northern climes
where they dress up our bleak winter interiors. These
plants do well in those offices and mall atria because they
are adapted to survive in the understory of their tropical
forest homes. Their leaves are a handsome, shiny, dark
green, and I'll use them to convey some principles about
leaf structure.

The green of foliage is a symbol of life on our planet,
an indicator of life's productivity. We all know that normal healthy leaves are green. Leaves vary in hue, however,
sometimes subtly and sometimes dramatically. We may be

Figure 6.1 Leaves of the umbrella tree (*Schefflera actinophylla*). It is widely planted as an ornamental tree in the tropics and used to decorate offices and homes at higher latitudes.

conscious of our ability to distinguish among the hues of green foliage even while driving full speed down an interstate highway.

In order to begin to understand the reasons for the differences among leaf colors in different plants, it is necessary to understand some basic principles of the functions of leaves, including photosynthesis. Then we'll look into the natural history of foliage colors in plants.

Leaves as Photosynthetic Organs

Leaves are expendable organs, and their function is nicely understood in economic terms. Leaves are investments of capital, allocations of energy, carbon, and expensive nitrogen. The result is an organ that captures light energy, converts it to electrical and then chemical energy, ultimately producing many carbon-based molecules. The return on the capital investment is the production of sugars and other molecules used by plants as a source of energy and as building blocks in constructing the entire organism. During their life spans, leaves recover the initial investment and accumulate considerable profit for the plant. Such profits continue as long as leaves are healthy, but decline as they age and wear out. Eventually the production (and profit) is reduced, and the plant discards leaves. Depending upon the environment, leaves may last just a few months or up to several years. In evergreen tropical rain forests, leaves generally last for a few months to a year, fall off, and are continually replaced by new leaves.

Let's consider that umbrella tree again (fig. 6.2). Its leaves are large and produced at the tips of thick branches on a sparingly branched tree (a fact that makes them easy to count). The "umbrella" name comes from the palmate leaflets extending out from a center point (like seven fingers radiating out from a palm). Its leaves last approximately a year, depending

Figure 6.2 Umbrella tree leaves at two scales. *Top*, the crown in a Miami neighborhood tree. *Bottom*, a transverse section of a fresh leaf.

upon their exposure. The umbrella tree in this photograph has about 600 leaves, totaling an area of 120 square meters for absorbing sunlight. During daylight hours the leaves capture carbon dioxide in the atmosphere (and produce oxygen molecules in equal measure) at a rate of some 3.6 milliliters of gas per hour, per square meter of leaf. This doesn't seem like much, but it quickly adds up for the tree over a single day. This means a capture of carbon dioxide of 1,200 liters (with the same amount of oxygen liberated), and the production of glucose sugar at around 1.4 kilograms (or 3 pounds) per day. At night, the leaves respire, reducing the net production of oxygen by a fraction. The excess production of photosynthesis goes to the growth of the tree, replacement of old leaves, and ultimately to the production of flowers, fruits, and seeds.

In contrast, how do we function energetically? The sugars fixed by this tree would provide the energy requirements of a normally active adult male of fairly large stature (myself), although these products are eventually assimilated by the tree as cellulose and secondary compounds and are not an appropriate diet for us. I breathe in some 15 cubic meters of air a day, taking in about 1,000 liters of oxygen gas (and breathe out a little less carbon dioxide, depending on the amount of fat in my diet). So this tree is a close metabolic match for me. My yard has the vegetation to match the activity of my family, but not if the enormous consumption of oxygen (and production of carbon dioxide) in driving our automobiles and consuming electricity is added. My economical Toyota Corolla generates 1,000 cubic meters of carbon dioxide per year (a little less than 2 metric tons) in local driving, not to consider the production of carbon dioxide in generating electricity, 308 liters per kilowatt-hour, or 11 tons total from a mixture of fossil and nuclear fuel. It would take a small forest of umbrella trees to match that consumption.

The umbrella tree also loses an enormous amount of water, as vapor from its leaves, some 300 molecules of water for every molecule of carbon dioxide diffusing into the leaf. Thus on a given day, this umbrella tree loses about three barrels of water. This loss of water helps the tree in two ways. First, it cools the leaf. Excessive heat can damage the photosynthetic tissue, and the evaporative cooling is essential for leaves to function well. Second, the cohesion of the water molecules rising through the trunk pulls a column of water up through the plant, via the xylem, and into the leaves. This column carries minerals in solution, including nitrogen, which is assimilated by the leaf as chlorophyll and the enzymes of photosynthesis. Consequently trees keep the surrounding environments cool, as a bystander on a tree-shaded city street can appreciate on a summer day. When water evaporates, it absorbs a large amount of energy, just as it takes the same amount of energy to boil water. Thus for any leaf to function optimally, it faces the trade-off between its requirements for carbon dioxide and the limited amounts of water available. Not surprisingly, photosynthesis is sensitive to many environmental factors. These include light intensity, temperature, carbon dioxide concentration, and water and nutrient availability. Photosynthesis increases with higher light intensity, up to a maximum, but may be inhibited by very high intensities. At very low intensities (particularly low for the umbrella tree), photosynthesis may equal the respiration by the leaf with no net carbon intake (the light compensation point). In darkness, leaves produce carbon dioxide through respiration. Light responses vary in plants adapted for different light environments, and in leaves within a single crown that have become acclimated to different light environments. Leaf color is important because it is the result of the amounts and wavelengths of radiation absorbed by the leaves, helping to balance their energy. Too much energy may damage the leaves, and too little may be insufficient for photosynthesis.

Inside the Leaf

To examine the process of photosynthesis in more detail, we shall explore the inside of an umbrella tree leaf. These leaves vary in size and thickness in different parts of the canopy; those more exposed to sunlight become smaller and thicker. They all have similar anatomical features, seen in transverse section (fig. 6.2). The upper surface is covered by a waxy layer (cuticle), secreted by the epidermal layer just beneath. Beneath this outer cell layer is an area of nonpigmented cells, the hypodermis. The photosynthetic cells come in two types and layers (fig. 6.3). Above is the palisade parenchyma (long and thin like the basalt rocks of the Palisades along the Hudson River), and below are the irregularly shaped cells of the spongy mesophyll. The cytoplasm of these cells is full of chloroplasts, surrounding the large central vacuoles (fig. 6.4). The lower epidermis of leaves contains special pores, the stomata (one is called a stoma), which open to the extensive air spaces connected to the surfaces of the photosynthetic cells. Stomata occur on the lower surface in high density, accounting for the high capacity for gas exchange and particularly the enormous loss of water through transpiration. In a typical umbrella tree leaf the density is 25,000 per square centimeter, or 15 billion for the entire tree.

Leaves are optical organs that absorb light energy for photosynthesis, and that light-capturing process takes place on the membranes (stained dark in fig. 6.4, left) within the chloroplasts. In these organelles, the absorbed light is funneled into two reaction centers (photosystems I and II) and converted into a flow of high-energy electrons, which drives a gradient

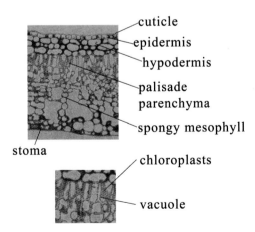

cuticle

epidermis

hypodermis

palisade
parenchyma

spongy mesophyll

stoma

chloroplasts

vacuole

Figure 6.3 The anatomy of an umbrella tree leaf: the entire transverse section, with major tissues identified, and a detail of palisade parenchyma cells.

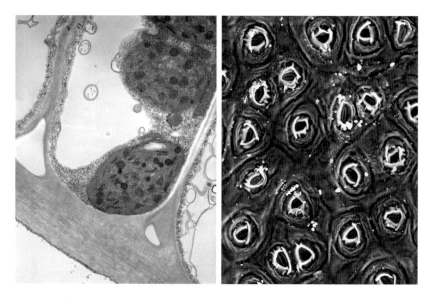

Figure 6.4 Electron-microscope photographs of an umbrella tree leaf. *Left*, a palisade parenchyma cell, showing chloroplasts with dark grana stacks and large vacuole. *Right*, the undersurface of the leaf, revealing the high density of openings (the stomata), each surrounded by two guard cells.

of hydrogen ions, which in turn is converted to chemical energy and finally into sugars. The high-energy compounds produced in the light reaction of photosynthesis feed the dark reaction, where carbon dioxide is fixed in simple organic acids and eventually glucose. The dark reaction takes place in the chloroplast, but free from the membranes.

In photosynthesis, carbon dioxide (CO_2) moves into leaves from the atmosphere, and water (H_2O) moves up the stems and into leaves via transport. We can consider the product of photosynthesis to be glucose $(C_6H_{12}O_6)$, although other molecules can be made. Oxygen, so important to all life on our planet, is a by-product of photosynthesis, harmful if retained in the leaves at high concentration.

The basic metabolism of photosynthesis is summarized as

$$6CO_2 + 6H_2O + light \rightarrow C_6H_{12}O_6 + 6O_2.$$

We "burn" carbohydrates to produce forms of chemical energy widely used in cells. The summary of the process makes this respiration look like the reverse of photosynthesis, but the details are not the reverse. Photosynthesis takes place in chloroplasts and respiration in mitochondria. The same process of respiration occurs in plants at all times, reducing the gross photosynthesis rate during daytime and consuming oxygen at night.

Energy and Photosynthesis

Plants are not very efficient in processing solar energy. The optical properties of leaves, and thus their colors, are a factor in this energy conversion. The optimal conversion of sunlight to sugars is around 1 percent, limited by the absorption of light and by the biochemical mechanisms of the process. Humans can burn up sugars with about the same efficiency as a large electricity-generating plant: a bit less than 40 percent. Furthermore, most trees, including the umbrella tree, can use only a fraction, about one-third, of the most intense sunlight during the day. That excess energy must be discarded without overly heating and damaging the leaves. Some radiation is reradiated at longer wavelengths, in the infrared region, as heat. Some radiation is reflected directly. Some is processed by mechanisms within the chloroplast organelles. Much of the excess radiation is absorbed by the evaporation of water in the air spaces within the leaves. Water vapor then moves out of the leaves via the stomata.

Water is a particularly important factor in photosynthesis. Gases move in and out of leaves through the stomata. The diffusion of these molecules results from the gradients of concentration inside and outside of the leaf. Carbon dioxide is a larger molecule than water and tends to diffuse more slowly. Stomata are regulated by the inflation (opening) and deflation (closing) of the two surrounding guard cells. Lack of water in the roots sends a signal to the leaves, causing the stomata to close—and shut off photosynthesis. Stomata are also sensitive to temperature and carbon dioxide concentration; increases in both may shut them.

As photosynthesis increases at higher light intensity, the limitation of carbon dioxide causes the process to level off. Higher carbon dioxide concentrations (think of global warming!) may promote increases in photosynthesis, temporarily at least. Some plants have evolved clever biochemical mechanisms to increase the concentration of carbon dioxide inside the cells, such as desert succulent plants, many herbs of the semiarid tropics, and even some aquatic plants and rain-forest epiphytes.

In leaves, photosynthesis is also limited by the amount of nitrogen absorbed by the roots and transported to the leaves. Most of the nitrogen is assimilated as chlorophyll *a* and *b* and the photosynthetic enzymes. The partitioning of nitrogen between pigments and enzymes is important. Plants adapted to intense light conditions, or leaves acclimated to these conditions, may produce less pigment and more enzymes, as the latter may limit the rate of photosynthesis. Plants adapted, or acclimated to, very shady conditions do the opposite, allocating more of their nitrogen to pigments and less to enzymes. The pigment concentrations affect light absorption, and color, but the distribution of pigments in organelles is equally important.

For the leaf to function properly in photosynthesis, all of these factors must be in balance. It is important that the leaves have the means to process the solar energy that they absorb. The bulk of this energy must be shed by the leaf to limit the increase of temperature. The light energy absorbed by the chloroplast must also be used. That absorbed in the light reaction must make the products that allow the dark reaction to occur. If not, the excess energy inhibits photosynthesis. Photoinhibition eventually may lead to damage of the chloroplast and bleaching of the leaf tissue.

The range of sizes and shapes we encounter among different plants also affects leaf temperature and gas exchange by leaves. Consider a palm frond 20 meters long as the upper end of that range and the tiny diaphanous leaf of a filmy fern a few millimeters long as the lower end. All objects attract a thin film of still atmosphere to their surfaces, known as the boundary layer. The thickness of the boundary layer is determined by the distance of a point on the surface to the nearest edge: The greater the distance the thicker the layer, over 3 millimeters in the center of a large leaf. Projections on the surface of a leaf, such as hairs or scales, can increase the layer's thickness, and turbulence from winds can reduce it. A large and highly lobed leaf may thus have the same boundary-layer thickness of a much smaller, nonlobed leaf. The boundary layer adds resistance to the diffusion of gases into and out of the leaf, and slows the transfer of heat energy. All of the above concepts help us understand the diversity of leaf sizes and shapes in different environments, and the range of colors in leaves—related to the amounts of light captured by leaves.

Leaf Color and Photosynthesis

This brief summary of physiology helps us understand the subtle differences in the color of green leaves we encounter in nature. The shifts in hue and absorbance are adjustments in the balance between the various environmental factors that determine photosynthesis and plant survival: temperature, water availability, nitrogen availability, and light exposure.

Let's go back to the umbrella tree again. Light strikes the upper surface of the leaves of this tree. It enters into the leaf and passes through the palisade layer with very little scatter. Since the chlorophyll is in packets (the chloroplasts), much of that light passes through this layer via the central vacuole. This provides the sieving effect discussed in chapter 4. The light then enters into the spongy mesophyll, where it is strongly scattered by reflection due to the differences in the refractive indexes of the cell walls and air spaces. This provides the path-lengthening effect discussed in chapter 4. Some light is backscattered through the palisade cells, adding to photosynthesis, and then passes out of the surface. This makes the leaves look dark green to us. Some light is scattered to pass through the lower epidermis of the leaf. If we hold the leaf up to the sun, we can see the green light transmitted through it (fig. 6.5). We turn the leaf over; the undersurface of the umbrella tree leaf is very pale green. On this side, light passing into the spongy mesophyll is immediately scattered by the air spaces and passes back out of the surface, seen by us as brighter and paler, less altered by interaction with the pigments. Leaf

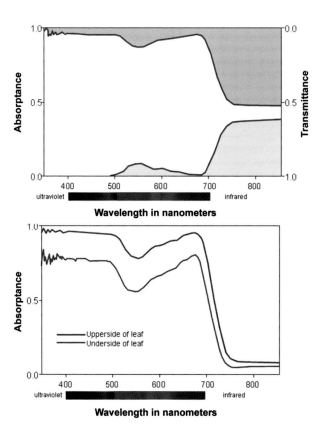

Figure 6.5 Optical characteristics of an umbrella tree leaf. *Top*, diffuse reflectance and transmittance of a leaf in a normal upright position. *Bottom*, comparison between the absorptions of leaves with light arriving on upper surface (*dark green line*) and on the undersurface (*yellow-green line*).

color, and the difference between the color of upper and lower leaf surfaces, is a consequence of this anatomy and the distribution of air spaces.

In any forest canopy, the subtly varying colors among individual tree crowns reflect the adaptations of individual species to the light environment. These are the results of the evolution of leaf traits within and among the different plant families that make up the diversity of the forest. Subtle shifts in color are caused by small differences in the composition of the chlorophyll and carotenoid pigments, as well as the distribution of chloroplasts within the leaves. Reduced chlorophyll concentrations (as caused by a deficiency in nitrogen) reveal the absorbance of carotenoids, and leaves appear light green or yellowish.

Some leaves differ very little in color between the upper and lower surfaces. In such leaves, there is a similar amount of spaces and stomata beneath the epidermis on both surfaces. As a contrast to the umbrella tree, buttonwood (a tree native to the mangrove swamps of south Florida and the American tropics, fig. 6.6) has stomata on both surfaces and is similar in color on both sides of its leaves.

Figure 6.6 Leaves of buttonwood (*Conocarpus erectus*), a mangrove tree of the New World tropics. *Left*, two forms are seen here, the normal green form and the silvery hairy variety (*argentea*). Note the small color difference between both sides of the leaf in the green form, because of the presence of stomata (and air spaces) on both sides of the leaf. *Right*, scanning-electron-microscope photograph of the hairs of the silvery form.

Many plants deviate strongly from the normal greens of foliage because they produce various structures that scatter light directly at the surface of the leaf. These structures include hairs, scales, and waxy coatings. Their optical effects were mentioned in chapter 4.

Hairs

Epidermal cells of leaves, and stems and flowers as well, may produce long projections that stick out from the surface. These may grow in a single direction, like neatly combed hair, or form a jumble of different directions (fig. 4.1). Hairs may be produced on the upper or lower surface, or over the entire leaf surface. Their functions appear to be twofold. Hairs interact with light directed toward the leaf surface. They intercept it, scatter it in all directions, and significantly reduce the amount of light captured by the leaf. Hairs also form an effective insulating barrier for the leaf. Such a barrier may reduce the flow of heat through the barrier, insulating the plant structure from effects of the air temperature, especially in small leaves. More importantly, they are barriers effectively preventing diffusion of water vapor, thus reducing the rate of water loss from leaves (and the intake of carbon dioxide, reducing photosynthesis). Hairiness, or pubescence, is commonly produced on leaves of plants exposed to extremes of temperature and to the stress of drought. The hairy form of buttonwood (fig. 6.6) is much appreciated by landscapers in south Florida. Occasionally such hairs are branched and form an even more complex network, as in castillo, a succulent shrub native to semideserts in Madagascar, and often planted in suitable climates in the subtropics. Hairs may also protect leaves against herbivores.

When hairs are produced exclusively on the leaf undersurface, they reduce water loss and increase the efficiency of light absorption by the leaves. Light that passes through the leaf, both through packaging and scattering effects, is backscattered into the leaf upon encountering the hairs on the undersurface. This increases the total absorption of the leaf, and foliage of such plants, as in satin leaf (fig. 4.1) and in snakewood (fig. 6.7) is dark green.

Scales

Plants also produce more complex multicellular structures on leaf surfaces. These scales are produced from single points on the leaf and spread out in various shapes. The most common scales are round, but may be quite ragged on their edges. The Jamaican caper (fig. 4.5) produces such scales on the leaf undersurface. These probably help reduce the loss of water vapor through transpiration, because of increased reflectance of radiation. Many species

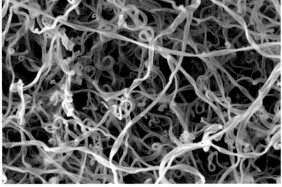

Figure 6.7 Leaves of the cecropia, or snake-wood, tree (*Cecropia peltata*), a fast-growing, early pioneer tree of New World tropical rain forests. *Top right,* note that the seedling leaves are smaller and variegated. *Top left,* adult leaves are larger, palmately lobed, and densely hairy on the undersurface. *Bottom,* scanning-electron-microscope photo-graph shows the curli-ness of the pubescence.

in the air plant family (Bromeliaceae) produce elegant scales on their leaf surfaces. Notable is the genus *Tillandsia*, which is represented by fourteen species in south Florida and is widespread throughout the New World tropics. Spanish moss is neither Spanish nor a moss. It is a small flowering plant whose gray color is due to the dense array of scales on the surface of the entire plant. Its somewhat larger cousin, ball moss, has particularly elegant scales, as viewed by scanning electron microscopy (fig. 6.8). In some bromeliads the scales may aid in the absorption of water by collecting drops of condensed fog through capillary attraction. Water trapped underneath the scale surface is absorbed by specialized cells at the base of the scale.

Wax

Leaves and other plant parts may increase reflectance of light, and reduce absorption and heating, through the production of waxes. Waxes are syn-

Figure 6.8 The elegant scales of the ball moss (*Tillandsia recurvata*), native to the live oak trees of my backyard in Miami. These scales both reflect excess sunlight and help capture and absorb water from rainfall.

thesized in the epidermal cells and deposited on the surfaces of aerial plant parts. These are long-chain compounds of fatty acids joined (esterified) with alcohols, and they are strongly water repellent. They form irregular bodies of different sizes that very effectively scatter light of all wavelengths away from the surface. Waxes can give leaves a glaucous or white color, the intensity depending on the amount of deposition. When particularly small particles are deposited, they are more effective at scattering shorter wavelengths of light, more blue than yellow or red. Some leaves, such as those of blue spruce, look powdery blue. In the case of the blue spruce, such a color seems to protect particularly against the damaging effects of ultraviolet radiation (see fig. 10.4). Those of the silver thatch palm, native to south Florida, produce a more whitish cast (fig. 6.9), as does the undersurface of the southern willow (fig. 4.4).

Silver Landscapes

In certain climatic conditions, of intense sunlight and/or lack of water, plants with hairs, scales, and waxy coatings are particularly common, and give the landscape a silvery or whitish cast. The correlation between hue and climate has been used as evidence for the adaptive function of these structures, and there is some direct evidence in desert and tropical alpine plants.

I grew up in such a landscape, the sagebrush steppe of the arid and cool regions of the western United States. My particular experience was the Columbia plateau of Washington State (fig. 6.10), but I find a familiar landscape in the San Luis Valley of southern Colorado, where we have some property. The silver cast is due to the dominance of a few shrubs with

Figure 6.9 The palmate frond of the silver thatch palm (*Coccothrinax argentata*), native to south Florida, is whitish on the undersurface due to the production of long strips of waxes, easily seen with a scanning electron microscope.

protected leaves: the hairy leaves of sagebrush (even the flower buds are pubescent), the hairy leaves of rabbit bush, and the slightly waxy leaves of greasewood. The whitish silvery appearance makes this plant community more efficient in water use, and slower growing, in an arid climate.

Similar changes in the vegetation occur with increase in elevation in mountains, and many high mountain species have a similar appearance. I once did research in the summit region of the highest mountain in Southeast Asia, Mount Kinabalu. The base of the mountain supported a diverse tropical rain forest, which gave way to montane forests and moss-covered cloud forests. The areas beneath the summit, over 3,500 meters in elevation, supported a dwarf montane forest of bizarre twisted trees, some found only on Kinabalu (fig. 6.11). These included a strange tropical conifer with phyllodes (petioles only) functioning as leaves. Also common was a glaucous conifer, *Dacrydium gibbsiae*, and the bluish cast of the leaf rosettes of *Schima wallichii* (which I later encountered in the Himalayan foothills in northeast India). However, the most common and remarkable tree is a small member of the eucalyptus family, *Leptospermum recurvum*, closely related to its more widespread cousin, *L. flavescens*, growing at lower elevations. I visited there in 1973–74 during the most severe drought of the twentieth

Figure 6.10 Northrop Canyon on the Columbia plateau, Washington, illustrates the gray landscapes of the sagebrush steppe. Two important plants in this community are the sagebrush (*Artemisia tridentata, bottom left*), and rabbit bush (*Chrysothamnus nauseosus, bottom right*).

century, and many shrubs had been killed by the drought. However, populations of individuals with a silvery (and hairy) upper leaf surface survived in much greater numbers than the more widespread individuals that were hairy only on the lower surface.

In such extreme environments of intense sunlight, not all plants are silvery. Other adaptations have appeared. The creosote bush is common in deserts of the Southwest and has a close relative growing in even more harsh deserts in South America. There are no scales, hairs, or waxes on its leaves, which are a shiny green. These plants have evolved a rich secondary chemistry, some of which is secreted on the leaf surface (fig. 6.12). These

Figure 6.11 On tropical mountains at high elevation, plants are frequently covered with hairs or waxes, means of protection against intense sunlight. Here on the upper slopes of Mount Kinabalu (*top left*), in Sabah on the island of Borneo, we see two such species. The dominant species of these forests is *Leptospermum recurvum* (found only on this one mountain), with the pubescent form (*top* and *bottom right*). A waxy-leaved plant (*Schima wallichii*) grows in the same forest (*bottom left* and *center*).

Figure 6.12 The secondary chemistry of the creosote bush (*Larrea divaricata* subspecies *tridentata*), native to deserts of the Southwest and closely related to another species from South American deserts, protects against being eaten and against damage by ultraviolet radiation.

compounds apparently serve two purposes. First, they deter animals from eating the leaves. Second, they protect the leaves from the damaging effects of ultraviolet radiation, yet allow more absorption at longer wavelengths than in the ultraviolet.

Foliage Color and Leaf Movement

Many plants produce strong differences in color, from the upper to lower surface of a leaf, one side green and the other side covered with waxes or hairs. In the trembling aspen, the twinkling shifts in leaf color in individual crowns happen when leaves twist to reveal a white and waxy undersurface; the mechanically weak petiole promotes such movement in a mild breeze. That is a passive movement. In a number of plants, particularly in certain families such as the legumes (Fabaceae) and prayer plants (Marantaceae), the petioles contain swollen structures that allow active movements during the day. In some desert plants, leaf movements may prevent too direct an absorption of sunlight—and therefore reduce water loss. In alpine plants, however, leaf movements may track the sun to increase absorption and maintain higher temperatures. In a few cases, such movements may affect the appearance of the entire plant community.

The mother-in-law's tongue (or father-in-law's) tree, of the seasonally dry forests of South Asia, is common on Miami streets (fig. 6.13). Its common name is inspired by the loud gossipy whispering sound its fruits

Figure 6.13 In the mother/father-in-law's tongue tree (*Albizia lebbeck*) growing in my neighbor's backyard, the leaves open in the day and close in the late afternoon, changing the crown's appearance. Such leaf movements are common in legume trees.

make when blown by the wind during the dry season. Its compound leaves begin to close in late afternoon, with leaflet undersurfaces pressed against each other, and they open early the next morning. Thus the tree takes on a dramatically different appearance at different times of the day, easy to spot from a distance. Another common example with similar behavior, in Miami and urban landscapes throughout the tropics, is the poinciana. In both of these legume trees, cells within the swollen areas (pulvini) at the base of the leaf and leaflets swell or deflate, thus causing the movements.

The pompano is a large herb common along forest edges and in pastures in Central America. The large banana-like leaves of this clumping plant spring from a rhizome. They are deep green on the upper surface and chalky white on the lower surface, due to waxes (fig. 6.14). The leaves of these plants move during the day, starting at a more flat and open angle in the morning and moving to a vertical position in midday, when the sun is brightest. Thus the leaf position is optimized for the minimal absorption of light energy at the hottest and brightest time of the day. Although such leaf movements had been observed by people living in the region, Tom Herbert, a colleague working at the nearby University of Miami, has shown the sophistication of these movements that include both twisting and bending of the blade. I measured the light absorption of the upper and lower surfaces of these leaves; the waxy surfaces absorb 30 percent less sunlight for photosynthesis. These leaf movements allow the plant to balance its requirements of light for photosynthesis in a range of temperatures and reduce its water use.

Leaf Colors in the Understory

Plants growing in the understory of a forest, or other closed canopy, encounter dramatically different conditions than plants growing in open sunlight. The canopies filter out most of the direct sunlight, reducing the amounts and shifting the spectral quality of the radiation (see fig. 12.6). Plant growth in these environments is generally limited by the amount of sunlight, and not by the concentration of carbon dioxide or water availability, which are more important in high-light environments. In shade, the intensity of sunlight may be 1 percent or less of full sunlight. Brief sun flecks move across the forest floor and may provide most daily light energy for the plants. Plants must ramp up their machinery to take advantage of these pulses of sunlight, and most seem inefficient at doing so. They are, for the most part, adapted to survive on the predominant low levels of

Figure 6.14 The banana-like leaves of pompano (*Calathea lutea*), native to Latin America, move to an upright position during the heat of the day. Such a position hides the light-absorbing upper surface and exposes the waxy and reflective undersurface to sunlight. The graph shows clearly the reduced absorption of the waxy undersurface (*light green line*) compared to the dark green upper surface (*dark green line*). Leaf movements are characteristic of the family, Marantaceae, to which this plant belongs.

light. Plants under these conditions have evolved different strategies for optimizing their efficiency of energy capture. Some of these have resulted in the production of colors and textures that are dramatically different than plants growing in more direct sunlight. These seem to be adaptations for survival in shade, although some of the commonly observed traits continue to puzzle us. The basic molecular machinery of photosynthesis, the light and dark reactions and the two reaction centers of the light reactions, is the same among plants adapted to shade and to sunlight. So perhaps the altered leaf colors are mechanisms of fine-tuning the light environments in leaves.

Generally, plants in deep shade allocate greater resources for harvesting sunlight (the light reaction), and less for producing the enzymes of photosynthesis (the dark reaction). The patterns of display of the light-harvesting pigments optimize energy capture and lower the rate of carbon fixation. Shade plants are generally more efficient at capturing sunlight partly because their structures are more flimsy (less tough tissues), with greater support due to the pressure of thin-walled cells, and their rates of dark respiration are consequently much smaller than in the leaves of sun plants. We see such adaptations in plants in tropical rain forests, and in the shade of temperate forests.

I use an example of a plant adapted to extreme shade, a rain-forest begonia, to contrast it with the leaves of the umbrella plant previously examined. The umbrella plant is capable of acclimating to fairly low light conditions, and can grow in interior environments, such as public areas of shopping malls in temperate regions. However, these plastic responses are limited, and it cannot grow in the extreme shade conditions where the begonia lives.

Begonia mazae grows in the dense shade of tropical rain forests in Brazil (fig. 6.15). It is a tuberous begonia, with a thickened underground stem that allows it to store energy and survive in the rare dry periods in the rain forest. It produces a dark green array of slightly lobed and lopsided leaves. The leaf blades are produced on upright petioles that grow from the tuber. The leaves are soft and velvety in touch and appearance. A look at the anatomy of the leaf reveals why. The leaves are thicker than those of the umbrella tree, principally because of the huge cuboidal cells beneath the surface. The water pressure (turgor) in these cells keeps the blade erect, much as an inflated air mattress remains stiff. There is little tissue with tough (lignified) and thick-walled cells, such as the fibers seen in the umbrella tree leaf. As a consequence, the begonia leaf is 93 percent water, compared to 67 percent

 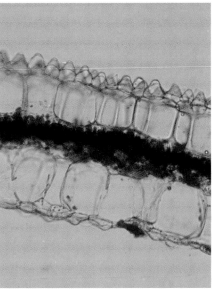

Figure 6.15 Left, the dark leaves of *Begonia mazae*, a begonia representative of plants growing in the understory of tropical rain forests. *Right*, a section of the leaf showing the
narrow distribution of chloroplasts and the pointed outer cells.

for the umbrella tree leaf. Similarly, the dry weight per unit area of the begonia leaf is about one-seventh of the umbrella tree leaf. The photosynthetic
properties of this plant have not been studied, but are probably similar to
those of other extreme shade plants, with a very low light-saturation point,
low maximum photosynthesis, and low light-compensation point. With
less enzymes, less production of lignified cells, and even less chlorophyll,
the cost of maintenance (partly seen as dark respiration) is substantially
reduced. The chloroplasts in *B. mazae* leaves are located in a dense band
in the middle of the leaf, almost entirely in very small palisade cells. This
concentration reduces the sieving effects of packets of pigments, because
they are stuck together. The anatomy also reduces path-lengthening effects
except for the intercellular spaces beneath the chloroplast layer that scatter
light back through the chloroplasts. The net result of this anatomy is that
less chlorophyll is required for light absorption in the begonia leaf, which
absorbs about 95 percent of sunlight. Two other features make the begonia
leaf striking. One is the dark appearance, which is partly due to the presence of anthocyanins in cell vacuoles of the lower leaf epidermis, and the
red leaf undersurface. The second is the velvety appearance, which is due

to the convexly shaped epidermal cells on the upper surface. We'll take up these characters later.

A sample of understory plants from a tropical rain forest reveals an amazing variety of leaf colors, patterns, and textures. Leaves are commonly patterned, or variegated, with red-purple undersides, with velvety texture, and even with an iridescent blue. The iridescent blue and red-undersurface leaves will be discussed in chapters 10 and 11. We'll discuss the rest of the phenomena here.

Variegation

There is certainly no more spectacularly variegated foliage than among the crotons (fig. 6.16). All forms of this cultivated plant are most likely derived from a single green-leaved understory shrub (*Codiaeum moluccanum*). The croton aficionado Frank Brown has pieced much of this history together by visiting the East Indies, particularly the Moluccas and Ambon, where these forms are found in high diversity. Apparently, a single mutation led to the production of leaves that are gold-flecked on a green background, and then more mutants were produced. These were attractive to natives on the islands and were selected and planted in their villages. Thus the famous natural historian Georg Everhard Rumph (latinized and better known as Rumphius) documented nine planted varieties on the island of Ambon in the mid–seventeenth century. These varieties were transported to other areas by the Dutch, perhaps first to China. Some were present on the Malabar coast of India late in the same century. Brown saw twenty-five varieties on

Figure 6.16 The variegated leaves of croton (*Codiaeum variegatum*), native to the islands of Indonesia, are produced through combinations of the three principal leaf pigments, chlorophyll, carotenoids, and anthocyanin. *Right*, transverse section of leaf.

Ambon near the end of the twentieth century, and he has described dozens and dozens of varieties. We can only speculate on the genetic basis of the pattern production; transposable elements are likely involved. None of the varieties breed true from seeds, and all must be propagated vegetatively through rooted cuttings.

I mention crotons to contrast them with the plants I discuss next, all of which produce color patterns in their native environments of forest shade. Variegation is produced in such extreme-shade plants through a variety of methods. The most obvious means is simply through the production of tissues without functioning chloroplasts and no chlorophyll. Those areas appear white from the scattered light. Examples of such patterns are found in many aroids. Some species of the New World dumb cane produce rather complex patterns in this way, very similar to patterns produced in their Old World relatives, *Aglaonema* (chapter 5). In certain Southeast Asian plants, as in *Sonerila margaritacea* (in the Melastomataceae family, fig. 5.10), the white blotches occur in a regular pattern at the same distance from the midrib, looking like the portholes of some understory airship. This pattern also occurs in begonias in the same forest. Another pattern is caused by the association of the non-chlorophyll-containing cells with the major veins on the leaf. This results in a fine-grained variegation, as seen in the silver nerved plant (fig. 5.6).

Silver patterning is produced by separation between the epidermis and palisade layer on the leaf blade. A common ornamental terrarium example is the aluminum plant, in the nettle family (fig. 6.17). Even some of the variegation in *Aglaonema* is caused by this separation (fig. 5.11), along with areas and layers without chlorophyll.

Some variegation in tropical understory plants is at least partly due to the production of anthocyanins in areas of the leaf blade. This may be joined by the absence of chlorophyll in other leaf areas, giving a variety of color patterns. Such a leaf may have dark brown areas (chlorophyll and anthocyanin), bright red areas (anthocyanin alone), white areas (no pigments), and normal green (chlorophyll alone). Caladiums produce patterns in this way, and the naturally variegated wild varieties have been exploited to produce the bewildering variety of modern cultivars. In Gabon, West Africa, I observed the juvenile leaves of a common ginger plant, *Aframomum longipetiolatum* (fig. 5.10). The developing juvenile leaves produce pinpricks of dark red color on the blade. As the leaves expand, the red color appears to "bleed" from this point. Ultimately, the color is lost in mature adult leaves.

The above examples of variegation are all from tropical understory plants, but such patterns are also occasionally seen in temperate-forest

Figure 6.17 Variegated-leaved plants, all from forest understories. *Top left*, prayer plant (*Maranta leuconeura*), of the New World tropics; detail of leaf surface at right. *Bottom left*, a trout lily (*Erythronium americanum*), common in temperate North American forests. *Bottom right*, the aluminum plant (*Pilea cadierei*), from the New World tropics.

shade plants. Two examples well-known to me are the trout lily, a spring wildflower of temperate forests in the United States (fig. 6.17), and some forms of the catbrier, *Smilax*. In trout lily, variegation is produced through the patterns of anthocyanin and chlorophyll production. In the catbrier, the white splotches are due to a separation between epidermis and palisade layers. In some temperate plants, variegation occurs in distinct patterns that are often correlated with geography. The best-studied example is clover, which occasionally produces silvery blotches (from tissue separation).

Since many plants may be strongly shade-adapted as small seedlings, and more sun-adapted as adults, they often produce variegated leaves as juveniles and normal green leaves as adults. Such patterns are observed in different developmental stages of vines. The seedling leaves of cecropia, which I used as an example of undersurface hairiness, are strongly variegated, due to sectors of tissue separation and anthocyanin production (fig. 6.7).

The function of variegation in understory plants is a puzzle that has received much speculation but little research. It makes no logical sense that a plant surviving in a low-energy, shady environment should produce variegated leaves, because such leaves reduce the plant's efficiency in capturing light energy. In evolutionary terms, such a trait as leaf variegation must be under strong selection pressure in these environments. Various arguments have been raised to explain the adaptive significance of leaf variegation, but only a single serious scientific study has been published. Variegation could reduce energy absorption in certain environments, such as open sunny locations, yet persist when the canopy closes over those environments. Variegation could make the leaves more apparent to a potential consumer, alerting it to some undesirable property such as concentrations of toxic molecules. In a similar manner, variegation could make the leaves look like they had already been colonized by a consumer, such as an insect. For instance, some species of passionflower produce outgrowths on their leaf petioles that look like the egg cases of butterflies. The butterflies are sometimes dissuaded from laying eggs and producing larvae that would consume the leaf. Finally, leaf variegation might disguise the leaf, and the consumer would not be able to locate it. Aposomatic coloration has been well-described in butterflies, in which related (but nontoxic) species appear similar to the species that have accumulated toxic molecules from the plants they eat. The monarch butterfly comes to mind. Perhaps such traits could be found in groups of plant species. Of course, it is quite possible that the patterns in leaves may be of no adaptive significance and merely be linked to some other trait that is adaptive.

Alan Smith, a colleague at the University of Miami and the Smithsonian Tropical Research Institute in Panama before his untimely death, conducted a classical study of the patterns of variegation in a rain-forest liana common in tropical rain forests in the New World: *Byttneria aculeata*. This plant, a member of the Malvaceae family, starts out as a small and compact shrub, and later grows into a subcanopy vine (fig. 6.18). Alan first showed

Figure 6.18 Variegated leaves of the understory plant *Byttneria aculeata*, of the New World tropics, may help it escape predation by leaf-mining insects and may reduce the temperature of leaves in intense sunlight.

that patterns of variegation in the juvenile plants, silvery flecks due to the separation of the epidermal layer near the midribs of leaves, are produced to a greater extent when plants are grown in sunny environments, and he found that the variegated forms were naturally prevalent in open forest environments, as in gaps. He also observed that the plant is attacked by a leaf-mining insect. Such insects lay eggs on the leaf, and the larvae burrow into the photosynthetic tissue in the middle of the leaf. Their trails of consumption are easily seen as areas of white or silver in a green leaf. Alan found that the leaf miners were significantly less common on the variegated leaves than on normal leaves. He thus concluded that this plant probably develops variegation in response to both the intense sunlight (reducing energy absorption) and insect damage (making it look to the potential herbivore as though it were already mined). These are interesting results, but it is unfortunate that they are the only solid data we have to discuss. The phenomenon is still very much a mystery, a very attractive one.

Leaves That Appear Dead, Variegated or Otherwise

When leaves produce both chlorophyll and anthocyanin in proper balance, the absorptions of the pigments combine to efficiently capture the visible wavelengths of light. The result is a brown, or almost black, color. Some leaves produce this color uniformly on the surface, or as variegation on a light background. Such color may actually camouflage a living leaf so that it appears dead to a potential herbivore. I was told about such a plant with dark brown leaves growing on the forest floor in dense tropical rain forests near Saül, in French Guiana, in South America (fig. 6.19). When I was led to these plants, *Psychotria ulviformis* (in the Rubiaceae family), it was impossible for me to find them, until their precise location was pointed out. In these plants the green photosynthetic tissue is overlain by an epidermis and

Figure 6.19 In *Psychotria ulviformis*, seen in a rain-forest understory in central French Guiana, a combination of anthocyanin and chlorophyll gives the leaves a brown—and dead—appearance, presumably a form of camouflage to defend against herbivory. The transverse leaf section shows the distribution of anthocyanin in hairs and superficial cells.

extensive curved hairs filled with anthocyanins; these provide the right balance of the two pigments to produce the brown color. Nearby we located an undetermined orchid with brown splotches that were a combination of the two pigments in areas of the leaves. Although camouflage is a reasonable hypothesis for such color production (which absorbs some of the light that could be used in photosynthesis), no one has journeyed to these obscure places to do the critical experiments.

Leaves and Cells as Lenses

Any curved surface on a plant organ can potentially refract light, focusing it in smaller areas and creating different textures and colors in plants, particularly in leaves. The basics of this phenomenon have already been discussed in chapter 4. Here we'll look specifically at the ways leaves refract light, producing color effects and enhancing their optical properties for survival in different environments.

Rarely a whole leaf may produce a curved surface, giving it a distinctive appearance and refracting light into the leaf to enhance photosynthesis in different environments, as in the living stones (*Lithops* spp. and other members of the Aizoaceae in South Africa). In many species the tips are also somewhat variegated, looking like the stones where they grow. Some succulent rain-forest epiphytes produce rounded and fleshy leaves, whose curvature refracts light into a photosynthetic layer in the interior of the leaf, as in *Peperomia rotundifolia*, fallen from branches of rain-forest trees in French Guiana (fig. 4.8).

When individual cells on the leaf surface are curved, light refracts into individual cells within the leaf. The degree of cell curvature determines the depth to which the beam of light is focused. In some extreme-shade plants, the level of concentration occurs at the position of chloroplasts, where light may focus on specially oriented chloroplasts to improve the efficiency of light capture. I turned my attention to this phenomenon almost twenty years ago, working with Richard Bone, a colleague in our physics department, and John Norman, now at the University of Wisconsin and a biophysicist who works on plant microclimates. Bone, Norman, and I developed a ray-tracing model to calculate the depths and degrees of concentration. An example is seen in the pothos vine, native to tropical rain forests in Southeast Asia (fig. 6.20). The velvet texture of the leaves is due to the

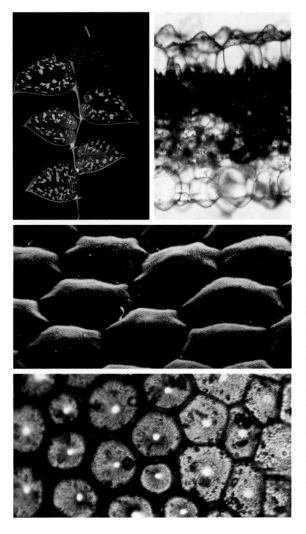

Figure 6.20 In the pothos vine (*Scindapsus pictus*), an understory vine from Asian tropical rain forests, the shape of the epidermal cells refracts light to the densest layer of chloroplasts, refraction seen by the surface photograph of the cells, *center,* and the scanning-electron-microscope photograph of the leaf surface, *bottom.*

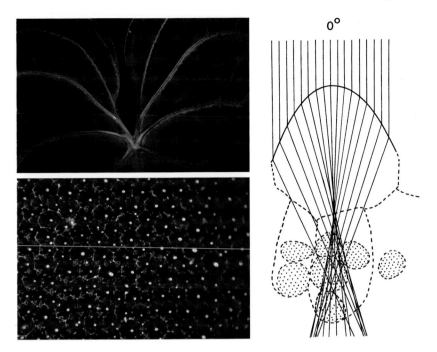

0°

Figure 6.21 In this velvet-leaved anthurium (Anthurium warocqueanum), the convexly curved surface cells focus light into the fine points, seen in both the microscope photograph of the leaf surface (bottom left) and in the model of cell refraction (right).

shape of the epidermal cells, and the silver variegation is caused by a tissue separation of the same layer. The shape of those cells concentrates light on the level of the chloroplasts, and we speculated that such concentration in the diffuse shade of leaves parallel to tree trunks would increase their capture of sunlight. Similar cell shapes and focusing effects also occur in other plants for which focusing may increase efficiency of light use (fig. 6.21).

One possible consequence of such focusing is that a fleck of direct sunlight could be directed into the leaf at such high intensities as to destroy the chloroplasts. So such a feature must occur in very shady habitats. A possible benefit of such epidermal cells is that they are more efficient in capturing oblique rays of diffuse sunlight, which would normally reflect off the leaf surface. A cell shape optimal for absorbing the oblique rays would be more conical and less rounded. Thus in the Begonia mazae leaf that was discussed earlier, the cells are more likely to function in diffuse shade to capture such rays. A consequence of such focusing effects is that the leaves have a velvety texture that affects the production of color in subtle ways.

Figure 6.22 The intense green color of these two cave-dwelling plants from West Malaysia is due to the focusing and refraction of light into specially oriented chloroplasts within the leaves. The large-leaved plant is *Monophyllaea patens*, and the smaller, luminous liverwort is *Cyathodium foetidissimum*.

Some of the most extreme examples of light collection and intense green color production occur in plants that grow in the twilight of caves. In Malaysia I often observed two such plants growing in the recesses of Batu Cave, not far from the capital city of Kuala Lumpur (fig. 6.22). One of these is a liverwort, *Cyathodium foetidissimum*, which grows as a thin crust on the rock. The other is a remarkable flowering plant, *Monophyllaea patens*, a member of the African violet family; it produces only a single leaf, one that ultimately produces flowers and breaks one of the basic rules of plant organization (that flowers are produced from buds on stems). In the photograph both plants seem almost to glow in the dim light. A change in observation to a more oblique angle would shift the color to dull green, and then to a light yellow. Only the green light is scattered intensely and directly out of the long palisade cells and so is observable perpendicular to the leaves.

Leaves That Quickly Change Color

Since green leaves produce color primarily by the absorption and scattering of light by chloroplasts, the positions of these organelles in the cells of a leaf can affect color production. In many plants, including algae and mosses, chloroplasts can quickly change position, in a few minutes, after exposure to direct light. The chloroplasts typically move from the bottom of the palisade cells to the sides of the cells. This shift in position makes the cells more transparent to light, as more light can be transmitted through the cells without encountering any chlorophyll molecules. This phenomenon has been observed in a variety of plants in many different families. An important factor regulating the movements (and change in color, which varies among different species) is the size of the palisade cells in which the chloroplasts are predominantly produced. Larger cells, particularly wider ones, allow for more chloroplast movement and more color change. The phenomenon is easy to demonstrate by covering a portion of the leaf of a shade-loving plant in direct sunlight. After a short period, remove the cover to see that this area is much darker than the rest of the leaf. These simple experiments can be performed with a variety of ornamental houseplants, such as begonias and the small aroid vine *Philodendron oxycardium*. Just thirty minutes of light exposure revealed the dramatic difference in color between the exposed and shady areas of the leaves (fig. 6.23).

Some time ago I received a thick envelope from B. L. Burtt of the Royal Botanical Gardens at Edinburgh, one of the experts on the Gesneriaceae of the tropical rain-forest understory, including the strange *Monophyllaea* just

Figure 6.23 Leaves of this shade-loving aroid vine (*Philodendron oxycardium*) were exposed to sunlight for 30 minutes, except for areas covered in foil. When the foil was removed and the leaves photographed, a dramatic difference in color due to chloroplast movements was visible.

discussed. He sent two photographic prints taken of a group of African violet plants collected by a plantation manager and plant hobbyist near Victoria Falls, Uganda. One view shows the plants at 11 a.m., and the second view shows the same plants at 7:15 p.m. on the same day (fig. 6.24). These are the same plants, but strikingly different in color, and this color change occurred every day. This shift in color is most likely due to the movement of chloroplasts. Chloroplasts move not only in response to sunlight, but also in a night-and-day (circadian) rhythm in many plants. Just recently I became aware of a little fern ally, *Selaginella serpens*, native to forests in the Caribbean, as in Jamaica (fig. 6.24). This little plant turns silvery white every night, and back to green during the day. During the day, its surface cells contain a single chloroplast, which covers the bottom of the cell like

Figure 6.24 Top, in this unnamed African violet (*Saintpaulia* sp.), collected near Victoria Falls, in Uganda, the leaves look dramatically different in the morning (*left*) compared to the evening (*right*), due to chloroplast movements. Photograph provided by B. L. Burtt. *Bottom*, photographs of the identical area of a fern ally, *Selaginella serpens*, *left* at 12 noon and *right* at 12 midnight.

a pancake. At night (and in response to stress) these flat chloroplasts turn into small spheres and make the cells largely transparent. This appears to be the explanation for the dramatic change in color, but begs the question of why the plants do this, that is, what is the selective advantage of such a rapid color change.

When Leaves Perform Other Functions

Johann Wolfgang von Goethe, the great German writer whose scientific work on color vision was mentioned in chapter 2, would have liked to be remembered most for his studies on plant form. Goethe viewed plants as repetitions of fundamental organic units, the leaf being particularly important. He saw the leaf transformed ("metamorphosed") into different organs. Similarly, in our contemporary molecular view of plant development and evolution, the flower is seen as a modified shoot in which leaflike units make up the parts of a flower. Thus it is not at all surprising that leaves would be transformed in evolution to take on different functions, and in those different functions to produce unusual displays of color.

Leaves that develop beneath flowers become intensely colorful in a few plants. They thus compensate for the lack of visual attractiveness of the flowers (fig. 6.25), and may supply visual signals to attract agents of pollination. A spectacular example is the poinsettia (of the Euphorbiaceae), in which the brilliant red coloration that graces our festive spaces of winter holidays is produced by leaflike structures (technically called bracts) just beneath the flowers. The red color is due to the presence of an anthocyanin, cyanidin glucoside. In the tropics, mussaenda (an Old World forest shrub) graces many a landscape with its branches of colorful bracts underneath the small tubular flowers. Colorful bracts in bougainvilleas provide brilliant colors through their production of the unusual betalain pigments (see fig. 7.13).

When shoot tips produce fruits, sometimes leaves near the tips change color to advertise the delicious items for would-be consumers and dispersers. The discolored shoots have been called "flags" to attract such animals. Early color change during the autumn in temperate forests (to be discussed in chapter 11) may be associated with ripe fruits and attract birds to eat and disperse the fruits; this may be the case for some species of sumac, common along forest margins and roadsides throughout North America.

Scales on young shoots, persistent or later falling, are leaflike organs that protect the young bud before it expands. These may produce an unusual

Figure 6.25 Leaves, or leaflike organs, in these plants take on functions other than photosynthesis. *Top left*, a spathe in *Heliconia collinsiana* protects flowers and fruits and attracts hummingbird pollinators. *Top right*, bracts on the inflorescence spike of an agave
protect the spike when it is young. *Bottom left*, *Warszewiczia coccinea*, with red-colored
bracts that attract pollinators. *Bottom right*, *Musa coccinea*, with bracts that protect and
then display male flowers.

pattern on the shoot when it expands, as in the inflorescence of a century
plant. The red color on these scales persists as the inflorescence rapidly
expands to a height of more than 3 meters (fig. 6.25).

Among the most remarkable modified and colorful leaves are those of
the carnivorous pitcher plants. Those in Southeast Asia grow in a vinelike
habit, with the pitchers produced at the tips of leaves. At least part of the
pitcher is red, and sometimes the sides are mottled. Such plants are com-

Figure 6.26 Insect-trapping leaves of *Sarracenia leucophylla*, a pitcher plant, photographed at a bog in southern Alabama. The photosynthetic function of these leaves is minimal, and the white with red anthocyanin patterning helps attract insects, which fall into the trap.

monly encountered below the summit of Mount Kinabalu, where the acid and organic soils of the cold subalpine environments are ideal places for plants to obtain alternative (and animal-based) sources of nitrogen. In acid bogs in North America, pitcher plants grow as hollow leaves from an underground rhizome base. Some species produce traps that are low and round, and others produce tall, thin, hollow traps that are a mottled red near their tips (fig. 6.26). Such a mottled red color is often associated with attracting insects that eat animal flesh or lay their eggs on it. In southeastern Alabama, I saw such plants raised above the shorter plants in the bog at Weeks Bay.

Reading the Landscape

Leaves provide subtle signals of color that we can use to learn about the function of plants in their landscapes. The more knowledgeable we become about the identity of individual plants in their landscapes, the more sensitive we become to slight shifts in hue, intensity, and saturation. Each species responds to its physical environment in a distinct pattern, and each "behaves" in a particular way, in its rhythm of flowering, fruiting, and producing and shedding leaves. We call such rhythms of activity the *phenology* of a particular species. Development causes these changes, which are much slower than the diurnal movements discussed above. In some cases the young leaves take on a coloring distinct from the adult leaves, and the senescing leaves may do the same (see chapter 11). In tropical forests, the greatest differences occur during development and vary dramatically among species. In temperate forests, the greatest changes in leaf color occur during autumn senescence, with most species turning red and many yellow. Individual species may also respond differently to stresses, such as a combination of intense light and drought. In some species, leaves lose chlorophyll and become yellowish, or chlorotic. In other species, leaves may turn brown and fall off. Becoming comfortable with the natural environment in which we live, as with learning the wildflowers and birds, also means becoming familiar with the subtle color changes in the landscape. In traveling across the country and through different climatic zones, the same sensitivity yields insights about the changes in the plant communities we encounter.

The same possibilities of interpretation exist at larger scales. We can interpret photographs of vegetation from above. A photograph taken by aircraft above the canopy of a tropical rain forest, as at Barro Colorado Island in Panama (fig. 6.27), reveals its topography, the identities of the crowns of trees that compose the canopy, and the activities (whether making new leaves or losing old ones) of individual species. Photographs by astronauts, or by automatic cameras from satellites, reveal subtle differences in colors and textures of vegetation at a planetary scale. They are routinely used now to classify forests and determine their extent. Perhaps some day these photographs will be used to determine the health of these ecosystems, as they are used for field crops today. In the Everglades in south Florida, a composite of satellite images reveals the patterns of vegetation and human habitation (fig. 6.28). The light areas on the right of the image are densely settled. The vegetated areas are indicated by the red color (actually the strong reflectance in the far-red just beyond our vision). The reds at the top (north) are the sugarcane fields of the Everglades Agricultural Area. The lighter pink near the bottom of the image is pinelands and tropical hardwood hammocks, which once occupied the land now urbanized. The main sources of color in all of these photographs are the leaves that give our planet a blue-green color over much of its surface. The newer cameras have greater sensitivity in the infrared regions of the spectrum, and using

Figure 6.27 Canopy trees on Barro Colorado Island, Panama. Photograph courtesy of Richard A. Grotefendt and Stephanie Bohlman.

Figure 6.28 A composite satellite photograph of southern Florida and the Everglades: my home for the past twenty-five years. The photograph was taken at many bandwidths, and the colors represent some of them. *Blues* represent an absence of vegetation, and *green* and *red* represent different types of vegetation. The image shows the flows of the historical Everglades and the human settlements impinging on the integrity of this ecosystem. Distance from east to west is 140 km.

those bandwidths allows us to discriminate among the vegetation types with greater sensitivity.

Many years ago the Princeton biologist Henry Horne wrote an interesting book, *The Adaptive Geometry of Trees*, about plant architecture and its relationship to the capture of light, something akin to the adaptations of leaves in light absorption but at a higher level of organization. He wrote that "trees are clever green strategists." Such strategies are obvious in the structures and functions of flowers, discussed in the next chapter. They are also evident in the patterns and colors of the foliage that surrounds us.

Chapter Seven Flowers

....

The flower is the poetry of reproduction. It is an example of
the eternal seductiveness of life.

JEAN GIRAUDOUX

Colours change: in the morning light, red shines out bright
and clear and the blues merge into their surroundings,
melting into the greens; but by the evening the reds lose their
piquancy, embracing a quieter tone and shifting toward the
blues in the rainbow. Yellow flowers remain bright, and white
ones become luminous, shining like ghostly figures against a
darkening green background.

ROSEMARY .VEREY

....

Flowers are brilliant and ephemeral producers of color
in our lives. They are magical, and amazingly diverse,
in form and color. In this chapter I cannot describe the
patterns of colors in flowers of the some 280,000 species
of flowering plants, or the millions of varieties produced
through breeding and horticultural practice. Instead I
chose a good example to establish principles and then
emphasized the natural history of flower colors. I initially
thought that roses could be that example, but succumbed
to my tropical biases, and chose the most spectacular of
orchids instead: the cattleyas. Not only do these plants
produce an amazing range of colors, but species of cattleya can cross with many other genera within the orchid
tribe, and intense interest in them has led to some understanding of how these flowers produce color. These
showy flowers remind me of my adolescence, presenting a

Figure 7.1 A detail of the orchid variety Blc 'Hillary Rodham Clinton'.

strange and exotic cattleya corsage to a date for a high school dance. As an example of how flower combinations are produced in cattleyas, I will use the example of a single variety, Blc 'Hillary Rodham Clinton'; the center of this spectacular blossom appears in figure 7.1.

Before we take up the colors of cattleya flowers, it is useful to compare orchid flowers with more general flower structures, such as those of the "orchid" tree, the chalice vine, and alstroemeria (fig. 7.2). The first two plants are in the division of flowering plants known as the Eudicotyledonae, or eudicots for short. Eudicots have two embryonic leaves, net-veined leaves, young stems with their conductive tissues in bundles near the periphery, and taproot systems. Their flowers usually produce parts in fours and fives, or in multiples of those numbers. All flowers evolved from the shoot system of an ancestral gymnosperm, and the flower parts are homologous to leaves on such a shoot. Flowers generally produce sets of accessory organs in two tiers, the sepals and petals. In the pink orchid tree flower, the sepals are fused together rather than separate as in the idealized eudicot flower (fig. 7.3). The showy pink petals are not identical to each other, and a bilateral

Figure 7.2 The flower morphology of three tropical plants. The top two are eudicots: *top left*, the chalice vine (*Solandra maxima*); *top right*, the orchid tree (*Bauhinia purpurea*). *Bottom*, a monocot, an *Alstroemeria* hybrid.

symmetry is established, reminiscent of orchid flowers. The sexual parts include five stamens (the male function) and a single pistil (the female function, which is formed from two leaflike organs, or carpels). The chalice vine flower is a further departure from the ideal eudicot flower type; the petals are fused together to form the chalice-shaped corolla. Again the sexual parts consist of five stamens and a single pistil.

A typical orchid flower has these basic flower parts, but arranged in a unique manner. Orchids are monocots (in the Monocotyledonae). Monocots produce embryos with a single cotyledon, leaves with parallel veins, scattered vascular bundles in the stems, and diffuse root systems. Their flowers produce parts in threes. Lilies are often described as generalized

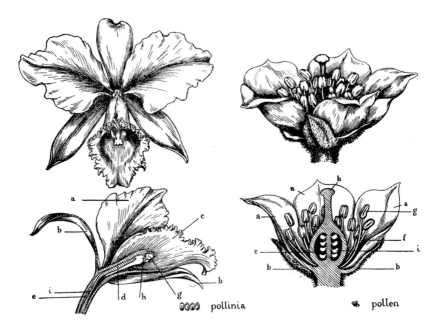

Figure 7.3 The parts of a typical dicot flower compared with a typical orchid flower. The parts are labeled *a*—petal; *b*—sepal; *c*—lip; *d*—column; *e*—ovary; *f*—filament; *g*—anther; *h*—stigma; *i*—ovule. Note the lip and column, which, along with the pollinia, make orchid flowers unique among all flowers. Drawing by Priscilla Fawcett, used with permission of Fairchild Tropical Botanic Garden.

monocot flowers, and we'll use the example of alstroemeria, a popular and inexpensive cut flower that comes to us from South America (fig. 7.2). Its flowers have accessory parts in threes, both sepals and petals (when identical they are all called tepals). In lilies these parts are arranged in radial symmetry, but in alstroemeria some tepals are slightly different, and the symmetry, as in orchids, is bilateral. They produce a single pistil, with a three-chambered ovary at the base, and six stamens.

In orchids the flower structure is modified in three ways. First, the petals are differentiated into two flags and a dramatically different lip. The lip serves as an attractant and landing platform for the pollinating agent, quite often a large bee. The sexual organs of the flower are fused to form a column, and the stamens are modified to produce pollen in sticky clumps called pollinia. The structure of the flower promotes the efficient transfer of pollen in pollinia from one flower to another, allowing fertilization of tens of thousands of tiny ovules in a flower from a single visit. Finally, the entire flower twists during development to end up upside down. It is particularly easy to observe these structures in cattleyas, because the flowers are so large.

Color in Cattleyas

Color in Blc 'Hillary Rodham Clinton'

I chose the variety 'Hillary Rodham Clinton' to explain color production because of its distinct colors and patterns, and because it was made available to me by a well-known orchid grower in Miami, Robert Fuchs. It is a spectacular flower, firm and large, with brilliant colors of purple and yellow on a white background. It is a recent production, registered by Christopher Chadwick in 1993. It is only a matter of time until it begins to accumulate awards of distinction as growers figure out how to produce the best flowers. I will simply call this spectacular flower by the first name of the public figure for whom it was named: 'Hillary'.

This flower has an anatomical feature common to many flowers: the pointed cells of the epidermis. On the purple portions of the lip, the epidermal cells contain a large central vacuole, which is filled with an anthocyanin pigment (fig. 7.4). The yellow portions of the lip contain a second water-soluble pigment, a flavonol. The inner veins of the lip contain yellow flavonols and red-purple anthocyanins together. Where both occur, the yel-

Figure 7.4 The anatomy of 'Hillary'. The dark purple areas have sharply pointed epidermal cells, filled with anthocyanin. The yellow areas have similar cells with yellow flavonols in the vacuoles, some condensing into bodies. Some areas have flavonols above and anthocyanins below. The white petals are devoid of pigments and full of light-scattering air spaces.

low pigment is located beneath the epidermis. The epidermal cells are particularly important in the production of flower color because they reduce the amount of light scatter on the lip surface, and increase light absorption and the intensity of color production. The white petals and sepals (collectively the tepals) contain no visual pigments, and the pure white color is the consequence of the efficient scatter and reflection of the entire visible spectrum of light by air spaces and cells in the interior of these organs. It is possible to see how the scattering properties of these air spaces contribute to color by infiltrating them with water under vacuum, and then to see the petal lose its whiteness and become translucent. The patterns and intensities of colors produced in 'Hillary' are the consequences of the collection and cultivation of wild species from Central and South America, going back to the beginning of the nineteenth century, and their subsequent hybridization in a bewildering variety of combinations. To see how this explains the color patterns in this beautiful flower requires some background information about orchids, and the cattleya group in particular.

Orchids, mainly tropical and epiphytic, are among the greatest of all plant families in the number of genera (788) and species (18,500). I expect that we have quite a ways to go before completing the global inventory of orchids, and some tiny rain-forest orchids will become extinct before we can study them. I became aware of the small and delicately beautiful terrestrial orchids of the conifer forests of western mountains from the backpacking trips of my youth: the small and beautiful lady's tresses and the delicately gorgeous orchis (fig. 7.5). In Europe, Carl von Linné (Linnaeus) observed common species of *Orchis* and named them in his *Species Plantarum*. *Orchis* then became the name after which Pierre Magnol created the entire family: the Orchidaceae. The name evokes the strongly sexual undercurrents of the family, the clever and devious means of practicing plant reproduction. *Orchis* is the Greek word for testicles, inspired by the shape of the underground tubers of the plant.

Later, as I began to pursue my studies as a botanist and studied in tropical rain forests, I became aware of the more typical habits of plants of this family. Orchids are, first and foremost, epiphytic plants of forests of the tropics, particularly rain forests. Despite all of the rain, their branch habitats are frequently dry, and most orchids have evolved the means to survive in such inhospitable environments. David Mabberley succinctly described them as "perennial *mycotrophic* epiphytes"; the seeds germinate only in the presence of root-inhabiting fungi, mycorrhizae, which help absorb nutrients for the plants. This great family has been divided into sixteen tribes,

Figure 7.5 My experiences with orchids. *Bottom center, Orchis rotundifolia*, encountered while hiking in the mountains of the Northwest. *Bottom right, Ophrys apifera*, encountered in the garrigue of southern France. *Top right, Cattleya aurantiaca*, growing in our yard in Miami. *Left*, an exhibit at the Miami Orchid Show, one of the largest orchid displays in the world.

and here I focus on one tribe, the Epidendreae, in which the cattleyas occur. In my adopted hometown of Miami, orchids are a big deal, and roses are not. Roses are difficult to grow in our semitropical climate; varieties often succumb to fungal diseases. On the other hand, it is an ideal place for raising orchids, with many commercial growers and thousands of hobbyists. The annual spring show of the Miami Orchid Society takes up a convention center, and was described in Susan Orlean's excellent book, *The Orchid Thief*, later made into a popular movie, *Adaptation* (fig. 7.5).

Cattleya and Allies

Arguably the most spectacular orchid flowers are in the tribe Epidendreae and the subtribe Laelinideae, including the genera *Cattleya, Encyclia, Laelia, Epidendrum, Brassavola, Rhyncholaelia*, and *Sobralia*. When I say arguably, there are many hobbyists convinced that vandaceous orchids (of the tribe Vandeae) are far more beautiful, and others would argue for *Paphiopedilum*, and so on. I suppose similar arguments could arise among rose fanciers. The genus *Cattleya* comprises about forty-five species, distributed in Central and South American tropical forests. They are all epiphytes. They produce thick leaves, one to three on a pseudobulb. They produce a remarkable range of flower colors and shapes, with considerable variation even within individual species.

Hybrid Cattleyas

The cattleya orchids evolved barriers to crossing between species by dramatic shifts in the shapes, colors, and fragrances of their flowers. These promoted some fidelity among the insect visitors. Nonetheless, some species could cross in the wild to produce hybrids that were named separate species, such as *Cattleya guatemalensis*, before taxonomists were aware of their hybrid origin. All members of *Cattleya* have twenty pairs of chromosomes in their cells, the same number as in the other members of the subtribe Laelinideae, such as *Brassavola* and *Laelia*. This makes it easier for members of the different genera to hybridize. All genera within the Laelinideae can cross-reproduce by hand pollination, yet the dramatically different flowers prevent this crossing in nature.

Cattleyas were introduced to Europe, particularly England, early in the nineteenth century. The generic name was established by the English taxonomist John Lindley, honoring William Cattley, an enthusiastic horticulturist and collector of orchids. The first successful orchid hybridist, John Dominy, crossed two *Cattleya* species—*C. guttata* and *C. loddigesii* most probably—in 1853. Hybrids between different genera in the Epidendreae appeared less than fifty years later. Among others, a British horticulturist, F. K. Sander, compiled lists of all of these successful hybridizations. This list, along with a formal registration process, is kept current by the Royal Horticultural Society. From it we can deduce the pedigrees of all registered orchid varieties.

Blc 'Hillary Rodham Clinton', 'Hillary', is the product of a complicated history of hybridization. The abbreviation *Blc* stands for the three genera that were crossed to form this variety: *Brassavola* (now *Rhyncholaelia*), *Laelia*, and *Cattleya* (fig. 7.6). Its genealogy includes twenty-two crosses, some going back six generations, involving eight species (some of which produce different varieties, used in separate crosses).

Judges, certified by the American Orchid Society and trained through apprenticeship and careful observation, can dissect any orchid hybrid and pick out the contributors to flower form and color in its ancestry. Robert

Figure 7.6 Three orchid species out of a total of eight making up the trigeneric cross of 'Hillary'. *Top, Laelia purpurata; bottom left, Cattleya labiata; bottom right, Rhyncholaelia digbyana.*

Fuchs is such a judge (but a grower of vandas; see fig. 5.6), and so is Robert Peters (and a commercial grower of cattleyas); both live in Miami and helped me tease out the contributions to color in this variety. 'Hillary' is the final product of two lines of hybrids, combining attractive flower size and form in one (such as flower size, a large frilly lip, and firmness) with color combinations in the other. The large frilly lip came from *Rhyncholaelia digbyana* and the tall inflorescence came from *Laelia purpurata*. The white background for the flower came from the albino forms of *Cattleya mossiae* and *C. warszewiczii*. The two-eyed yellow throat pattern came from *C. warszewiczii*, and the dark purple from *C. mossiae*.

This precise analysis of the orchid pedigree contrasts with our ignorance about the actual inheritance of the genes that control the color patterns. Mendel's rules of inheritance, of segregation, independent assortment, dominance, recessiveness, and so on (chapter 5), were worked out with the garden pea. Forty years after his discoveries, when breeders became aware of these patterns of inheritance, it was natural to look for such patterns in the organisms they studied, including orchids. Although the new science of genetics produced a revolution in biology, its influences were slowly felt for organisms that grew slowly and took time to produce results from crosses of long-lived organisms, such as trees . . . and orchids. It may take six to seven years before the orchid breeder can see the results of a particular cross. Animal breeders referred to the characters they were trying to breed into new combinations as "blood," a throwback to the earlier humoral theory of blending inheritance. Orchid breeders often refer to the "carriage" or "penetration" of a character (such as flower color) expressed in a hybrid cross.

What little we know about the inheritance of color in orchid flowers was established by the English geneticist C. C. Hurst, working around the time of the rediscovery of Mendel's laws. He established the genetic rules for purple color inheritance in cattleyas. He found that two color factors (loci C and R) determined the expression of purple anthocyanic colors in many orchids, including cattleyas. Genes C and R were dominant over c and r for such color in the petals. Purple varieties could be RRCC, RrCc, but not RRcc, rrCC, or rrcc. Thus the albino *Cattleya warszewiczii*, used in the production of 'Hillary', is rrCC. Additionally, Hurst discovered a third factor, P, necessary for the expression of R and C throughout the flower. When the recessive gene is present (pp) with R or C, then only the lip is pigmented. For 'Hillary' this meant that parents of the final cross were heterozygous for the gene controlling lip pigmentation, and an individual with white petals and a purple lip could be selected from products of that cross.

Although inheritance of purple petal color is dominant in *Cattleya*, it is recessive in *Laelia*. Conversely, the character for yellow flower color (probably from the production of flavonols) in *Cattleya* appears to be recessive but in *Laelia* is dominant. Lip color is controlled by different genes. Thus an orchid such as 'Hillary' can be produced with colorless petals and a purple lip.

Given the fascination and occasional obsession for orchids, such as I observe at the South Florida Orchid Show every spring, we can expect more and more new hybrids and color combinations. Perhaps the techniques of plant biotechnology, of introducing foreign genes into orchids, will yield color combinations unseen up to now, such as a true blue cattleya, or a patriotic red, white, and blue plant. The will is certainly there.

Back to the Roses

At age ninety-three, my mother still tends her roses. One of her favorite varieties is *Rosa* 'Double Delight' (fig. 7.7). This is a cross of hybrid teas, made in 1977 of two other hybrid tea roses, 'Grenada' and 'Garden Party'. The paternity of ancestral crosses that added to these beautiful cream and red flowers can be traced back through more ancient crosses to the origin of the tea roses, ultimately from crosses of varieties of wild species from central and east Asia. Most of these modern varieties have their beginnings as sports of *Rosa gallica* (perhaps the earliest in cultivation over 3,000 years ago), *R. gigantea*, and *R. chinensis*. Because roses can hybridize freely with each other, the number of species of wild roses can only be guessed: 100–150 species, mostly in Europe and Asia. My mother's garden includes the brilliant red colors of 'Mr. Lincoln', the mauve-lavender of 'Paradise', the yellow-orange of 'Caribbean', and the raspberry pink of 'Friendship'. Each of these varieties has a record of paternity that can easily be traced farther back than the roots of my mother's family. The reds and pinks of roses are produced by accumulating anthocyanins, particularly cyanidin-3- and -3,5-glucosides. Uniquely for roses, a single enzyme catalyzes the attachment of both sugars on the anthocyanidin molecule.

Outside of my hometown in Washington State, I encountered one of the few wild roses native to North America. The dwarf rose (fig. 3.16) grows along watercourses in the springs of this arid region, and in the mountains farther west and east. Its delicate pink flowers are more fragrant than my mother's hybrids, but pale in comparison to the grandeur of her rose garden. From the wild species, tens of thousands of hybrids have been pro-

Figure 7.7 My mother's favorite hybrid tea rose, *Rosa* 'Double Delight'.

duced over the centuries (I even read an estimate of a quarter of a million), creating a wild array of colors and blossom shapes among these plants.

Remember, the revolution of genetics, culminating in the biotechnology revolution today, began with Gregor Mendel's simple crosses of flowers. After Mendel, much early progress in genetics was made by crossing flowers. Thus all of the cultivated flowers, many featured in this chapter, have some history of breeding and some fundamental genetics behind that. Specialists in universities and plant-breeding companies throughout the world work on their favorite varieties, whether they are orchids, roses, tulips, or other beautiful flowers. Lately, as alstroemeria has become more popular as a cut flower, we've also learned much about it. Curiously, we know more about the popular but less economically valuable summer ornamental flowers, petunias and snapdragons, than all the others.

The Subtleties of Flower Color

The colors of flowers vary in a spectacular manner, within a single plant, among varieties within a species, and among the 280,000 species of the flowering plants. We know the combinations of pigments and their genetic controls in very few of them. Our best understanding comes from the few annuals that have been subjected to some intensive research, such as snapdragons and petunias, which are easily and quickly grown, and a few ornamental woody plants, such as roses. I can't very well cover the chemical and genetic basis for flower color in all these plants, but I can discuss some of the major processes contributing to the subtleties of flower color. They include the properties of individual pigments, mixtures of pigments, alterations of pigments, and optical properties of flower tissue.

The Flower Palette

In a few cases flower colors are primarily caused by a single pigment. After Willstätter determined the structure of the first anthocyanin, named cyanidin, the subsequent structures were named for the flowers from which they were extracted. It is pretty simple to determine that these colors are produced by anthocyanins, or another flavonoid pigment, by pressing the petal against some blotting paper and placing it in a glass so that some rubbing alcohol (with a few drops of vinegar added) will move up the paper (see appendix B). Anthocyanins move up the paper with the solvent. We'll see a bit later that it takes a different solvent to move the carotenoid pigments. Willstätter extracted cyanidin from the petals of red roses (figs. 3.16,

7.7, A.39). The orange-red color of geranium (*Pelargonium*) yielded the anthocyanidin pelargonidin (figs. 7.8, A.41). Mallows (*Malva* sp.) yielded malvidin (figs. 7.9, A.42), and peonies (*Paeonia* sp.) yielded peonidin (figs. 7.10, A.43). These last two pigments produce colors rather similar to cyanidin. The colors of violet-blue flowers are due to the presence of delphinidin, extracted from *Delphinium* (figs. 7.11, A.44). These pigments absorb in the blue-green region of the visual spectrum, and transmit and reflect blue and red wavelengths. The small differences in the absorbance ranges of these

Figure 7.8 The anthocyanin pelargonidin is primarily responsible for the color produced by this hybrid geranium cultivar. Photograph courtesy of Ball Horticultural Company.

Figure 7.9 The anthocyanin malvidin is primarily responsible for the red flower colors in hibiscus and its relatives, such as this tropical shrub *Pavonia multiflora*.

Figure 7.10 The antho-cyanin peonidin is primarily responsible for the red flower colors in this variety of *Paeonia officinalis.*

Figure 7.11 The anthocyanin delphinidin is primarily responsible for the violet-blue flower colors in this western desert species, *Delphinium nuttallianum.*

pigments alter the amount of blues and reds reflected—and dramatically change the hues. Thus the slight shift in the absorbance of delphinidin to longer wavelengths adds more blue wavelengths to the reds, producing the violet colors we are accustomed to in many flowers (fig. 7.12).

The red-purple varieties of *Bougainvillea*, so important in tropical landscapes, are produced by quite different pigments, the betalains (see fig. 3.17). The red color is primarily due to the pigment first extracted and studied in the red-purple roots of the common beet (*Beta vulgaris*): betanidin. This pigment also produces red flowers in the four-o'clock plant and in some cacti (fig. 7.13).

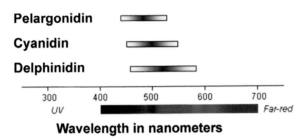

Pelargonidin

Cyanidin

Delphinidin

300   400   500   600   700

UV   Far-red

Wavelength in nanometers

Figure 7.12 A comparison of the principal absorbances of three anthocyanin pigments. Slight alterations in structure can shift the absorbance slightly to longer wavelengths, changing the hue.

Figure 7.13 Flowers colored by betalain pigments. Betacyanin pigments produce the red-purple floral bracts of *Bougainvillea* 'La Jolla' (*top left*), and the western desert cactus (*Pediocactus simpsonii, right*). Betaxanthin pigments produce the yellow-orange colors of *Bougainvillea* 'California Gold', *bottom left*.

Carotenoid pigments also vary in their cut-off of visible light absorbance, producing yellow through red colors as the cutoff wavelengths increase (see fig. 3.9). β-Carotene produces yellow to orange colors in a variety of flowers, as in the sunflower and other members of its family, the Asteraceae (figs. 3.8, 7.14). Since carotenoids are bound to lipid membranes and are water repellent, they can be distinguished from the flavonoids by their solubility in paint thinner or acetone, and they migrate in filter paper in this solvent (appendix B). Orange colors in flowers of the California poppy are produced by an unusual pigment, eschscholzxanthin (figs. 7.15, A.10). Although few red flowers are produced by carotene pigments, those

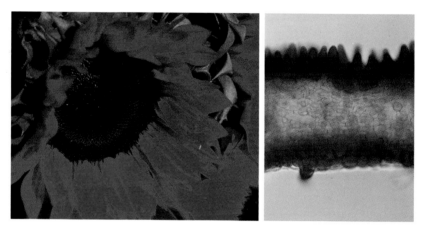

Figure 7.14 *Left*, color in the ray flowers of the sunflower (*Helianthus annuus*); *right*, transverse section of a ray.

Figure 7.15 Unusual carotenoid flower pigment. The brilliant orange flowers of the California poppy (*Eschscholzia californica*) are due to the accumulation of eschscholzxanthin.

of annatto are produced by small amounts of the same pigment that coats the seeds, bixin (fig. A.11), used by Native Americans for decoration in tropical rain forests of the New World (chapter 1; see fig. 1.3).

Yellow flowers are produced by quite different pigments. In addition to the yellows produced by carotenoid pigments, quite common in sunflowers and other members of the aster family as well as in yellow tulips and pansies, yellows are also produced by individual flavonoid pigments, such as the flavonols and aurones (see chapter 3). The bright yellow color of the yellow shower tree, native to Asia and grown throughout the tropics, is caused by a water-soluble flavonol pigment (figs. 7.16, A.35). Yellow-orange colors in snapdragons are produced by an aurone (fig. A.32), and orange flowers in the flame of the forest tree by a chalcone (fig. A.31). The yellow colors of the flower of cotton and other members of its family, such as the seaside mahoe so common in tropical landscapes, are due to accumulations

Figure 7.16 Flavonol pigments in the yellow shower tree (*Cassia fistula*). *Top*, the tree; *bottom*, the flowers.

of the flavonols gossypetin and quercetin. An anthocyanin-like pigment, 3-deoxyanthocyanidin, produced by an alternation in the flavonoid pathway, is responsible for the brilliant orange-red flowers in the African violet family (figs. A.36, 5.1), such as episcia.

Cacti and their allies make yellow nitrogen-containing pigments, betaxanthins. These produce the brilliant yellow flowers of a prickly pear cactus, and the dark yellow varieties of *Bougainvillea* (figs. 3.17, 7.13, A.48).

Mixing Pigments

Most flower colors are produced by mixtures of pigments. Thus hues can be altered by small changes in the proportions of pigments. A mixture of anthocyanins produces subtle colors in a variety of flowers, from pink to purple. Yellow to orange colors may be produced by mixtures of carotenoids, and by mixtures of flavonols and aurones, along with anthocyanins. Although plants can draw on both carotenoid and flavonoid pigments to produce yellow and orange, quite often their combination produces a copper brown color, as in the French marigold (fig. 7.17). The ranges of color bred for many popular flowers, such as roses, result from genetically manipulating the proportions of different flavonoid pigments. Yellow varieties are produced by high concentrations of quercetin and kaempferol (yellow flavonols; fig. A.35). Orange varieties contain a predominance of kaempferol and pelargonidin (fig. A.41), and deep red varieties (such as my mother's 'Mr. Lincoln') by a mixture of kaempferol and very high concentrations of cyanidin (fig. A.39). In snapdragons, orange flowers result from a combination of aurone and pelargonidin. Orange tulip varieties are produced by pelargonidin, cyanidin, kaempferol, and quercetin. The possibilities are virtually limitless, and result in different colors among closely related species, varieties within species, and even individual plants growing in different conditions.

The poinciana, native to Madagascar, is one of the most beautiful flowering trees in the tropics (figs. 5.9, 7.18). Just prior to the wet and humid summers of Miami, it fills our streets with leafless branches loaded with orange-red flowers. We even hold a small festival in celebration of flowering trees, especially commemorating the poinciana. The color of poinciana flowers is a balance between anthocyanin pigments in the petal epidermal cells and cells beneath loaded with large carotenoid-carrying chromoplasts (fig. 7.18).

At the opposite ends of this color spectrum, we have cream colors produced in certain flower varieties and in many wild species, and the dark purple-blacks of the hearts of some flowers and in certain cultivated flow-

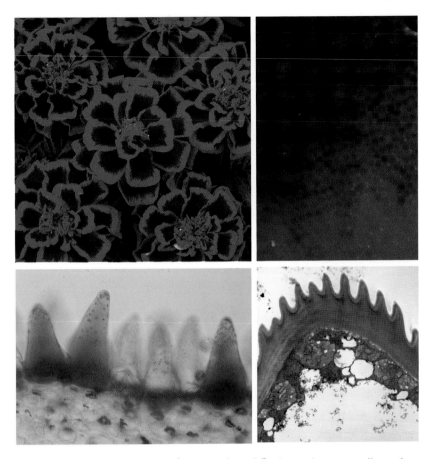

Figure 7.17 In the French marigold (*Tagetes patula, top left*), the predominant yellow color is produced by the accumulation of carotenoid pigments in the ray flowers. These water-insoluble pigments are located in chromoplasts, which pack the cytoplasm at the tip of one of the surface cells. The copper tinge in the ray flowers is due to the accumulation of anthocyanins in the vacuoles of these surface cells (*top right* and *bottom right*). Flower photograph courtesy of Ball Horticultural Company.

ers, such as tulips. The cream colors are produced with flavones and small concentrations of flavonols. The blackish colors are simply the result of the accumulation of really high concentrations of anthocyanin, particularly delphinidin.

Modifying Pigments

Although the numbers of anthocyanins, flavonols, and carotenoids are not great, particularly compared to the range and subtlety of flower colors, they

Figure 7.18 The orange-red flowers of the poinciana (*Delonix regia*), perhaps the most spectacular tropical flowering tree, are produced by a combination of anthocyanin and carotenoid pigments. Their locations are seen in the light-microscope photographs at the bottom.

can be chemically modified to alter their color properties in new directions.

Most of the flavonoid pigments are rendered soluble (and accumulate in the solution of the cell vacuole) because sugars are attached to them. In anthocyanins, sugars are primarily attached at two ring positions. Different sugars, in addition to the common glucose, may be present, and often more than one sugar is attached. These sugars may slightly shorten the wavelength of absorbance, producing a shift toward the orange. Other molecules may be attached on the sugars, through a process called acylation. Often these are simple organic acids or aromatic acids. In petunias, anthocyanins may be acylated with caffeic acid. Pigments acylated with *p*-coumaric acid dramatically increase their absorbance, although there is little shift in their peak wavelength of absorbance. Acylation may also increase the stability of the pigments in solution. Carotenoid pigments in flowers also may be complexed with long-chain fatty acids to form more stable pigments.

THE PIGMENT ENVIRONMENT · The flavonoid pigments, most important in producing flower colors, are in solution in the vacuoles of cells. Their production of color is affected by interactions with other molecules in solution (fig. 7.19). They are flat molecules, which easily associate with each other, forming stacks of stable complexes. Individual molecules, such as anthocyanins, may also associate with other flavonoid pigments, to increase stability, increase absorbance, and alter color properties. Even for pigments with chemically added molecules, these molecules may be positioned between the flat pigment molecules to alter their color properties (fig. 7.19). When another flavonoid, such as a flavonol, is mixed with an anthocyanin, the two molecules interact to form loose complexes that shift the wavelength of absorbance toward longer wavelengths (and produce colors toward blue).

Anthocyanins are also affected by the acidity of the vacuole solution; they are unstable at more alkaline conditions. Even mild shifts in acidity may alter their absorbance and color properties. The charge in the middle

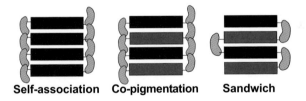

Self-association   Co-pigmentation   Sandwich

Figure 7.19 How flavonoid molecules may form stacks. *Red disks* are anthocyanins; *yellow disks* are flavonols and other flavonoid pigments; and *kidney-bean shapes* are sugars. *Left,* anthocyanins may form stacks of pigments at high concentration, deepening color intensity. *Center,* anthocyanins may form stacks with other flavonoid molecules, deepening color intensity as well as changing the hue. *Right,* anthocyanins and other flavonoids form organized stacks because, through acylation, they are attached to each other.

ring of anthocyanins promotes association with ions in solution, such as the positively charged metal ions magnesium and aluminum. These ions may add to stability and alter color production. It should be clear by now that flower color, at least its chemical basis, can be very complicated.

THE PROBLEM OF BLUE FLOWERS · We have seen a variety of pigments producing yellow, orange, red, and purple, but none is truly blue. Yet we know many blue flowers: bluebells, hydrangea, blue morning glories, blue agapanthus, and so on (fig. 7.20). There is additional interest in blueness in flowers, because many of our favorite garden and cut flowers, such as roses, orchids, tulips, and snapdragons, do not produce blues. There is intense commercial interest in the chemical bases for blue colors, so that we can transfer this capability to other flowers.

Several mechanisms, all involving the alteration of anthocyanin pigments, produce brilliant blues. In plumbago, the predominant delphinidin pigment is slightly modified, and its absorbance shifts to longer wavelengths in the presence of the flavonol azalein, a slight modification of quercetin,

Figure 7.20 Four examples of clear blue flowers, each produced by different modifications of anthocyanins, as described in the text. *Top left*, hydrangea (*Hydrangea macrophylla*); *top right*, dayflower (*Commelina communis*); *bottom left*, lead plant (*Plumbago auriculata*); *bottom right*, cornflower (*Centaurea cyanus*).

probably something like the model in figure 7.19 (center). In the dayflower, the blue pigment, commelinin, complexes with a charged magnesium atom (Mg^{2+}). In the morning glory (the heavenly blue cultivar), the blue pigment is an anthocyanidin complexed with caffeic acid residues. This complex produces blue color in special cells that are less acidic than other layers in the petal. Blue agapanthus flowers produce similar pigments. In sky blue hydrangea, the blue color is due to a complex of aluminum ion (Al^{3+}), delphinidin, and caffeoylquinic acid. If aluminum is removed, the flower color changes to red. Perhaps the blue flowers that are modified with the greatest complexity are the cornflowers. Varieties vary in color, from red, to maroon, to bright blue. The blue flowers contain the common anthocyanin, cyanidin-3-glucoside. Three of these molecules are stacked with alternating molecules of the flavone apigenin. This complex is further stabilized by the metal ions iron, magnesium, and calcium. Thus there are a variety of mechanisms for modifying anthocyanin pigments to produce blue colors.

With the growing knowledge of the mechanisms of production of blue colors, it is only natural that plant biologists turn their attention to the artificial production of blue flower varieties through biotechnology. Given the complex mechanisms that have evolved to produce flowers in nature, however, it is not surprising that success is not easily obtained. Yet the rewards in producing a truly blue rose, or other flower, would be great. Australian scientists originally working for their government (in the Commonwealth Scientific and Industrial Research Organisation, or CSIRO) established a biotechnology company, Florigene, for that purpose. They have used two discoveries, the cloning of the gene that makes the anthocyanidin pigment delphinidin in petunias and a technique to silence specific genes, to introduce foreign genes into carnations and produce flowers that are not exactly blue, but purple, and growing in popularity. They have teamed up with a Japanese company, Suntory, to apply the same genetic tricks to producing a blue rose; they announced this achievement in 2005. At present, the blue rose owes its color as much to the alteration of the digital photograph as to the production of delphinidin. The wavelength shift in color production by this pigment is not very substantial (fig. 7.12). Other alterations, such as a more neutral vacuole pH or copigmentation, will be required to produce a true blue rose. Certainly such success will eventually arrive; the will is there and the rewards are great. When I talk to my chemical colleagues about the difficulties for organisms in producing blue colors, they express puzzlement. Chemists have no trouble producing a variety of blue pigments from different precursors. A Norwegian chemist, Hans-Richard Sliwka, routinely tweaks carotenoid molecules to make them blue. But a clever chemist is not the same as the template of biodiversity and the process of natural selection.

GREEN FLOWERS · Flowers are seldom green. Colorful flowers attract pollinators (more on that later), and green flowers do not stand out from foliage. Small flowers that may not be pollinated by animals often lack color, and such flowers may be self-compatible and self-pollinating. Because of its rarity, green color has been exploited in the breeding of certain flowers, particularly orchids. The pigment in this case is chlorophyll, sequestered in chloroplasts in the petals. A remarkable example of green flowers (and an exception to the use of chlorophyll in producing it) is the bright blue-green corolla of the jade vine (fig. 7.21), originally from New Guinea and now a popular ornamental plant in the tropics. The pigment of this flower has not been identified, but it appears to be water soluble and possibly the mixture of a blue pigment (a modified anthocyanin?) and some other pigment: a real mystery.

Figure 7.21 Green flowers. *Left*, the jade vine (*Strongylodon macrobotrys*). *Top right*, the pigments are soluble in water or rubbing alcohol, which shifts the pigment color toward blue. Perhaps the color is due to a mixture of flavonoid pigments. *Bottom right*, the green petals of an orchid (*Dendrobium* 'Peng Seng').

FLOWERS THAT CHANGE COLOR · Some flowers shift through a range of colors in the process of attracting pollinators and being fertilized. In the seaside mahoe (fig. 7.22) the flowers open a brilliant yellow due to flavonols, with a dark shiny maroon anthocyanin base. When the flowers age and prepare to fall from the tree, they turn progressively redder, starting from the petal veins and expanding to the entire flower, due to the synthesis of the anthocyanidin malvidin. Its relative, *Hibiscus mutabilis*, changes from yellow to red in a single day, due to the rapid accumulation of anthocyanin. In the pansy variety 'Yesterday, Today, and Tomorrow' (*Viola tricolor*), flowers change from white to purple, due to the synthesis of its principal anthocy-

Figure 7.22 Changing flower colors in seaside mahoe (*Hibiscus tiliaceus, top left*). Anthocyanins appear sequestered as bodies within the vacuole (*top right*). Flowers of yesterday, today, and tomorrow (*Brunfelsia australis, bottom*), change from blue to violet to white over the course of three days.

anidin, malvidin, and made more purple by copigmentation with myricetin. Increase in cellular acidity as petals age may also change colors in some flowers. Fuchsias change from blue-violet to purple-red with age, because increasing acidity affects colors produced by malvidin and a copigment. The "yesterday, today, and tomorrow" flowers most familiar to me are produced by shrubs in the genus *Brunfelsia* (such as *B. australis*) in the nightshade family (fig. 7.22). In these attractive shrubs, the flowers open as a deep purple, turn lavender, and then white—all in the course of several days. I expect that this color shift may be due to a decrease in acidity, the opposite of the fuchsias. Certainly we can observe subtle shifts in color during the aging of many common flowers, and this phenomenon is much more widespread than commonly realized. Its presence in flowers of many families suggests that there may be floral guilds where color change attracts certain pollinators.

DIFFERENT COLORS IN THE HEART OF THE FLOWER · As a rule, the colors of the sexual parts of the flower, the stigma of the pistil and the anthers of the stamens, contrast dramatically with those of the accessory parts, such as petals. Anthers are generally yellow, as can be seen in many flowers illustrated in this book, and particularly in *Schima wallichii*, seen on Mount Kinabalu (see fig. 6.11). In general, both carotenoids and flavonols are responsible for this color in different plants, and the function of such compounds is not understood, although some sort of photoprotection is the most likely explanation. In a few plants, other colors are produced. In the annatto (fig. 1.3) the anthers are a deep red, due to the accumulation of the carotenoid pigment bixin (fig. A.11). Very rarely, anthers are blue, as in the Mediterranean anemone (fig. 7.23), due to their accumulation of delphinidin. Their anthers contrast with the red petals. In rare cases, the natural pigments contributing to this color, as in horse chestnut and chamomile, may be azulenes—very similar to commercial dyes of the same name.

We know more about the function of pigmentation in the stigma of flowers. They are quite frequently pink or even red, due to the accumulation of different flavonoid pigments, as in the female flowers of the castor bean (see fig. 9.3). These pigments play important roles in the mechanisms that establish compatibility between the pollen grains and the stigma, and promote the germination of the pollen tube. The most economically important color from stigmas is that produced in the saffron crocus (fig. 7.23). The chemistry of its stigmas is complex, involving flavoring agents and pigments useful as dyes. The principal contributor to color is the carotenoid pigment crocetin (fig. A.12)

Figure 7.23 Flowers with strong contrasts in color between petals and sexual parts. In the Mediterranean anemone (*Anemone coronaria, top left*), the dark blue anthers contrast strongly with the intense red petals. In the saffron crocus (*Crocus sativus, top right*) the orange-red stigma lobes contrast with the light purple petals, due to the synthesis of the carotenoid crocetin (fig. A.12). Photograph courtesy of Philippe Latour. The stigma lobes of the strawberry hedgehog cactus (*Echinocereus engelmannii, bottom left*) are bright green from chlorophyll production. The stamens of an unidentified *Medinilla* species (*bottom right*) are a bicolored blue and yellow.

Petal Optics and Plant Colors

Just as in leaves, the petal surface and distribution of intercellular spaces in the interior strongly affect the intensity and efficiency of floral coloration. The distribution of pigments within flowers may also be important. Most flowers produce convexly curved surfaces that effectively capture light, and most produce pigments in the outermost layer (the epidermis), where they are most likely to receive light. In the orchid hybrid 'Hillary', the epidermal cells are pointed (fig. 7.4), as they are in the French marigold (figs. 7.17, 7.24). However, in another orchid hybrid, *Ascocenda* 'Princess Mikasa', the

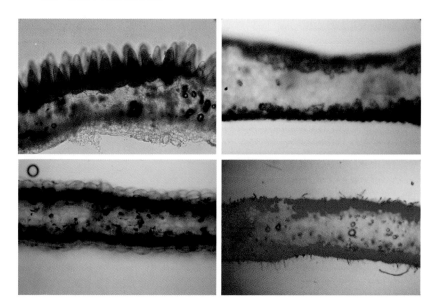

Figure 7.24 Transverse sections of flower petals, revealing differences in cell shapes. *Top left*, French marigold (*Tagetes patula*); *top right*, purple orchid tree (*Bauhinia purpurea*); *bottom left*, *Ascocenda* 'Princess Mikasa'; *bottom right*, flame of the forest (*Butea monosperma*).

cells are more round, as they are in the purple orchid tree (fig. 7.24). Pointy, or papillose, cells are more efficient at capturing light by reducing reflection. The rounded cells are more efficient at focusing light into the petal (see chapter 4). Petals vary in their surface structure, and some (as in flame of the forest, fig. 7.24) are smooth-surfaced, or even slightly hairy. Sometimes a flower is smooth on a dark and shiny interior, as in the seaside mahoe (fig. 7.22). What are the trade-offs that might favor different petals in different conditions? Perhaps the papillose cells have more surface and lose more water through evaporation than the flat ones: the orchids are rain-forest epiphytes, and the flame of the forest is a drought-tolerant tree. Flowers also vary in the layers where pigments are produced, and different colors may be produced in different layers, as in the lip of 'Hillary' (fig. 7.4).

We know something of the genetic control of the papillose cells of petals, so much so that mutants can be produced with abnormal flat cells. Snapdragons produce such cells, and mutants without the cone shapes produce duller, paler colors. These mutants are less effective in attracting bee pollinators than are normal flowers. Although more intense floral color may be the principal effect of these bulging cells, a more slippery petal surface in flat cells may also discourage pollinators.

All petals have a loose arrangement of the cells in the interior that scatter light back through the pigmented layer to promote the intense color production. This dramatic effect is revealed by the simple treatment of the petals of 'Hillary'. When the petal was immersed in water and put under partial vacuum, water infiltrated the tissue and filled up the air spaces. The result was a translucent cellophane-like transformation of the white petal. A pigmented petal treated in this way would produce very little color because most of the light would be transmitted through the petal and not backscattered through the surface.

Finally, just as leaf colors are affected by packaging of chlorophyll pigments in plastids, similar packaging may affect flower colors. Pigments in discrete bodies increase the likelihood that they will avoid absorbing part of the light penetrating a petal, although this is partly offset by the high degree of scatter within (and path-lengthening; see chapter 4). Most flowers make their yellow-orange and red-purple pigments water-soluble by attaching one or more sugars to the molecules. These are then shunted into the large vacuoles of the pigmented cells, filling some 90 percent of the cell volume with pigments. This location increases the likelihood of their absorbing light, and makes for more efficient color production.

However, pigments may be present as small particles. Flowers with yellow-orange carotenoid pigments (such as the French marigold, fig. 7.17) produce them in small plastids in the periphery of the cells, around the vacuole. All of these pigments are water-insoluble and are associated with the membranes in the plastids, or chromoplasts. Even flavonoid pigments may be condensed as smaller bodies within the water solution of the vacuoles. In the yellow lip spots of 'Hillary', the color is seen partly as diffused throughout the cell, and partly as the condensed bodies that some have called "crystals." Anthocyanins form condensed reddish bodies in many flower petals, well-studied in lisianthus (fig. 4.15). In the darker shiny areas in the throat of the flowers, the anthocyanins are highly condensed, and less so in the outer parts of the petals of lisianthus. Such particles are also seen in the aging flowers of the seaside mahoe (fig. 7.22). The functional significance of this pigment condensing is not worked out, although I expect it has something to do with the petal structure and optics. Packaging and path-lengthening effects have a tendency to broaden and flatten the peak of color production.

This is a brief summary of a complex and rich topic. There are virtually limitless combinations of pigments and modifications to produce every color perceivable by us. I have focused on the details in the few flowers that are bred for display in our homes and businesses, or to grow in our gardens. However, the driving force for the production of such color variation in nature has not been human intervention, but natural selection.

The Natural History of Flower Color

Colorful flowers attract pollinators. We have been piecing together the relationship between the pollinating animals and flowers during our history of observing and using plants. We have put this study on a more scientific footing in the past couple of centuries, and can look to the writings of Darwin to appreciate our degree of understanding in the nineteenth century. Volumes could be written about this topic, but I can only summarize our general understanding of flower color and pollination, embellished with a few examples of the sophistication of these relationships.

Pollinators are attracted to flowers in "anticipation" of a reward, a solution of sugars (and lipids, perhaps a few amino acids and vitamins) in water, the rich complex food of the pollen itself, or sometimes other rewards such as resins and secondary compounds. These pollinators are attracted to specific flowers through perception of shape, odor, and color. The importance of color depends upon the color sensitivity of the visiting animals and the time of day that the animals are active. There are several syndromes of floral characteristics and pollinator visitors, and I provide examples from a few flowers.

Small white-yellow, open, and sweet-smelling flowers are generalists. Many insects may visit such flowers: bees of different sizes, flies, and an occasional butterfly. In such flowers there is little fidelity between the pollinator and the flower shape. Such generalist flowers illustrated in this book include the paradise tree (fig. 11.14) and the false tamarind (fig. 9.8). Sometimes, however, such flowers may attract a very specific pollinator.

Bats are attracted by large white (or brown) musty-smelling flowers displayed at the edge of the plant. Bat-pollinated flowers are typically pale. The flowers of the durian (a famous fruit tree of tropical Asia, fig. 7.25) are large and pale ivory in color (probably pigmented with flavones and flavonols), open in the evening, smell like sour butter, and provide a primary reward of pollen. There is now evidence for visual sensitivity by bats well into the ultraviolet (fig. 2.13), and some flowers may effectively reflect or absorb in these wavelengths to help attract bats at dusk. Most bat-pollinated flowers don't smell particularly nice to us, but they must be attractive to bats.

Moths are attracted by tubular, very sweet-smelling, white flowers open-

Figure 7.25 Two bat-pollinated flowers and a pollinating bat. *Top left*, the durian (*Durio zibethinus*); *right*, the sausage tree (*Kigelia pinnata*); *bottom left*, *Eonycteris spelea*, the bat pollinator of the durian.

ing in the evening. Many of our sweetest-smelling garden plants, such as gardenia, tuberose, and jasmine, are naturally pollinated by moths. Because of their heavy sweetness, many are important perfume-oil plants. The large and deeply tubular flowers of the chalice vine (fig. 7.2) are pollinated by a large sphingid moth with a proboscis long enough to reach the base of the tube for the nectar reward. A few orchids are also pollinated by moths, such as the white and fragrant flowers of *Rhyncholaelia digbyana* (fig. 7.6).

Butterflies are attracted by colorful (yellow to purple) tubular flowers without fragrance that are open during the day. Many attractive garden plants are pollinated by butterflies, and it is easy to design a home or school garden to attract them. The purple tubular flowers of the golden dewdrop (fig. 5.8) are pollinated by butterflies, as are those of the bougainvillea (fig. 7.13). In this case, the flowers are not very showy (they are tubular and provide a nectar reward), but the bracts are colored by betalain pigments of varying hues. The "yesterday, today, and tomorrow" (fig. 7.22) is another butterfly-pollinated plant, but it has close relatives with longer tubular and exceedingly fragrant flowers pollinated by moths. Petunias are pollinated by butterflies in nature (fig. 5.13).

Bees are attracted by colorful flowers that are often sweet-smelling and open during the day. Bees generally use both the nectar and pollen rewards of flowers. Most orchid flowers are pollinated by bees of many different sizes. The cattleya group is generally pollinated by bees. The lip is a landing platform for the insects, and the size of the opening is optimized for the particular species that visit the flowers (fig. 7.26). Roses are bee-pollinated, but it is difficult to see this in the highly altered varieties we grow in our gardens (figs. 3.16, 7.7). Other bee-pollinated flowers illustrated here include the purple orchid tree (fig. 7.2), seaside mahoe (fig. 7.22), sunflower (fig. 7.14), delphinium (fig. 7.11), snapdragon (figs. 3.13, 5.13), and many others. Bees are generally not sensitive to the red wavelengths (fig. 2.13) and rarely visit red flowers. Bees use color to choose flowers to visit, so much so that they have been shown to select flowers in paintings—and preferred Van Gogh's sunflowers most of all!

Birds are attracted to bright red and tubular flowers, open during the day, that reward visitors with lots of nectar. Some of the most spectacular flowers displayed in this book are pollinated by birds: flame of the forest (fig. 3.13), heliconia (fig. 6.25), and coral tree (fig. 8.10). Birds are very sensitive to the red wavelengths reflected by these flowers (fig. 2.13). Families

Figure 7.26 The diagram (*left*) shows how the floral traits of a cattleya attract and orient bee visitors, so that they deposit and remove the pollinia as they move from flower to flower. *Right*, a bee-pollinated cattleya, *Cattleya bowringiana*. Drawing by Priscilla Fawcett, used with permission of Fairchild Tropical Botanic Garden.

of birds, such as the honeycreepers of the Old World and the humming-birds of the New World, have evolved in association with the flowers they pollinate. Birds that perch on flowers may be attracted to colors other than red. Finally, bird-pollinated flowers are not fragrant, as birds normally do not have a well-developed sense of smell.

Carrion flies are attracted by flowers that mimic rotting flesh. Through a combination of red anthocyanin pigments mottled with white areas (like fatty meat) and the fetid smell of the same chemicals produced during the rotting of meat, flowers have evolved that attract flies that lay eggs in "flesh" and that pollinate the flowers (fig. 7.27). Flowers of *Stapelia* have such an appearance and smell, and attract flies. Different species have fine-tuned odors for fecal, carcass, or urine mimicry. The inflorescence of *Amorphophallus* species (with many small male and female flowers inside of its spathe) is of similar, but darker, color.

In assessing the relationships between these pollinating animals and the flowers they visit, it is important to consider what the animals see, and not to look at flowers from our own visual perspective (and limitation, fig. 2.13). Bees and butterflies are generally not attracted to red flowers because

Figure 7.27 Stinky flesh-simulating flowers. *Top, Stapelia gigantea*, with flies on its petals; *bottom, Amorphophallus paeoniifolius* from my front yard in Miami.

they lack visual sensitivity to that color. Moths and bats are attracted to creamy white flowers not because of appearance, but because of odor. The pigments are a metabolic cost that such flowers need not expend for night pollination. Although many animals may lack some of the color sensitivity we have (in the red wavelengths), they make up for it with their greater sensitivity in the ultraviolet. They have ultraviolet-sensitive cones and can see pigmentation not visible to us, particularly the flavonoid pigments, flavones and flavonols, which barely absorb in wavelengths visible to us. Quite often, flowers visited by pollinators (such as bees and butterflies) produce lines and blotches that function as guides, helping to direct the pollinators toward the nectar rewards. These can be observed by using photography that is sensitive in the ultraviolet range. These days, digital cameras with filters that block out visible wavelengths can be used for this purpose, to give us the additional information that might be available to these pollinators. A colleague at Fairchild Tropical Botanic Garden, Scott Zona, has developed this photographic technique and supplied me with some photographs that illustrate these floral marks (fig. 7.28). In the butterfly ginger,

Figure 7.28 Flowers of the butterfly ginger (*Hedychium coronarium*), photographed in color as visible to us (*left*), and with an ultraviolet filter showing shorter wavelengths that would be detectable by an insect or a bird (*right*). *Bottom*, color and ultraviolet photographs of *Acmella pilosa*. All photographs courtesy of Scott Zona.

the white blossoms have a light yellow strip to our eyes, which would be a dark contrast to an insect visitor. The heart of a flowering head of *Acmella* looks yellow to us, but would be darker in color to an insect.

We have much to learn about the relationship between flower structure, including color, and the fidelity and consequences of visits by pollinators. This research is moving from a background of natural history observation into experimentation and molecular biology, leading to a much more fine-tuned understanding of these interactions. A group of gingers, the genus *Costus*, inhabits rain-forest environments throughout the tropics. Closely related species vary in the colors of flowers and bracts, and are pollinated by different animal guilds. In Central American rain forests, species with more tubular flowers and a strong red display from flowers and bracts (e.g., *Costus barbatus*, fig. 7.29) are pollinated by hummingbirds. The more open, more pale flowers and green inflorescence bracts of other flowers (*Costus malortieanus*) are pollinated by bees. It will be interesting to figure out the evolutionary relationships (the genealogy, or family tree) of all these species, using the techniques of sequencing the DNA of certain genes. Then the characters, including the different pigments produced in the flowers, can be placed on this evolutionary tree to learn more about evolutionary

Figure 7.29 Contrasting colors of the floral bracts of inflorescences of gingers help attract different guilds of pollinators, as in the genus *Costus*. *Left*, green bracts and open flowers of *Costus malortieanus* attract bee pollinators. *Right*, red bracts and tubular flowers of *Costus barbatus* attract hummingbirds.

Figure 7.30 Illustrations of flower color and pollination fidelity. *Top left*, scarlet gilia (*Ipomopsis aggregata*), this form pollinated by hummingbirds. *Top right* and *bottom left*, wild radish (*Raphanus sativus*), in which flower colors in a population affect pollinator fidelity. *Bottom right*, *Mimulus luteus*, in which pollinator fidelity is affected by the dark pattern in the corolla.

change. Perhaps the genus started out as a small group pollinated by bees, and then evolved into new species and took advantage of the availability of new pollinators.

The scarlet gilia (fig. 7.30) is a western mountain flower well-studied by population biologists. Although the most common floral color is described by its common name, other floral colors occasionally occur in natural populations. Does color change shift the visits of pollinators? There is now quite a bit of research accumulated on these little wildflowers. The scarlet gilia, with normal red flowers pigmented by pelargonidin, is pollinated by hummingbirds. Related species, pigmented with purple delphinidin or lighter in color with little pigmentation, are pollinated by butterflies. Even within the scarlet gilia, variations in color change the fidelity of pollination, and pale flowers are visited by butterflies rather than by hummingbirds; such colors and pollinators may even change within a growing season.

In the Cascade Range and the Sierra Nevada, another mountain wild-flower, the monkey flower, varies in color and morphology among different species. These species are visited by different pollinators, just as in scarlet gilia. Two species, *Mimulus cardinalis* and *M. lewisii* (fig. 7.31), are closely related and primarily differ in flower color and corolla shape. Flowers of the former species are a brilliant scarlet (pigmented with pelargonidin) with a tubu-lar corolla; these are faithfully visited by hummingbirds. The latter species produces light pink flowers (very low concentrations of the same anthocy-

Figure 7.31 Two Sierra Nevada monkey flowers, in which flower color and shape have experimentally been shown to affect pollinator fidelity. *Top, Mimulus cardinalis,* which attracts hummingbirds; *bottom, M. lewisii,* pollinated by bees.

anin) and a much broader corolla; these are faithfully visited by bumble-bees. What genetic changes would result in the differences that changed the flower structure, and the visits by different pollinators, of these two species? Douglas Schemske, now at Michigan State University, and his graduate students have been answering these questions by conducting a series of elegant experiments. They have artificially crossed the species and obtained plants with varying characteristics of the two parents, yet these combinations are never seen where the two species live together. Furthermore, they have discovered that the genetic changes that determine the differences between the two species (including the levels of anthocyanin) are mostly due to few genes with very large effects. This means that selection leading to the different flower forms of these two species could occur quickly. This is very elegant research, which Schemske once modestly described as "mopping up after Darwin"; it leads us in the direction of understanding how changes in flower color are involved in the evolution of flowering plants.

A yellow monkey flower, *Mimulus luteus*, grows in the Andes Mountains of South America. Chilean botanists have studied the effect of flower shape and dark color patterns in the throat of the flower on visitation by birds or bees (fig. 7.30). They have shown that the shape of the color patterns, the "nectar guides," influence the visits of these two pollinators. Bees are more attracted by long marks and hummingbirds by broader marks. So even the patterns of color can influence choices by pollinators.

Sometimes flower color is correlated with other characters that influence the seed production of the plants. The wild radish, escaped from Europe and the progenitor of the radishes in our salads, grows throughout North America as different-colored flower forms (fig. 7.30). Flower color is controlled by two loci; at one locus a gene dominant for anthocyanin production produces pink flowers, and at another locus a gene dominant for carotenoid production produces yellow flowers. When both genes are expressed, the flowers are bronze in color. The predominant bee pollinators prefer the yellow and white flowers. Yet some insects that attack this plant, eating flowers and leaves, prefer the plants with yellow and white flowers. The anthocyanins are somehow associated with chemicals that reduce the attack on the plants. Quite often, floral pigments are also produced in leaves, and what is observed as an effect of floral color may be due to their effect on leaves, as has been shown in the blue morning glory. Such effects may limit the color combinations of flowers.

We can see that the phenomenon of flower color is exceedingly complex. For most of the flowers in nature, we do not even know the pigments of flower color, let alone how they are modified to produce the subtle coloration that marks them apart. I hope you can add an appreciation of this complexity to the sensual delight of flower color. We will see similar complexity in the colors of fruits and seeds.

Chapter Eight Fruits and Seeds

....

I advise students on the subject of color as follows: "If it looks good enough to eat, use it."

ABE AJAY

One classic American landscape haunts all of American literature. It is a picture of Eden, perceived at the instant of history when corruption has begun to set in. The serpent has shown his scaly head in the undergrowth. The apple gleams on the tree. The old drama of the Fall is ready to start all over again.

JONATHAN RABAN

....

The apple is certainly the fruit of my childhood, for I grew up in Washington, "the apple state." I was born in Wenatchee, "the apple capital of the world," where my mother graduated from high school. My dad's father moved his family from Illinois to Chelan in 1915, one of the fruit-growing areas, so he could start an apple orchard. The young lad on the apple box label (fig. 8.1) bears a strong resemblance to my informal childhood photographs, with reddish hair sticking out at different angles, holding that red and prized apple. Fruit growing came to central Washington in the 1880s, and expanded rapidly with the arrival of the Great Northern and Northern Pacific railroads. The most popular variety, the 'Red Delicious', was discovered by a Quaker orchardist, Jess Hiatt, in Peru, Iowa, in 1872. That apple was a seedling sport of a now antique variety, the 'Yellow Bellflower' (fig. 8.2). The 'Red Delicious' was eventually purchased by the Stark

Figure 8.1 An apple-box label used in the 1940s for the Red Head brand, distributed by Northwest Wholesale, Inc., a grower's cooperative in Wenatchee, Washington. Courtesy of Dale Easley.

Figure 8.2 A collection of mainly antique apples collected in central Massachusetts, from Tower Hill Botanic Garden. The dark red apples are 'Red Delicious'. See notes for a guide to the identity of other apples.

Brothers Nursery 1894, and was made widely available. It soon became the single most popular variety of Washington apple, presently accounts for about 40 percent of the state's crop, and is still the most popular apple in the United States.

During my childhood years, it was feast or famine with apples. During and after the harvest, my mom made all sorts of things from apples, and what was put up lasted until the next harvest. Only much later, in the 1970s, were apples made available throughout most of the year with the development of controlled-atmosphere storage. Storage and marketing steadily reduced the availability of "antique" varieties to the consumer. Although we have some new and very tasty varieties of apples on the market, such as 'Jonagold', 'Gala', and 'Fuji', the number of varieties available in the produce section or fruit stand is small in comparison to what my parents ate when they were children.

I was exposed to the symbolism of the apple: the forbidden fruit of the Garden of Eden, the poisoned fruit (sure looked like a 'Red Delicious' to me!) that put Snow White into a deep sleep, and the polished apple on the teacher's desk. So the apple was a symbol of knowledge, and the dangers of knowledge, as well as a symbol of fertility. It entered into the myths of all cultures in the Indo-European language family, and beyond.

We all know at least part of the story of Johnny Appleseed (John Chapman), who grew up in Massachusetts, not far from where I worked on this chapter. Johnny collected seeds of the antique varieties of apples around Leominster and along the Mohawk Trail as he traveled west in the late eighteenth century, and began to spread apple seeds among settlers, ultimately in the Midwest. The antique varieties popular in the region were reproduced (cloned) from stem cuttings. The apple seeds that he collected did not breed true for the traits of the parents, because they were the product of sexual reproduction, not cloning. So the apples that settlers throughout the Midwest obtained from Johnny were a variety of shapes and degrees of sweetness. That was not a problem, since a predominant use for these apples was for making hard cider. That cider softened the hard edges of pioneer life, as Michael Pollan has revealed to us. Figure 8.2 depicts those antique varieties (I've added a 'Red Delicious' variety to that mixture). I collected lumpy, small fruits from wild trees in the forests of central Massachusetts, left over from the farms that were abandoned in the nineteenth century. Those fruits are quite similar to those of the ancestral trees from central Asia.

The 'Red Delicious' apple makes a good example for the discussion of fruit color and natural history to which I will devote this chapter, along with seeds. Fruits are ripened ovaries (see fig. 5.3), formed after their ovules are fertilized by the sperm cells traveling down the pollen tubes in the flower's pistil. The ovary wall develops into the fruit (or part of the fruit), and the ovules develop into seeds. An apple fruit, or pome, is the composite structure of the ripened ovary largely covered by the receptacle on which the ovary develops. So the fleshy part of the fruit is the receptacle, and the less palatable inner part that encloses the seeds is the actual ovary. Fruits, especially fleshy and colorful ones like the 'Red Delicious', are short-lived organs. Ripening is under developmental control, further modified by interaction with the environment. The 'Red Delicious' gets its first flush of red when the fruit begins to form. It turns green during development, and then red just before maturity. Fruit growth is mediated by different growth-regulating substances; one of particular importance is ethylene gas. Our discovery of this molecule's role in promoting the ripening of fruits, and ultimately their senescence, led to the development of controlled atmospheric storage, not only of apples but also of bananas.

The intensity of color of the 'Red Delicious' is promoted by the bright early autumn sunshine and crisp weather of central Washington. In nature, fruit color is timed by plants as a cue to attract animals to consume fruits and disperse seeds. The ancestors of the modern apple were small reddish fruits 2–3 centimeters across, which were eaten and spread by a variety of animals. The apple was domesticated from a complex of species present in Europe and central Asia; there are about thirty species of apples (the genus *Malus*) worldwide, including some wild crab apples in North America. The small reddish fruits of these species are a valuable food source for many animals, who then efficiently disperse the seeds packaged in a most nutritious medium of excretion. The concentrations of wild species and varieties of the domesticated apple in central Asia, as in Kazakhstan, have led us to conclude that the first apple varieties were produced there. The most direct ancestor may be *Malus pumila*, among a closely related complex of species. The first apple varieties produced very small fruits, such as the garland of dried apples produced in Ur, in Mesopotamia, some five millennia ago. From them, a bewildering variety of cultivars arose, perhaps 7,500 worldwide.

Apple varieties vary in intensity and patterns of color. The 'Red Delicious' is one of the most spectacular because of its deep red coloration. Depending upon the variety, apples employ anthocyanins to produce reds, carotenoid pigments to produce yellows, and chlorophyll pigments in chlo-

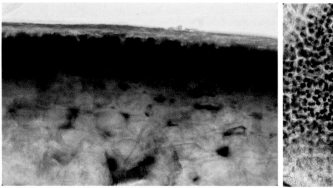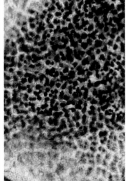

Figure 8.3 Details of the skin color of the 'Red Delicious' apple. *Left*, a transverse section shows the intensity of anthocyanin production in the skin (epidermis) and cell layers immediately beneath. *Right*, a view through the surface reveals that the anthocyanins are condensed into structures within the cell vacuole, like those described for flower petals (fig. 4.15).

roplasts to produce greens. All of these pigments may be mixed together, sometimes in patterns that are generated early in the development of the fruit, much like variegation in a leaf.

The 'Red Delicious' is a solid red, caused by the accumulation of a single anthocyanin, cyanidin-3-glucoside, in the epidermis and layers immediately beneath (fig. 8.3). The pigment appears to be sequestered in large bodies within the cell vacuoles. Light passing through these pigmented layers is scattered by the white fruit tissue beneath, and reflected back out of the surface. In apples, as in other fruits and flowers, the synthesis of anthocyanin is under genetic control. The genes controlling apple color appear at a single locus. At this locus, the gene for red color is dominant over all other genes, including yellow and green.

There is certainly something attractive about the brilliant red color of an apple. Many of us will select that color over its flavor. Apparently the first 'Red Delicious' apples were less intensely colored and were more tasty, and some taste was lost when more intense color was selected. I wonder if the same fate awaits some of the new varieties.

Color and Domestication

Most of our food is fruits and seeds, or the flesh of animals partly fed on them. We have selected desirable traits from the wild species that gave rise to these foods since the beginning of agriculture, as far back as 10,000 years ago. We call this process domestication. Color production in fruits, as in flowers, has been affected by our manipulation of plants over time. The flower examples of chapter 7 describe modifications only during the past three centuries. In the case of fruits (and seeds), we also selected traits in these plants, and inadvertently hybridized some varieties, to improve food quality, increase yield and ability to harvest, and improve resistance to pests. The apple is a relative newcomer in this history. As far as we can tell, the first domestic varieties were known only to the Greeks; Theophrastus wrote about apple cultivation. Some of the other fruits and seeds described in this chapter, and elsewhere in the book, were domesticated much earlier in different parts of the world: the squash in Mesoamerica about 10,000 years ago; beans and maize a bit more recently; the eggplant (or brinjal) in India and of unknown antiquity; citrus in China at least 4,000 years ago; and mango in India at about the same time. The domestication of these important crops was associated with the rise of agriculture and the first civilizations. Other fruits, such as the tomato and strawberry, were domesticated fairly recently. Part of the process of domestication was selecting for colors that were attractive or culturally important to the consumers, and some of those pigments are of nutritional value. In fact, these days there is a dramatic increase in interest about the principal color pigments of fruits, both carotenoids and flavonoids, because of their antioxidant activities and alleged value in protecting against cancer and aging. Rats fed extracts of blueberries and strawberries (both with high concentrations of anthocyanins) retained their abilities to remember how to run a maze better than those not fed in this way. The effectiveness of blueberries on the maze-running memories and on reduced aging symptoms in the actual brain tissue of such animals has led researchers to conclude that the antioxidant properties of these fruits (particularly because of the anthocyanins) halt memory loss associated with aging. Hence, there is a greater demand for blueberries in local supermarkets, thanks to the marketing efforts of the blueberry growers' cooperatives.

The more we learn about antiaging chemistry, the more we see that these common supermarket fruits are not only nutritious, but also essential to good health. Now we hear about the benefits of plant colors in our diets, some even advocating a mixture of plant colors to promote good health. We are even told that similar antioxidant and health-promoting molecules are found in high concentrations in the seeds of coffee and cacao. So our daily addictions may be good for us. It may be that fruit colors were selected by traditional people in the domestication of their foods, and that the early civilizations of Central America knew about the health-promoting benefits of chocolate, and the highland people of Ethiopia knew those of the coffee bean.

Perhaps flowers, sometimes added to salads in upscale restaurants, provide the same sorts of nutritional benefits. Scientists at Chrysantis, a subsidiary of the Ball Horticultural Company, are developing varieties of flowers with high carotenoid concentrations, as a source of antioxidant dietary supplements.

Given the association of fruit color with nutrition and health, along with our enjoyment of bright colors, humans most likely have been selecting fruits for bright colors for our entire history. With the rediscovery of Mendel's laws of inheritance and the establishment of the modern science of genetics, it was natural for us to focus our intention on inheritance in crop plants to create new varieties. Thus it is not surprising that we know quite a bit about the inheritance of fruit color in different fruit and seed crops.

Chemistry and Color in Fruit Ripening

Eventually even the most remarkably attractive and brilliantly colored fleshy fruits turn soft and brown-black. I bite into the apple flesh, put it down, and return later. The apple bite is now a discolored brown. I forget to return for a few days, and the entire fruit turns brown and eventually shrivels. The chemical process of browning during senescence involves the production of molecules that further react with each other as well as other molecules in the fruit tissue, the tannins (figs. A.25, A.26). Both sorts of tannins are produced, soluble and condensed, and both derive from aromatic compounds involved in the production and degradation of the flavonoid molecules. When plant cells are broken open, as when we chew on an apple, these ringed compounds are released to react with other molecules in the tissue to produce the brown color noticeable when we return to finish off that partly consumed fruit. The same occurs when we eat other fruits, such as bananas. Larger complexes of molecules constitute the condensed tannins. These aggregations of molecules are extremely complex because of the variety of precursors that may be available to join together.

Condensed tannins produce the brown or reddish brown color in the tissues (see chapter 3 for this and other modes of formation of brown and black). The short wavelengths absorbed by these aromatic molecules are extended into the visible wavelengths from the joining together of carbon rings and additions to them. Tannins may be most visible in decaying fleshy fruits, but they are also produced in other decaying organs (such as flowers and leaves) and in response to mechanical damage or attack by insects and disease. Tannins in high concentration are an effective defense against browsing by herbivorous animals. The tannins of damaged cells cross-link proteins and make the plant material indigestible, and can also bind to the animals' digestive enzymes. Ethylene produced from the damaged portion of a plant can induce production of tannins in another part of the plant, or even in a neighboring plant.

Both carotenoid and flavonoid pigments are important in producing color in fruits. Carotenoids are particularly important in producing the yellows and oranges in fruits. Here we'll see how and where these pigments are produced to create the colors of some of our most common fruits.

Peaches

Peaches are of the same family as apples, the Rosaceae. The peach also originated in Asia (from China). Peaches are the second most popular fruit, after apples. They are stone fruits, in which the ovary wall develops into an inner stone (with one seed in it) and an outer fleshy layer. The pale to yellow colors of fresh peaches are caused by mixtures of carotenoid pigments, principally β-carotene (fig. A.8). The reddish brown burnish on the shoulder, most exposed to the sun, is due to the addition of anthocyanin pigment, cyanidin-3-glucoside (fig. A.40).

Strawberries

Strawberries are also members of the rose family, with a dramatically different fruit structure. In the strawberry, individual pistils form on a receptacle (fig. 8.4). After fertilization the receptacle expands, reddens, and becomes the delicious "fruit" and popular flavoring. The fruits are actually the small ripened ovaries that individually form a single seed with a thin wall adhering to the outside (fig. 8.4). Wild strawberries are intensely delicious, but tiny. The commercial varieties have been established in several ways; a common means is hybridization between two wild species (*Fragaria chiloensis* and *F. virginiana*), and is given the name *Fragaria ×ananassa*. The brilliant red of the fruit is the simple consequence of the accumulation of many anthocyanins, but especially pelargonidin-3-glucoside (fig. A.41), in the epidermis and layers beneath.

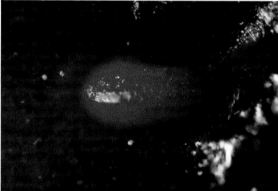

Figure 8.4 Fresh strawberries (*Fragaria* ×*ananassa*) are an aggregate fruit, in which the flesh is the receptacle alone. The actual fruits are small achenes on the surface of the red and fleshy structure, seen in detail in the bottom photograph.

Peppers

Sweet bell peppers and hot chili peppers originated in the New World and are closely related to each other. The bell peppers are members of one species (*Capsicum annuum*) in the nightshade family (Solanaceae) from Mexico and Central America, among some five species that produce the great range of color, size, and "heat" of all the peppers. This and at least four other species were domesticated in Mexico and Central and South America, and moved to the rest of the world during colonization in the sixteenth century. Different varieties of bell pepper mature as green, yellow, orange, red, or purple fruits. The green fruits produce chloroplasts in their outer layers, giving the brilliant green color of these varieties (fig. 8.5). In both the yellow and red varieties, carotenoid pigments, sequestered in chromoplasts, are responsible for the colors. In the red bell peppers, the pigments lycopene (fig. A.7) and capsanthin (fig. A.13) are responsible for the brilliant red color.

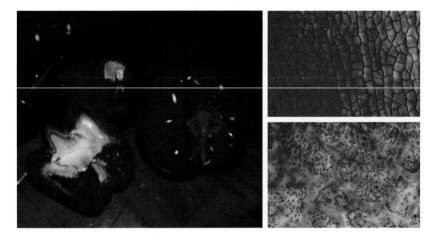

Figure 8.5 Bell peppers of different colors are all varieties of a single species (*Capsicum annuum*), which includes both sweet and hot peppers. *Top right*, outer cell layers of green peppers contain chloroplasts. Those of red peppers (*bottom right*) contain lycopene and another carotenoid pigment, capsanthin.

Tomatoes

Bright red tomatoes, fruits in the technical sense but vegetables by habit, are members of the nightshade family and were domesticated in the New World. We consume relatively few varieties produced commercially today, and many of these look much better than they taste. In certain markets, it is possible to purchase older heirloom varieties that give some idea of the shapes and colors in tomatoes that were available for further domestication (fig. 8.6). As I revised these chapters, I was fortunate to select such heirloom tomatoes from a stand down the road. We have learned quite a bit about the regulation of carotenoid synthesis in tomatoes, because of their importance in making fruits red. These controls are complex and involve genes at many positions. The principal red pigment, lycopene (fig. A.7), gets its name from the original genus of the tomato: *Lycopersicon esculentum*, now placed in *Solanum*, as *S. lycopersicum*. The outer layer contains high concentrations of this pigment in chromoplasts, and the pigment decreases in concentration toward the center of the fruit.

Eggplants

Eggplants were domesticated in India, and my wife and I became very fond of eating the small local varieties, called brinjal, when we lived in Southeast Asia. These fruits/vegetables are members of the nightshade

Figure 8.6 Left, antique varieties of tomatoes (*Solanum lycopersicum*). *Right,* view through the outer cell layer, showing the lycopene-bearing chromoplasts.

family and receive their scientific epithet from the dark colors of their fruits: *Solanum melongena*. Although there are white and purple varieties of eggplants in our markets, most varieties are a dark purple-black (fig. 8.7). The pigment mostly responsible for this intense dark color is the antho-cyanin delphinidin-3-rhamnoside (fig. A.44), with a residue of coumaric acid added to the sugar to shift the absorbance toward purple. The intense color is promoted by sunlight. It is interesting that the genes selected for desirable fruit traits in eggplant, including color, are located in similar po-sitions on the chromosomes of tomato and bell pepper, indicating very similar genetic mechanism for domestication in these three crops. This is not surprising, since tomato and eggplant are now both placed in the same genus of the same family.

Citrus

The colors of the orange and its numerous relatives are primarily due to the production of carotenoid pigments in the rind. Carotenoids, particularly β-carotene, are largely responsible for the yellow-orange of the rind and the pulp (and juice) of oranges and other fruits (fig. 8.8). The red navel orange's tint is changed by the addition of lycopene (fig. A.7). However, flavonoid pigments embellish the colors of citrus fruits in two ways. The yellow of

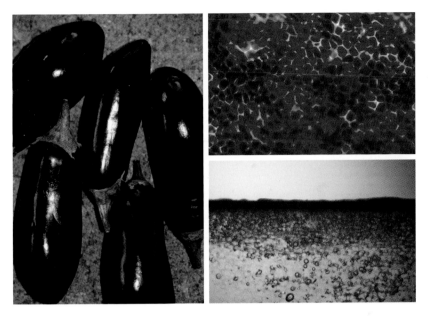

Figure 8.7 Left, a variety of eggplant (*Solanum melongena*), purchased from a Miami grocery store. *Right,* views through the skin of the eggplant, showing the purple anthocyanin pigments in the cell vacuoles.

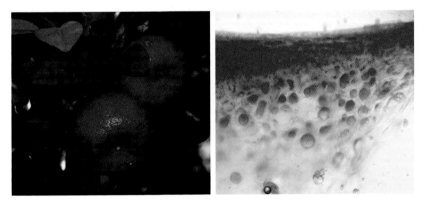

Figure 8.8 Left, the citrus variety 'Honeybelle', a tangerine (*Citrus reticulata*) in a Miami yard. *Right,* distribution of pigments in the rind, including carotenoids, soluble in the oil drops, and flavonoids.

lemon peel is partly due to the presence of a hydroxyflavonol, gossypetin (similar to dihydrokaempferol, A.33). This and other flavonoids are important in imparting flavors to various citrus varieties, and are alleged to have beneficial health-promoting effects (the "bioflavonoids"). The red surface color and red pulp of the blood oranges is produced by the anthocyanin cyanidin-3-glucoside, and not by lycopene. This pattern of color is similar to that in the pink grapefruit.

Mango

I can't resist adding mango, a member of the family that includes poison ivy and cashews, to the fruits I'm describing, because it is so delicious and so abundant in my yard and adopted city of Miami. The colors of mango fruits vary from green to yellow-orange, and are caused by an accumulation of carotenoid pigments, especially β-carotene in the outer wall of the fruit, and in the flesh as well (see fig. 8.15). In some varieties, the shoulder or even the entire fruit may turn reddish from the production of anthocyanins. In others, the fruits stay green from the persistence of chloroplasts in the outer layers.

Squashes

Squashes have the greatest variety of shapes, colors, and patterns of any fruit seen in the marketplace. These differences are the legacy of the long history of domestication, and preservation of different types by indigenous people in the New World. All squashes belong to the genus *Cucurbita*. A single species, *Cucurbita pepo* (in the Cucurbitaceae, or cucumber family), gave rise to an amazing variety of domesticated fruit/vegetables: pumpkins, gourds, squashes. Deena Decker-Walters, an expert on the domestication of cucurbits and formerly a colleague in Miami, provided me with a photograph illustrating the range of variation in squashes, in two of the five species that have been domesticated (fig. 8.9). It is difficult to gain evidence about the early history of domestication of the fruits, except from the hard parts that persist in some archaeological sites. The tough rinds of squashes, along with their seeds, are evidence for the very ancient domestication of this species in Mexico. This diversity and the importance of squashes and pumpkins in human nutrition have been an incentive for us to learn much about the control of color in these plants.

Although research on the inheritance of fruit color in squashes began soon after the rediscovery of Mendel's laws, early in the twentieth century, much of our understanding of these patterns is due to the pioneering research of a single scientist, Oved Shifriss, a hero in horticulture and one of my mentors during graduate training. Shifriss, best known as the originator of the Burpee Big Boy tomato, turned his attention to color inheritance in squashes intermittently throughout his career.

There are three interconnected factors explaining the patterns of colors in squash. First is the color itself. Squash colors vary from green (chlorophyll in chloroplasts) to orange (carotenoids in chromoplasts). Color

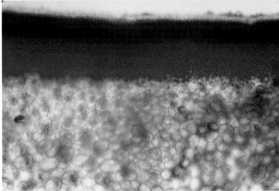

Figure 8.9 Top, squashes collected from a market. See notes for a guide to the identity of different squashes. *Bottom*, a section of a summer squash rind, showing the high concentration of carotenoids. The same pigments are also distributed in the flesh of the fruit, adding to their nutritional value. Courtesy of Deena Decker-Walters.

production is controlled at thirteen chromosome positions (loci) in squash, but three of those are of greatest importance. The second factor is the timing of pigment production; in some squashes the coloration is determined right after flowering, and in others initial coloration gives way to another color later in development, as in crookneck squash. In some squashes, such as the 'Turk's Turban', the change in color is accompanied by a change in surface texture. The third factor is differences in the production of stripes in squash fruits, as in zucchini. These stripes are associated with veins that develop beneath the fruit wall. Veins vary in width and color, depending upon the color genes that are present. The gene promoting veins is at the same locus as those promoting color. The intense-color gene suppresses expression of the stripes, and it is expressed over the light-color gene at the same position. It is difficult to think that early farmers were selecting all of these traits just for nutrition; they must also have enjoyed the amazing variety of form and selected for these as well, just as we continue to do today.

Colors in Seeds

Seeds, including some of commercial and nutritional importance, vary in color. Pigmented and nonpigmented varieties of maize (fig. 5.12) and rice were selected by ancient cultivators in Central America and Asia. However, these grain crops are not really seeds but are very simple fruits with thin walls adhering to the seed coats. Although most seeds are fairly nondescript in color, as in the apple and the other fruits just described, some are very brightly and attractively colored.

Structurally, seeds develop from ovules. The fertilized egg divides once, and one of the cells develops into the embryo—the future plant. In flowering plants, the nutritive endosperm tissue develops around the embryo, and the wall of the ovule develops into the seed coat (fig. 5.3). Seeds develop inside the ovary, which matures into the fruit. In gymnosperms (conifers and cycads, among others), the ovule develops into a seed that the embryo fills, without endosperm tissue and without an ovary wall to protect the seeds. In most cases, the basis for color in the seed is the hard seed coat, or testa.

Four of my favorite—and strikingly colorful—seeds are from the coral tree, coral bean, rosary pea, and castor bean (fig. 8.10). In the first three plants, intense red colors are deposited in the hard seed coat. The reds of the rosary pea and coral bean are due to partly condensed and highly stable anthocyanins, condensed tannins. The black end of the rosary pea (also present in some varieties of the coral bean growing in more humid climates) is probably due to the production of the plant equivalent of melanin. These phytomelanins (fig. A.29) are polymers of aromatic alcohols, really tannins that form during the development of the seed. The colors of the castor bean are produced by oxidized and polymerized phenolic compounds as well (tannins), but how the patterns form is a mystery. Each of these seeds is attractively colored, and very poisonous. The coral bean and coral tree produce alkaloids and sugar-binding proteins that interfere with protein breakdown and digestion. The castor bean produces lots of oils. Castor oil is used to lubricate fine machinery and to "lubricate" our digestive tracks when necessary. The protein-rich seed meal left when the oil is removed contains a sugar-binding protein named ricin. Ricin is one of the most toxic molecules known, one that we worry might be used as a terrorist weapon. Once assimilated, ricin destroys the cells of the digestive tract and capillaries of the circulatory system. This leads to death, and there is no antidote. To add insult to injury, castor beans also contain a toxic alkaloid molecule, ricinine. The rosary pea contains a protein toxin very much

Figure 8.10 Some of my favorite colorful seeds. *Top*, the red seeds of the coral tree (*Erythrina herbacea*), shown with the brilliant red tubular flowers. *Bottom*, from *left* to *right*: rosary pea (*Abrus precatorius*), castor bean (*Ricinus communis*), and coral bean (*Adenanthera pavonina*).

like ricin, but less toxic. One or two seeds of each of these plants can kill a human being, but the seeds must be well chewed so that they do not pass through the digestive tract intact. Seeds are often toxic; apple seeds produce cyanogenic glucosides. These compounds release free cyanide, a potent respiratory poison to small insects attacking the seeds but not likely to affect us if we swallow a few of them.

In addition to the shiny coats, seeds produce colors in other ways. In some seeds, the outer seed coat is fleshy, a sarcotesta, and may be colored with anthocyanin or carotenoid pigments. Then the inner layer becomes hard and protective. The red color of the seeds of the cardboard zamia (fig. 8.11) is due to such a layer, as is the bright red of the individual seeds of the lignum vitae (fig. 8.12). The blue color of seeds in heliconia, an herb common in the American tropics and a popular ornamental plant in tropical latitudes, is also caused by a sarcotesta (fig. 8.11).

A fleshy outgrowth may also develop from the outer layer at the point

Figure 8.11 Colorful and fleshy seed coats (sarcotestae) are found in a small minority of seeds. The fleshy seed coats of heliconia (such as *Heliconia stricta* seen at *top* and *H. mariae* seeds at *bottom left*) are a brilliant blue, due to the production of anthocyanins. In the absence of fruits, the red sarcotestae of the cardboard zamia (*Zamia furfuracea, bottom right*), a gymnosperm, attract dispersal agents. These seeds sprout all over my yard.

of attachment of the seed to the ovary wall, or funiculus. Such a structure is called an aril. An aril may cover most of the seed, as in the case of the traveler's palm (fig. 8.13), or only part of the seed, as in the reddish aril of the nutmeg seed, which is the familiar spice called mace. The tissue that grows from the funiculus at the base of the seed may also be colored and elongated, to produce an elaiosome. The ear-leaf acacia, a fast-growing tree common throughout the tropics and subtropics, produces twisted brown pods that snap open to reveal shiny black seeds dangling from bright or-

Figure 8.12 The beautiful blue flowers and red seeds breaking through the yellow fruit wall of the lignum vitae (*Guiacum officinale*), native to south Florida and the Caribbean. This is another example of a seed with a sarcotesta.

ange elaiosomes (fig. 8.13). Conspicuous structures may develop from the opposite end of the seed, around the micropyle of the ovule. In the case of the castor bean, a little of that tissue develops into the swelling known as a caruncle, which helps give the castor bean the appearance of a scarab beetle (fig. 8.10). The scarab was an important little animal to the ancient Egyptians, often depicted in their arts and crafts, and castor beans were stored in King Tut's tomb.

Along with maize and squash, beans were domesticated in the New World at a very early period. This domestication included selection of colors of the seed coats. We are familiar with this color variation today because many beans that are important in our diet, kidney, navy, pinto, and black, have dramatically different colors (fig. 8.14). These modern varieties were bred from ancient races of beans throughout the New World. As in the rosary pea, the wild species also contained sugar-binding and antidigestive proteins, but were only mildly toxic. Improved digestibility and nutritional value of beans was achieved during domestication. Beans collected in a market in Ecuador, and representing a wild species at the earliest stage of domestication (fig. 8.14), illustrate the dramatic variations in color and pattern produced in a single crop on a traditional farm. This variation was selected by Native American cultivators in northern Mexico

Figure 8.13 Colorful accessory seed parts. Arils are seen in blackbead (*Pithecellobium keyense*), *top left;* nutmeg (*Myristica fragrans*), *bottom left;* traveler's palm (*Ravenala madagascariensis*), *bottom right;* and the ear-leaf acacia (*Acacia auriculiformis*), *top right.*

and the southwestern United States into distinct color patterns, flavors, and growth responses. Just as in apples and squashes, this variation was the raw material for geneticists to work out the inheritance of color production. This research began very soon after the rediscovery of Mendel's laws, by some of the founders of genetic research in the United States and Europe. After all, beans are related to the sweet peas that Mendel chose as his model organism.

The colors of bean seeds are all produced by flavonoid pigments. The yellow colors are caused by flavonols, such as quercetin glucoside. The purple and black colors are produced by combinations of three common anthocyanins, delphinidin-3-glucoside (fig. A.44), petunidin-3-glucoside, and malvidin-3-glucoside (fig. A.42). The genetic controls of the pigment production are complex. There are three loci for color that can be expressed

Figure 8.14 Varieties of the common bean. *Top*, a land race collected by Paul Gepts at a traditional market in Ecuador, showing spectacular variation in color, shape, and size in semiwild cultivated plants. *Bottom right*, the common edible bean varieties: white, kidney, navy, black, and pinto. *Bottom left*, traditional bean races maintained by Native Americans in the southwest United States and northern Mexico.

only in the presence of a gene at another locus. Additionally there are four modifying genes that darken colors from the color genes. Such complex controls on color production makes one think that such patterns have some biological importance, as in dispersal. There is much more interest in the genetics of bean seed colors these days because of their influence on the synthesis of flavonoids. Bean varieties with high levels of these pigments may be nutritionally advantageous for us (just like the blueberries and strawberries we eat, hoping to retard aging).

Patterns of color are also genetically controlled. The Anasazi pattern, seen in some of the Native American varieties (fig. 8.14, top and middle of lower left photograph), is caused by a recessive gene expressed at one of the color loci, which produces the partial coloration in these seeds. Bean seed colors and patterns are certainly complex and beautiful.

Bean color has also become a controversial issue in the world of biotechnology. A yellow bean was patented by a U.S. scientist, which made it technically illegal for traditional cultivators to grow their own preserved cultivars with the same color and appearance. Very recently, Paul Gepts showed that the "engineered" bean was genetically virtually the same as one of the traditional varieties, maintained by Native Americans for centuries, invalidating the patent.

Fruits, Seeds, and Dispersal

Flowering plants, which originated about 125 million years ago and became dominant after the extinction of dinosaurs 60 million years later, have two advantages over their gymnosperm ancestors. First, the visibility of flowers, partly through color, helps attract animals to move pollen from plant to plant. Second, fruits provided new means of dispersal. The evolution of wings and parachutes promoted dispersal by wind, and the evolution of hooks and glues promoted attachment to motile animals. Many plants promote seed dispersal by attracting animals to consume the fruit as food and to disperse seeds by expelling them in their feces. Such fruits provide nutritive rewards to such animals, often some fleshy material rich in sugars and starches, and rarely also oil and protein. To advertise for the services of certain animals, syndromes of size, fleshiness, nutrition, and color evolved. Color became an important part of the attractiveness to animals, and a range of color and color combinations appeared. Color also became an important signal of readiness of seeds for dispersal, and the ripeness of the reward.

Among the fleshy fruits, two general types evolved. The first type is large, with a tough outer husk and fleshy tissue inside. Such fruits are usually dull in color, including yellows, greens, and browns (fig. 8.15). These fruits attract large terrestrial animals. The durian tree, of Southeast Asia, produces an enormous spiny fruit, larger than a football, and large flowers (fig. 7.25). Once the tough outer covering is split open, the nutritious and delicious fleshy arils surrounding the large seeds are revealed to the consumer: elephant, tapir, human, even tiger. The mamey of Central America (fig. 8.15) is dull brown, about 15 centimeters long. Its fibrous cover is relatively easy to break open to reveal the inner salmon pink and sweet flesh

Figure 8.15 Examples of large fruits, not strongly colored and dispersed by larger animals. *Left,* mango (*Mangifera indica*); *top right,* mamey colorado (*Pouteria sapota*); *bottom right,* durian (*Durio zibethinus*).

surrounding the very hard, shiny, black seeds. These fruits (and seeds) are ingested by numerous terrestrial animals, including agouti and tapir, in Latin America. It is a prized item in human diets in the region, as well. The mango is also a large fruit, mostly dull in color in nature, and with a single large and protected seed. It is consumed and dispersed by many animals in its Asian tropical forest home.

The majority of fruits in temperate and tropical forests are small (less than 2.5 centimeters across), colorful, and thin-skinned (fig. 8.16). The fruits are easily opened to reveal nutritious flesh and small seeds. They are consumed and dispersed by birds and, in the tropics, by primates. These fruits are visually much more attractive than those consumed by terrestrial animals. The latter animals have color vision in the blue and green wavelengths, not in the red. Birds (and some primates, including us) are sensitive to red wavelengths and can distinguish among blues, greens, yellows, oranges, and reds. Surveys have revealed that most bird-dispersed fruits are red to black. Wild roses produce fruits as "hips," red, rich in vitamin C, and consumed by birds (fig. 8.16). Figs are frequently dull red to violet and are displayed in masses on trunks and large branches. The fleshy red

Figure 8.16 Examples of fleshy fruits with strong visual attraction, either through color or contrast. *Top left*, Brazilian pepper (*Schinus terebinthifolius*); *top right*, *Lasianthus attenuatus*; *bottom left*, wild rose (*Rosa gymnocarpa*); *bottom right*, Kashmir mountain ash (*Sorbus cashmiriana*).

seeds of lignum vitae are also a vivid display (fig. 8.12). White is the next most important color, among the most important foods for birds in tropical forests. White fruits may stand out because of their contrast with the green of foliage, as in the snowberry from conifer forests in the Northwest, or the Kashmir mountain ash, whose white berries stand out from the dark branches at their tips (fig. 8.16). These fruit colors are followed in importance by orange, blue, brown, yellow, and green. The yellow fruits of the golden dewdrop are also dispersed by birds (fig. 5.8). These color patterns are seen in the tropical rain forests of South America and West Africa.

In a temperate forest in Illinois, blue-black fruits are most common, yet in southeast Alaska, red fruits predominate. These fruits mature much earlier in the higher latitudes in Alaska, compared to later autumn maturity in Illinois. The interactions between animals and plants, at the heart of these color differences, are a mystery. It will take some time to sort out the ecological reason for these differences. Although we see broad patterns of fruit characters in tropical forests, the interactions that explain color and dispersal are even more complex and less understood.

Plants that are invasive often have very efficient means of animal dispersal, which allow them to quickly colonize new sites. In south Florida, one of our most invasive trees is the Brazilian pepper, or Florida holly (fig. 8.16). It was introduced to our region as an attractive ornamental, for the glossy foliage and brilliant red berries produced during the autumn months. Song birds avidly seek out these trees, gorge on the berries, and distribute them everywhere.

What particularly stand out among the fruits of the tropics are the combinations of color that are so striking in many plants. Red is often associated with black, usually as black seeds on a red background. Cola nuts (yes, formerly an ingredient in the flavor formulae in cola drinks) are hidden beneath a black sarcotesta and are displayed in contrast to the brilliant red background of the interior fruit wall (fig. 8.17). This contrast is com-

Figure 8.17 Tropical fruits with strongly contrasting color displays. *Top left, Tabernaemontana orientalis; top right, Aechmea woronowii; middle left,* bamboo palm (*Chamaedorea seifrizii*); *middle right,* cola (*Cola digitata*); *bottom left, Dillenia suffruticosa; bottom right, Polygala oreotrephes.*

mon in this plant's family and it occurs in fruits of tulipwood, in the family Sapindaceae. There are additional variations on this theme, often adding creamy yellow to the red and black. A well-appreciated example in the West Indies is the fruit of akee, an African tree brought by slaves to Jamaica. Akee fruits ripen to a pink-red and split open to reveal shiny black seeds suspended beneath a creamy aril (fig. 8.18). The aril is toxic when green, but edible and sought after by Jamaicans (in Miami as well) when ripe. Francis Hallé encountered and drew the fruit of a vine of the family Connaraceae in West Africa, its black seeds suspended from a creamy aril and red fruit wall (fig. 8.18).

Any color combination of good contrast in our vision, and apparently in that of birds as well, seems fair game and is encountered among tropical rain-forest plants (figs. 8.17, 8.18). I add here, for illustration, blue fruits on red pedicels in a bromeliad, *Aechmea woronowii*; red fruits from the sarcotesta on an orange fruit wall in *Tabernaemontana orientalis* from the Asian tropics; black fruits on red pedicels in *Clerodendron speciosissimum* (or blue fruits on red pedicels in *C. minahassae*, fig. 8.19); black seeds on a brown capsule sus-

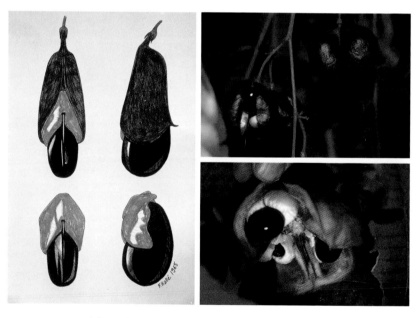

Figure 8.18 Tropical fruits from different families offer a display of red walls, seeds with black sarcotesta, and often a creamy to yellow aril. *Left, Manotes* sp. (of the family Connaraceae), drawn by Francis Hallé; *top right,* tulipwood (*Harpullia pendula*); *bottom right,* akee (*Blighia sapida*).

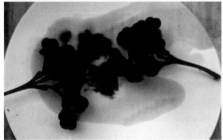

Figure 8.19 Clerodendrons, tropical Asian shrubs of ornamental importance, often produce color-contrasted fruit displays. *Top, Clerodendron minahassae; bottom, C. speciosissimum,* anthocyanin pigments extracted with rubbing alcohol (see appendix B).

pended from a red aril in the blackbead, a tree of south Florida and the Caribbean (fig. 8.13); black seeds on orange pedicels in *Chamaedorea seifrizii* from Central America (fig. 8.17); blue seeds peeping out of the red spathe in different species of *Heliconia* from Central America (fig. 8.11); a brilliant red sarcotesta against a yellow fruit wall in the lignum vitae (fig. 8.12); and seeds surrounded by a dark red pulp on a salmon pink fruit wall in the Jamaican caper (fig. 4.5). These color combinations are not limited to the tropics. In doll's eyes, a perennial wildflower in northeastern United States, the white fruits with a black spot are poised on a dark red pedicel, and in sassafras we have dark blue fruits on a bright red pedicel (fig. 8.20).

In most cases, we do not know what pigments produce these colors. We can get a good idea by performing experiments like those for flowers, using filter paper and rubbing alcohol or paint thinner (appendix B). So

Figure 8.20 Two temperate species with color-contrasted fruits reminiscent of tropical plants. *Left*, doll's eyes (*Actaea pachypoda*); *right*, sassafras (*Sassafras albidum*). Courtesy Ken Robertson.

in the case of the black fruits of *Clerodendron speciosissimum*, the fruit pigment migrates on the filter paper with the rubbing alcohol. The retention of the dark blue color suggests an anthocyanin and flavonoid copigment linked together to a sugar (fig. 8.19).

These contrasting color combinations in fruits bring us back to the color patterns in seeds, specifically the red black combination in the rosary pea and numerous other seeds in the humid tropics. This color combination might visually attract birds, but there is no fleshy and nutritious reward for the visitor. The hard seed coat, if broken, could even be lethal to the visitor. Other bright red seeds such as the scarlet bean and the coral bean contrast with the dark colors of the pod wall. Many of the seeds with such bright and contrasting colors are poisonous, so the color could be viewed as a warning . . . but that would keep animals away and the seeds undispersed. It is possible that such color combinations mimic those in fruits, where the dispersing animals obtain a reward. If the seed color combinations occur at very low frequency, such a ruse would work, just as it does in the warning coloration of the toxic monarch butterfly (which obtains its poisons by eating milkweeds as a caterpillar). Not only is that butterfly protected, but also a few species that escape being eaten by mimicking the wing patterns of the monarch. In a similar way, attractive seeds might be picked up and dispersed by birds.

We actually know rather little about the factors that contribute to fruit choice by animals, particularly in more remote locations as tropical rain forests. Just as in the pollination of flowers by animals, these interactions are likely to be complex and fine-tuned between these very different organisms, each obtaining some advantage. Brightly colored fruits are also attractive to us. This is not just some modern merchandizing ploy to get us to purchase stuff in the supermarket, but part of the ancient relationships we have with plants, just like that 'Red Delicious' apple.

Chapter Nine Stems and Roots

....

Garden writing is often very tame, a real waste when you
think how opinionated, inquisitive, irreverent and lascivious
gardeners themselves tend to be. Nobody talks much about
the muscular limbs, dark, swollen buds, strip-tease trees and
unholy beauty that have made us all slaves of the Goddess
Flora.

KETZEL LEVINE

The tree is more than first a seed, then a stem, then a living
trunk, and then dead timber. The tree is a slow, enduring force
straining to win the sky.

ANTOINE DE SAINT-EXUPÉRY

....

In most cases the colors of stems and trunks are less
flashy than those of flowers and fruits, but are more
durable. Flowers and fruits are ephemeral, and the trunks
of trees last for centuries. Thus the colors and textures
of trunks and stems are permanent features of the landscape. As trees grow, they are threatened by the physical
harshness of their environments and must defend against
predators (herbivores) and diseases. The trunks show evidence of these continuing battles of survival, something
like frozen combat. Because they are long-lived and because they form the essential links between the roots and
leaves, they are particularly vulnerable. Just beneath the
surface of each stem the conductive tissues of the plant
(xylem and phloem) are positioned to move the vital fluids of the plant. Water and dissolved minerals are transported via the xylem to the growing tips of the tree. Sugars

and some amino acids are translocated via the phloem from their sites of production to their sites of use. Generally this means moving from the leaves to the roots, or to developing tissues in the aerial parts of the plant. These tissues, wet and nutritious, are a valuable source of food for all sorts of organisms, especially insects and fungi. Outside of these vulnerable tissues in both roots and branches, plants have evolved elaborate chemical and structural defenses against such enemies. The metabolic pathways that produce the vivid colors of fruits and flowers also produce these chemical defenses, and a consequence of this production is the colors of many trunks and branches. As a result, trees produce many of the color patterns in the landscapes where we live and visit.

The trunks of the ponderosa pine are a good example. Ponderosa pine is a familiar tree of the conifer forests on dry slopes of the Cascade Range, and along moist creek beds near my hometown (fig. 9.1). The handsome, thick bark is partly explained by the ecology of this tree and these forests. The bark may bear the burn marks of an earlier fire, one that the tree easily survived, for this thick bark gives good protection against fire. The resin pushing out in cracks and holes reminds me of the numerous pests of this

Figure 9.1 The ponderosa, or yellow, pine (*Pinus ponderosa*), photographed in eastern Washington.

Figure 9.2 Trunks of two gum trees, or eucalyptus, of Australia and New Guinea. *Left*, the rainbow eucalyptus (*Eucalyptus deglupta*); *right*, cadaghi (*E. torelliana*).

tree, and the tree's sophisticated chemical defenses against attack. Pine-bark beetles enter the trunk, assimilate some of the defensive compounds present in the bark, and signal other beetles to enter and reproduce. It gets much more complicated, and some of the details will be discussed toward the end of the chapter.

Dramatically different in appearance, and no less spectacular, are the multicolored trunks of a tropical gum tree from the forests of New Guinea, *Eucalyptus deglupta* (fig. 9.2). Some call this tree the rainbow eucalyptus because of the multiple bark colors, but my students call it the GI-Joe tree, from the bark's similarity to army camouflage. Another eucalyptus, *E. torelliana*, starts out with a smooth chalky blue bark and then begins to establish a more conventional flaking bark as the trunk expands. This color production has both a chemical and a structural basis. To understand how such different colors can be produced in these trees, it is necessary to cover some basic plant anatomy and development. Trunk color may be less important than other features, such as leaf and fruit color, in the natural history of plants, and I'll discuss this toward the end of the chapter.

Shoot and Root Development

Stems (and roots) are derived from the meristematic regions at their tips; these ultimately come from the embryo of the seed (chapter 5). The shoot meristem produces the stem, and makes leaves at intervals down its length. The lateral shoots or branches then form from buds at the base of those leaves. The organization of the root meristem is less complicated, but it produces cells that make up the body of the root. The cells formed from these meristems elongate and then specialize into the different cells that compose the tissues of the stem or root. Castor beans, also used to illustrate the importance of meristems in producing patterns (fig. 5.4), are ideal for explaining the development of the stem. They start out as an herb (non-woody) but are capable of growing into a small tree, with a thick bark (fig. 9.3).

All stems produce an epidermal layer around their outside. In flowering eudicot plants, the internal tissue of the stem is divided into cortex and pith because of the circular arrangement of the tissues (or vascular bundles) that conduct water and nutrients (up the stem) and translocate sugars and some amino acids (generally down the stem, but from source to sink). All of this growth, derived from the tip, lengthens the shoot or root, and is called primary growth. In eudicots, as stems grow older and thicker, additional growth is derived from another meristematic region, the cambium. The cambium is composed of dividing cells within the vascular bundles; these cells expand laterally, eventually producing a complete circle, seen nicely in the growth of the castor bean plant. In longitudinal section (fig. 5.4), the meristem produces leaf primordia that are placed on the stem as it expands. As the seedling grows, the shoot extends and initially thickens, all due to cell divisions at the tip. In cross section the stem is typical of any eudicot seedling (fig. 9.3). In the tip of the shoot we have primary growth, which forms the ring of vascular bundles. Those bundles expand to establish a continuous ring, with the cambium tissue in the middle. The cambium produces cells that produce xylem to the inside (for transport of water and minerals) and phloem to the outside (for translocation of sugars and amino acids). Eventually the castor bean can grow into a small tree, with a trunk as thick as my arm, and with corky bark.

In their early growth, stems produce a remarkable variety of colors and textures. The stem of the prickly pear cactus is the pad itself; the spines are modified leaves. Thus the stem is the bright green photosynthetic organ of

Figure 9.3 The anatomy of shoot growth in the castor bean plant (*Ricinus communis*). *Top left*, the young shoot with male and female flowers; *top right*, the young shoot in transverse section, showing primary growth; *bottom right*, older shoots making the transition to secondary growth; *bottom left*, an old trunk with thick bark covering the stem.

the cactus (fig. 3.17). In *Didierea trollii*, a succulent desert plant from South Africa, the reddish stems (due to a betalain pigment) are armed with sharp spines and produce whorls of small leaves (fig. 9.4). The smooth stems of the black bamboo accumulate high concentrations of condensed tannins in the epidermis; most other bamboos are green (fig. 9.5). The orange-red shoots of the sealing wax palm are produced by carotenoid pigments in the leaf-base sheaths. When the leaf dies and falls off, new leaves grow to replace the lost color. The remarkable petioles (not really stems) of the massive leaves of *Amorphophallus* species, fairly common in the shade of tropical

Figure 9.4 Left, the massive shootlike petiole of an *Amorphophallus* plant; *right*, the well-armed stem of *Didierea trollii,* a desert plant of South Africa.

Figure 9.5 Stems of monocots are mostly derived from special meristems at the tip, and do not increase in thickness as they grow. These stems have an outer layer of epidermis. *Left*, a smooth, black stem of a bamboo (*Gigantochloa atroviolacea*); *right*, a smooth stem of the sealing wax palm (*Cyrtostachys renda*), where the red color is produced by the sheaths at the bases of leaves.

forests in Asia, are colored and splotched in a ways unique to each species (fig. 9.4).

In monocots, the vascular bundles are scattered throughout the stem. Secondary growth, generally not very common, takes place by different mechanisms. The most common examples of woody monocot growth are among the palms. In palms, no cambial layer is established, and the thick and woody palm trunk is formed by a special thickening meristem, shaped like a doughnut beneath the shoot tip. The appearances of palm trunks vary, from smooth in the royal palm to very rough and thick in the Canary Island date palm (fig. 9.6), and the mechanisms by which the tough surface forms are distinct. In palms, marks on stems are typically the scars where leaves were once attached, such as the ringed scars on the otherwise smooth trunk in the buccaneer palm. Projections on stems are derived from the bases of leaves. The packed leaf bases of the Canary Island date palm give its trunk a very distinct appearance. The prickly spines on the stems of the rattan palms are also derived from leaf bases. Even their leaves are armed with sharp and recurved hooks. In Malaysia we called these plants lawyer

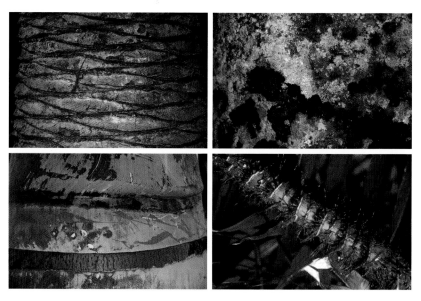

Figure 9.6 Trunk surfaces of four palms. *Top left*, Canary Island date palm (*Phoenix canariensis*); *top right*, royal palm (*Roystonea regia*); *bottom left*, wine palm (*Pseudophoenix vinifera*); *bottom right*, a rattan (*Daemonorops ochrolepis*).

vines. It is hard to imagine how these lethal-looking stems end up as lawn furniture, but the native collectors in Southeast Asia easily strip the spines away before they take them out of the forest and sell them to traders. The spines on the trunks of erect palms are also derived from leaf bases.

Protecting Stems and Trunks

As plants grow, most of the products of photosynthesis from the foliage (and even from young, green stems) are translocated along the stems via the phloem tissue. The conductive cells of the phloem, the sieve-tube elements, are loaded with sugars and some amino acids. Tissues vital to the function and survival of a woody plant are thus located very close to the surface of the stem, particularly among the eudicots. These tissues are vulnerable to physical damage, as well as to attack by insects and disease and even by parasitic plants. Initially, cells in the cortex may develop into tough fibers or stone cells to provide protection, and the epidermal layer may secrete a thick coating of waxes and cutin on the outside. Even then, the delicate stylet of an insect, such as an aphid, can penetrate to suck out sap from the stem.

In most plants, exposed organs develop photosynthetic tissue near their surfaces, are green, and are capable of fixing carbon. Young stems actually produce stomata to aid in gas exchange. Thus twigs normally are some shade of green, but may be altered by the production of other pigments, such as anthocyanin (already seen in castor beans, fig. 9.3). In some older plants with thick branches, the stems may retain photosynthetic tissues, as in the gumbo-limbo (fig. 9.7) and the silk cotton tree (fig. 9.8). An important protective layer subsequently forms in the older stems. As the stem thickens, the pressure of expansion distorts the cells of the cortex, although cell division may allow the continuous covering of epidermis to persist. In most woody plants, a new meristematic layer is established that produces more protective tissue to the outside (bark). Such protection becomes more important as the trunk becomes the Achilles' heel for the growth and survival of such large organisms; most of the products of photosynthesis from the thousands of leaves of the crown move through this vulnerable layer. Destruction of cambium and phloem in a small section around the base of the tree will eventually kill it. The cambium and phloem may be vulnerable to extremes in temperature (a sudden rapid freeze or exposure to fire) as well as damage by insects and disease.

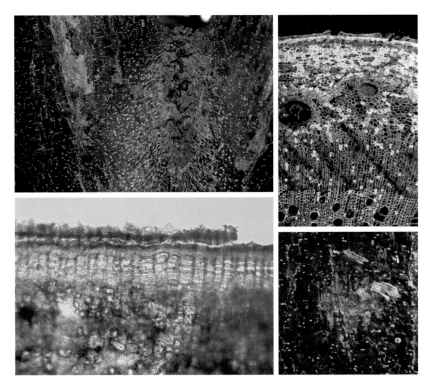

Figure 9.7 Bark formation in gumbo-limbo, also known as tourist tree (*Bursera simaruba*). *Top left*, a tree trunk about 40 centimeters across; *top right*, a light-microscope photograph of the stem, showing bark on the outside; *bottom left*, bark formation in a living stem, showing the thin layers produced and the photosynthetic tissue beneath; *bottom right*, a detail of the bark, showing the photosynthetic tissue underneath the surface.

The Formation of Bark

In very few trees, the epidermis and cortex may persist and expand to provide a continuous layer around the trunk, with a very smooth and even surface. However, most trees produce protective bark tissue. This tissue derives from a new meristem layer, phellogen, which develops from cells in the cortex, or perhaps deeper in the stem from phloem tissue. This new layer produces specialized cells to the inside, the phelloderm, and to the outside, the periderm, which we call bark. The outer bark cells are typically dead at maturity and often loaded with secondary compounds. The primary phellogen layer produces the thin barks that we see in young branches of many plants and that may persist in much larger trunks of others, as in the trembling aspen (fig. 9.9). An example of phellogen and bark forma-

Figure 9.8 Some examples of tree bark. *Top left*, pink silk cotton tree (*Ceiba insignis*); *top right*, false tamarind (*Lysiloma sabicu*); *middle right*, paperbark tree (*Melaleuca quinquenervia*); *bottom left*, live oak (*Quercus virginiana*) with resurrection fern (*Polypodium polypodioides*) growing on it; *bottom right*, large trunk of a madrone tree (*Arbutus menziesii*).

tion commonly used in textbooks is the American linden. Its young stems quickly establish the cambium layer from the fusion of vascular bundles, and they thicken. This thickening requires the establishment of a protective layer, produced by the phellogen (fig. 9.10). As the trunk grows even larger, the bark thickens and can be pulled away in strips.

The barks of most large trees are ultimately produced from the establishment of secondary phellogen layers beneath that first continuous one. The thickness, duration of activity, and types of cells produced determine the appearance of tree bark. Sometimes the transition between the first surface and mature bark is quite abrupt and remarkable, as in *Eucalyptus*

Figure 9.9 Tree barks in forests along creeks in the San Luis Valley at Crestone in southern Colorado. *Bottom left,* the mainly smooth bark of trembling aspen (*Populus tremuloides*); *bottom right,* the thick and ridged bark of the cottonwood (*Populus trichocarpa*).

torelliana (fig. 9.2) and in the madrone (fig. 9.8). In thin and smooth barks, wide bands of phellogen produce thin layers of tissue, which may peel. A remarkable example of such bark is seen in the gumbo-limbo, native to south Florida and the Caribbean region. A careful look at the surface reveals lenticels (fig. 9.7), which provide a means of gas exchange between the bark tissue and the outside atmosphere. Peeling away the paperlike layers of dead cells exposes the dense green of the photosynthetic tissue beneath. The peeling and reddish bark give rise to another popular local name, the tourist tree, for the hordes of sunburned tourists that visit south Florida to escape the northern winters. In temperate landscapes, the white bark of the paper birch is also produced in very thin layers, each deposited with wax particles that scatter light effectively. In a similar manner, paper layers of thin white birch bark are easily pulled from the trunks. The thicker layers of paperbark (also known as melaleuca or cajeput), a very invasive exotic tree

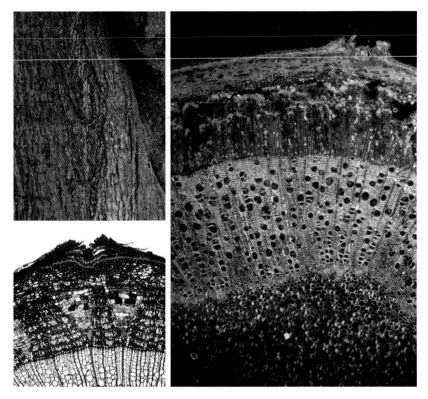

Figure 9.10 The formation of bark in the young branch of a linden tree (*Tilia americana*). *Top left*, the smooth bark of a young trunk, with a twig and buds against it; *right*, a transverse section of a young stem, showing the establishment of secondary wood to the inside and the formation of bark to the outside; *bottom left*, a detail of bark formation, with a lenticel at the top.

in south Florida (fig. 9.8), can be pulled from the trunk in large sheets. In the false tamarind tree, the plates of bark are thicker, and peel off in more narrow strips. In the pink silk cotton tree, the green tissue derives from the first continuous phellogen. Branches of such trees are highly photosynthetic. Then secondary phellogen is established to form the sharp spines on its branches (fig. 9.8). Spines of various sorts are produced by many trees.

In most cases, trunk expansion requires the production of plates of protective bark, from those regions of phellogen establishment. Trees vary dramatically in the types of bark they form, and these are often good characters for identification, particularly in temperate regions. The bark of the

Figure 9.11 Tree barks in large cone-bearing trees on the west slope of the Cascade Range in Washington. The two trees, left to right, are western red cedar (*Thuja plicata*) and Douglas fir (*Pseudotsuga menziesii*). *Top right*, bark details of Douglas fir conspicuously covered with wolf lichen; *bottom right*, bark details of western red cedar.

cottonwood is among the thickest of temperate trees (fig. 9.9). Bark that is produced in plates, as in ponderosa pine and a more moisture-loving conifer of the Northwest, the Douglas fir (fig. 9.11), is produced by such intermittent layers of secondary phellogen. Narrower and longer layers of phellogen produce a more fluted surface, as in the live oak (fig. 9.8). Other barks can be peeled from the trunk in long strands, as in the western red cedar (fig. 9.11) and the false tamarind (fig. 9.8). The phellogen layer may be more continuous over larger areas, producing mature and very smooth trunks, as in the allspice tree or the smooth and oddly rippled trunk of the baobab tree. These examples can only hint at the diversity of colors and textures in tree bark, and the photographs in this book reflect my own experiences in the landscapes where I have visited and lived.

Bark is produced in a similar way in roots. Bark may not protect the root from the same threats as the trunk. However, roots may be particularly vulnerable to attack by animals and disease. Root barks, in particular, may accumulate toxic compounds as defenses against those organisms. Consequently, root barks are often used medicinally or in tanning leather.

Color and the Complex Chemistry of Stems

When it comes to chemical production, plants pull out all stops in filling bark and wood with an amazing diversity of secondary compounds. These help to explain the great variety of bark colors, although such colors may be incidental to the actual defensive functions of barks in the natural environment. The aromatic-ringed molecules that are important precursors in the formation of the flavonoid pigments, such as anthocyanin, are building blocks for other molecules that are important for the mechanical strength of stems, the production of bark, and their patterns of color. The isoprene pathway, fundamental in making the yellow-orange carotenoid pigments and in producing the ringed terpenoid compounds, is also active in many barks. Finally, barks frequently accumulate toxic compounds, including alkaloids. Many are medicinally important; think of quinine collected from the bark of an Andean Forest tree, used as protection against malaria.

The principal chemical unit of a plant stem, whether herbaceous or woody, is the simple 6-carbon sugar, glucose (fig. 3.2). Molecules of glucose are condensed end on end to form two important plant polymers, starch and cellulose. The difference between these polymers is the orientation of alternating members of the growing chain: all "facing" up produces starch, and up and down produces cellulose. These different links profoundly affect their use and reuse. Starches are easily digested back into their single glucose units by plants and animals; we do it effectively with our saliva. Cellulose is not broken down by plants or animals. Animals that live on plants, such as ruminants like cows and deer, have specialized digestive tracts with fermentation tanks that culture bacteria with the enzymes required to digest cellulose. Cockroaches and termites also house protozoans that shelter bacteria doing this work.

Lignin

Cellulose is not very strong by itself (think of toilet tissue), but can be made dramatically stronger with the addition of an intercellular cement: lignin. Lignin is another polymer, consisting of different kinds of aromatic alcohols, monolignols (figs. A.23, A.24). The aromatic alcohols, such as coniferyl alcohol, are joined together into extremely complex structures. Different alcohols may be involved, and they may join at different locations on their molecules, but we do not know very much about how lignin is synthesized, other than that it is a complex and biologically regulated process.

Lignin does not color wood or bark significantly by itself; it is mainly a cement that makes the walls of the conductive vessels and fibers terrifically
strong. Smaller complexes of molecules in wood give it color, the lignans.
For instance, the reddish color of red cedar is due to lignan, plicatic acid.
Many molecules that color heartwoods, discussed below, are lignans.

Suberin

A parallel and spatially more complex process makes suberin, polymers in
bark. Suberin is produced in the corky portion of bark, the phellum. It is
also produced in response to the wounding of plant organs, even in potatoes. Suberin consists of layers of oily or waxy (water-repelling) material
alternating with layers rather like lignin. These chemical structures contribute to the strength of the cell walls of the tissue, and make them waterproof.
The oily layers are secreted from living cells and consist of long-chain fatty
acids. The aromatic portion of suberin is produced first from the living cells
of young phellum tissue. The building blocks are primarily a few aromatic
acids derived from hydroxycinnamic acid, with long water-repelling carbon
chains. These join together in complex ways similar to lignin.

Other joined aromatic compounds produce color in stems, in wood and
bark. Condensed tannins (mentioned in chapter 3) may accumulate in bark
and produce reddish or darker colors; these may also be present in the
heartwood. The color partly depends upon the shapes and distributions of
cells in the phellum. Dark corky barks are produced by phellum heavily suberized throughout the tissue. Lighter-colored barks are suberized, but air
spaces in the tissue or between separate layers allow for some backscattering
of light. Phellum produced in thin papery layers impregnated with suberin
may allow light to penetrate and be reflected out through the surface, producing more orange-red colors, like the thin layers in the gumbo-limbo.
Similar colors are produced in the thin peeling bark of ornamental cherries.
Alternatively, the peeling bark of the paper birch have waxes deposited on
each layer. The net result is strong reflectance of light by the wax particles
and layers, making the bark nearly white.

Secretions

Stems and other plant parts may produce color from fluids that are ex-
uded when tissues are wounded. Many tropical plant families produce latex, usually white. A distinguishing characteristic of the figs (in the family
Moraceae) is white latex (fig. A.5). White latex is also produced by members of the spurge family (Euphorbiaceae), the sapote family (Sapotaceae),

and the dogbane family (Apocynaceae). The elastic latexes are polymers of five-carbon isoprene groups (chapter 3). Secretory cells lining specialized vessels, laticifers, secrete the latex, which accumulates under pressure. When the laticifers are broken, the latex gushes out. Latex is suspended as tiny spheres in a water solution; those spheres scatter light to produce the white color. The economically most important latex is that collected from the rubber tree, native to Brazilian rain forests and now cultivated throughout the tropics (fig. 9.12). Many latexes are quite toxic, not because of their basic structure but because other molecules are incorporated.

At least one family produces latex with a distinct yellow color: the Clusiaceae (fig. 9.12). In the pitch apple native to south Florida and the Caribbean (also called the autograph tree because people write on its long-lived leaves), the yellow residue on the trunks and fruits is caused by a unique pigment, mangostin (fig. A.55). Yellow latex also exudes from the surface of that delicious tropical fruit, the mangosteen.

Other plant exudates may also be strongly colored. In the temperate woods of North America, bloodroot (fig. 9.13) produces an orange-red sap. The sap of almost all members of the nutmeg family, mainly trees in

Figure 9.12 Three tropical trees representing families that produce latex. *Left,* a fig tree (*Ficus microcarpa*), which produces white latex throughout all parts of the plant. *Top right* a pitch apple (*Clusia rosea*) produces latex with a yellow tinge, due to mangostin. *Bottom right,* a rubber tree (*Hevea brasiliensis*), being tapped for rubber latex in Malaysia.

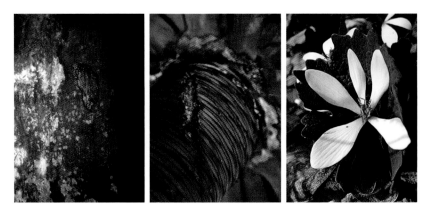

Figure 9.13 Plants that exude colored saps or resins. *Left*, members of the nutmeg family (Myristicaceae) can be identified by the red sap, colored with virolin and similar compounds, which exudes when the trunk is cut. *Center*, the red resinous sap of the dragon tree (*Dracaena draco*) from the Canary Islands has long been used medicinally. *Right*, a flowering plant of bloodroot (*Sanguinaria canadensis*). Bloodroot photograph courtesy of Jeff Abbas.

rain forests throughout the tropics, is bright red. The machete slash of the cruising forester easily reveals the family identity of such trees (fig. 9.13). The reddish colors are produced by complex aromatic molecules, including virolin (fig. A.27), common in this family. Quite often these saps are used medicinally by indigenous people, whether efficacious or not. Red saps became famous in Europe many centuries ago and are being rediscovered in the health stores of Europe and the United States. The dragon tree is a monocot native to the Canary Islands, distantly related to the Jericho tree of the deserts of the southwestern United States (fig. 9.13). This plant was well-known in Europe, and dried blood red extracts were traditionally used for a variety of medicinal purposes. The principal pigment molecule is dracorubin (fig. A.50). Today, dragon's blood medicine from a South American plant is sold in health food stores. This pigment is found in many dracaenas, and such plants have been sources of medicines in the Chinese pharmacopoeia for millennia.

Still other plants produce gums and resins that may discolor the surface of stems. Water-soluble gums are produced by plants of many families, and the precious incense of the frankincense tree is a resin. Its south Florida cousin, the gumbo-limbo, also produces similar resins, and the resin canals are visible in sections of the stem (fig. 9.7, top right). The terpenoid and water-insoluble resins of conifers (related in their mode of synthesis to carotenoids but condensing to produce ring compounds) dry to produce colored resins on their trunks. Yellow colors from dried resins color the bark of the ponderosa pine (fig. 9.1).

Members of the poison ivy family (Anacardiaceae) often produce long-chain aromatic compounds with economic value. Many of us are quite sensitive to some of these compounds; the skin reaction we get from handling poison ivy or poison oak is caused by such a molecule, urushiol (fig. A.56). These compounds are produced in a network of ducts beneath the surface of stems, and in leaves and fruits (fig. 9.14). When the tissue is damaged, as by an insect, the resins are pushed out under pressure. Some of these resins are toxic, and they dry quickly to produce a hard coating. When exposed to air, these same compounds are oxidized and produce black stains. Saps on trunks in the poison ivy family produce black blotches, one of the means of identifying poisonwood, a tree native to south Florida and the Caribbean (fig. 9.15). It causes skin rashes similar to poison ivy. Japanese lacquerware is produced from the resins of a Japanese tree, *Rhus verniciflua*, and related species are used for lacquerware in Vietnam and Thailand. These lacquers

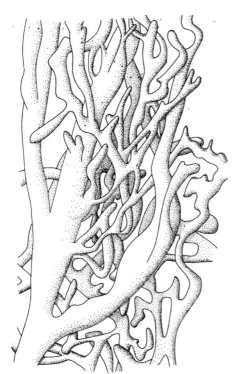

Figure 9.14 The network of resin ducts that protect the mango fruit against attack by insects. The distance across this reconstruction is about 2 millimeters. Such networks are found in plants that produce latexes, mucilage gums, or resins. Courtesy of *Journal of Experimental Botany.*

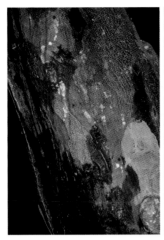

Figure 9.15 Plants that produce resins that oxidize to black. *Left*, such resins discolor patches of bark of the poisonwood tree (*Metopium toxiferum*). *Right*, resins of other trees from Asia are used to produce fine lacquerware, such as this bowl handcrafted in Japan.

are different from those produced by the lac insect (shellac), which is collected in India. Similar compounds from plants in the Anacardiaceae can be complexed with iron to produce superior black inks. These were previously used in Asia and the United States, but have been replaced by cheaper synthetic dyes. When I first went to India in 1975, my laundry was marked by such an ink produced from the marking nut tree. Richard Howard, former director of the Arnold Arboretum of Harvard University and an eminent botanist, provided training materials to the U.S. military during World War II for teaching survival skills in the "jungle." His expertise was also occasionally used by the military to solve specific problems. In the Pacific, a severe form of skin rash appeared on the rumps of U.S. military personnel. You may have already guessed the answer; the soldiers were sensitive to a plant-derived lacquer that locals had used to varnish the toilet seats.

The Color of Wood

The attractiveness of the wood of many trees is due to the way that the secondary xylem, or conductive tissue, is formed, as well as the chemicals that are deposited in the old tissue, principally in the center part of the trunk, the heartwood. The heartwood of the Brazilwood tree (fig. 1.14), from Central and South America, yielded the important red dye of com-

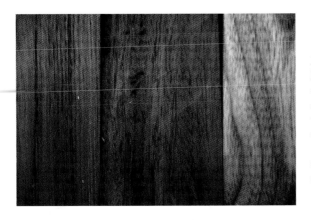

Figure 9.16 Three hardwoods with extremely attractive color and figure. *Left*, teak (*Tectona grandis*); *center*, Indian rosewood (*Dalbergia sissoo*); and *right*, walnut (*Juglans nigra*).

merce in the eighteenth and nineteenth centuries: hematoxylin (fig. A.52). Woods from Southeast Asia also provided raw materials for dyes later used in Europe, such as the heartwood of the red sanders tree (*Pterocarpus santalinus*), which accumulates a pigment, santalin (fig. A.51).

Many other woods have become valuable for their attractive appearance, or "figure," which is partly due to the production of coloring agents in the heartwood (fig. 9.16). Such woods are familiar to us because they are important in the production of furniture and musical instruments. In teak, perhaps the most important tropical hardwood used in making furniture, the dark heartwood accumulates a complex of pigments, including lapachol (fig. A.21). A valuable domestic hardwood used in furniture manufacture is walnut, whose dark heartwood contains juglone (fig. A.20). Rosewood is used to make fine furniture and musical instruments (such as the fret boards of expensive guitars). Dalbergiones are molecules that produce the warm colors of rosewood's heartwood (fig. A.22). Perhaps the most remarkable color in a tropical hardwood is found in purpleheart, a tree growing in New World tropical forests. Its heartwood is an intense rose red, produced by an anthocyanin-like molecule, mopanol (fig. A.28). Woodworkers know that tracking any sawdust from this wood into their house will permanently stain their carpets a bright red. Many woods do not have unusual chemistry that produces intense colors in their heartwood. A good example is the pale wood of the linden tree.

The Colors of Roots

The bright colors of roots seem counterintuitive. Why would a subterranean organ produce pigments? The answer is most likely because we have selected such colors for their attractiveness and nutritional properties. The wild ancestors of the domesticated carrot (fig. 9.17) produced nonpigmented, narrow roots. The modern carrot is loaded with β-carotene. The pigment of red beets, betanin (fig. A.47), was also selected for its attractive color and perhaps for the antioxidant properties of the pigments (fig. 9.17). Other varieties of the beet include the less-pigmented sugar beets. Potatoes have been selected with purple anthocyanin pigments, both standard supermarket varieties and heirloom varieties with purple throughout. I include potatoes here because they are subterranean; they are actually not roots, but swollen (usually underground) stems, or tubers. The Maori took sweet potato varieties with them on their long ocean journey to New Zealand. Sweet potatoes continue to be an important part of their diet. Many of those varieties are a deep purple due to the high concentrations of anthocyanins, which may partially explain the low incidence of colon cancer among the Maori. Young aerial roots, such as those of the banyan fig, may turn red from the production of anthocyanins when exposed to light. The roots of some plants among the composites, including the ragweed

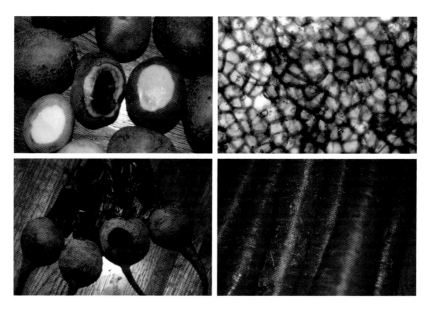

Figure 9.17 Roots, or rootlike stems, that produce colors. *Top left*, heirloom potatoes (*Solanum tuberosum*), whose tubers are purple; *top right*, tuber cells with anthocyanins present. *Bottom left*, beets (*Beta vulgaris*) have been selected to accumulate high concentrations of the betalain pigment, betanin; *bottom right*, the bright orange roots of the carrot (*Daucus carota*) with accumulation of β-carotene.

Figure 9.18 Turmeric (*Curcuma longa*) is a spice in the ginger family, also used as a dye plant. *Left*, workers in Cochin, India, are drying the roots and bagging them for sale. *Right*, a turmeric plant revealing its rhizomes.

so irritating to those who suffer from allergies to it, produce a remarkable sulfur-containing pigment called thiarubrine (fig. A.53). The color is probably much less important to these plants than the strong antimicrobial activity of this molecule. The red color of roots containing this pigment, as in ragweed, is at least partly due to the presence of anthocyanins in cells around the thiarubrine-containing tissues. The underground stems, or rhizomes, of the turmeric plant produce a pigment molecule, curcumin (fig. A.54), which colors the interior and even the surface of the rhizome. It has long been used in cooking, as well as being a dye and medicinal plant, in Southeast Asia (fig. 9.18).

The Natural History of Bark

Bark is an important defense produced by plants to protect their conductive tissues against damage from the physical environment, as well as by animals and disease. Bark may also protect plants from colonization by lianas in tropical forests, and may even attract certain organisms to grow on them, and repel others. In time, the bark surface may become a complex community of bacteria, algae, fungi, lichens, and other plants. How might the nature of bark affect these defenses and interactions, and how does color play a role in these defenses?

Although barks vary in texture and thickness, even among species in a single family, it is not clear how this variation correlates with the ecology of individual species. We might expect thicker bark among species that are exposed to fire. Such thick bark may protect large conifers in Northwest forests from damage by occasional fires, as in the exceedingly thick bark of the coastal redwood (fig. 9.19). The plates of bark in the ponderosa pine also provide some protection against fire, as for the many pines that have evolved in ecosystems where fire is a regular occurrence. However, much of the variation in bark characters seen in large tropical groups, such as the dipterocarps in Southeast Asia and the eucalypts of Australia and New Guinea (fig. 9.2), cannot be easily explained. Barks may protect the cambium and phloem tissues of stems from extremes in temperature. Some trees are sensitive to sun scald, when lesions form from exposure to the sun; this is more likely a response to the high temperatures from that exposure. However, thin barks seem to work about as well as thick barks.

Barks also vary dramatically in chemistry and, consequently, in color. We know most about the chemistry of the barks of economically important trees, especially sources of chemicals for tanning leathers and the barks of softwood conifer trees, such as the ponderosa pine and Douglas fir. Their barks have been studied to find uses for a by-product of timber production and to learn about bark as a potential defense against the economically

Figure 9.19 Two fire-resistant conifers native to various forests in northwestern United States. *Top*, trunk of the coastal redwood (*Sequoia sempervirens*); *bottom*, trunk of the ponderosa pine (*Pinus ponderosa*). Both trunks reveal evidence of surviving fires.

damaging beetles that bore through it to feed on the phloem and cambium tissues. The chemical compositions of the bark and resin of ponderosa pine and Douglas fir are exceedingly complex, consisting of numerous terpenoid, phenolic, and flavonoid compounds and complexes of these compounds.

In the ponderosa pine, these compounds are part of a complicated story of the tree's long evolutionary relationship with bark-boring beetles. The bark is a mixture of mechanical and chemical defenses against these beetles, and phenolic compounds (which contribute to the reddish color) are one line of defense. The terpenoids, produced by special cells lining ducts, are a line of defense that the beetles have partially evaded and even used to their own advantage. In the western United States, two principal beetles damage this pine, the California five-spined ips and the western pine beetle. Female western pine beetles are attracted to trees producing terpenes. They bore into the bark and produce a male-attracting hormone (a pheromone) from the raw material of the tree terpenoids. This attracts an infestation of both male and female beetles. In dense beetle aggregations, another pheromone is synthesized that repels additional invasions. Sex ratios of the attracted beetles are also controlled by the different pheromones produced. These compounds are also effective defenses against entry by fungal diseases, but the beetles carry a disease that decomposes wood tissues and provides places for beetle larval development. Pines die primarily from the fungal infections, which cause breaks in the xylem tissue that allow air bubbles to establish in conducting cells, breaking the negative pressure that pulls water up into the needles. The infections may also destroy the cambial layers and phloem translocating cells, starving the roots. We are still learning about the complex chemistry of bark and its defenses against bark-boring beetles. Perhaps bark color plays a role in defense or in recognition of the tree by the beetle.

The networks of ducts that store gums, resins, and latex are formidable defenses against attack by insects. When the insect bites the plant tissue, these materials are forced out and interfere with the chewing activity of the insects. The leaf-cutting ants of the American tropics, which denude trees of their foliage by cutting up the leaves and carrying them back to their fungal gardens, avoid the leaves of trees with latex. Cutting or picking an unripe mango fruit causes a sticky resinous fluid to gush out, perhaps an effective defense against insect attack (figs. 8.15, 9.14). These fluids may also contain certain toxins. However, insects that feed on the leaves can cut a vein to release the pressure, wait, and then consume the blade. Such evasions of effective plant defenses are evidence of the evolution of insects that can circumvent the effects of molecules in latexes and resins, evidence of a coevolutionary race between plants and animals. For any plant that produces a highly toxic chemical, there is invariably an insect that has evolved a mechanism or strategy avoiding the toxic effects.

Barks also vary in attracting or allowing organisms to grow on their surfaces. In Miami, the trunks of the Caribbean mahogany are frequently black with colonies of cyanobacteria, and those of the live oak are gray with lichens. The live oak also attracts epiphytic plants, such as the resurrection fern (fig. 9.8), but the gumbo-limbo trunk is free of them. Trunks of trembling aspen attract different lichens than trunks of Douglas fir. Trees of both species accumulate these organisms as the trunks age, but those of Douglas fir may become particularly well covered by lichens and bryophytes (fig. 9.11).

In tropical forests, bark is not only inhabited by lichens and bryophytes, but also by epiphytes. Thicker bark may permit the colonization of bromeliad plants that differ in their drought tolerances, some plants with roots in the furrows and others with roots on the ridges. Barks may also promote or discourage the establishment of lianas, woody mechanical "parasites" that are particularly important in these forests. Intuitively, we should expect that trees with smooth trunks and rapidly peeling bark might fend off lianas, but gumbo-limbo trees in forests in Central America seem to have as many lianas as other trees. In a rain-forest study in Borneo, the opposite was found. Trees with rough bark were less likely to be colonized by lianas than trees with smooth bark.

The colors of tree trunks make a much less coherent and satisfying story than those of leaves, flowers, and fruits. I suppose that is because stem color is more a by-product of plant chemistry and may not have any direct ecological function, as in energy balance in leaves, pollinator attraction in flowers, and dispersal of fruits. Certainly, the occasionally bright colors of heartwood do not provide any visual signal to invading organisms. We generally know less about stems, at least their surfaces, than about other plant features. However, I wouldn't be surprised if bark color plays a role in trees' interactions with invading animals.

From our perspective, the color of stems is an important part of the story of plant color. Stems dress up our landscapes, provide colorful wood for our furniture, and produce dyes for our foods and fabrics. Stems give us the aesthetic clues by which we recognize those landscapes.

Chapter Ten Iridescent Plants

....

We float through space. Days pass.
Sometimes we know we are part of a crystal
where light is sorted and stored,
sharing an iridescence
cobbled and million-featured.
Oh, tiny beacon in the hurting dark.
Oh, soft blue glow.

DAVID YOUNG

Now that these colours are onely fantastical ones, that is; such
as arise immediately from the refractions of the light, I found
by this, that water wetting their coloured parts, destroyed
their colours, which seemed to proceed from the alterations
of the reflection and the refraction.

ROBERT HOOKE

....

The peacock feather (fig. 2.1), along with other
metallic-looking feathers and insects, was an object
of fascination to scientists during the Renaissance. Robert
Hooke's observations on the iridescent colors produced
by peacock feathers were the first tentative steps toward
an understanding of the wave nature of light. They pre-
ceded that revolution in understanding, which came a
century later. My observations of plants from the understory of tropical rain forests in Malaysia that produce a
similar iridescent blue certainly did not precipitate any
similar scientific revolution, but it led me on the trail of
an intriguing mystery.

Soon after getting married, my wife and I moved to
Malaysia in 1973. I started my first teaching position at

the University of Malaya in Kuala Lumpur, and Carol taught printmaking to art students at a local institute of technology. I met Ben Stone, who had helped me obtain the position, and Brian Lowry, who was a plant natural-products chemist from New Zealand. They took me on my first explorations of tropical rain forests in the mountain valleys outside of the city. Ben was one of the experts on plant diversity in the Old World tropics, and Brian knew his plants pretty well and their secondary chemistry even better. So they were in a good position to guide me on these walks. The forest environment was bewildering. I was intellectually, but not emotionally, prepared for the bewildering diversity, the majesty of lofty trees festooned with epiphytes and lianas, the iridescent green-on-velvet wings of the Rajah Brooke birdwing butterfly (fig. 10.1)—I had read about all that. However, I was astonished by the leaf color of understory plants, especially the shimmering electric blue of *Selaginella* (fig. 10.2). We walked through understory glens where this plant covered the ground in bright blue. I asked Ben and Brian, "What is this?" They told me that it was a plant allied to the ferns and a relative of the ground pines (*Lycopodium*), that it was an extremely common

Figure 10.1 Top, a Rajah Brooke birdwing butterfly rests on a branch at a butterfly farm in the Cameron Highlands of Malaysia. *Bottom*, a microscope view of the wing scales, showing the points of green iridescence.

Figure 10.2 A small lizard rests on a shimmering blue branch of the peacock fern, *Selaginella willdenowii*. Although this scene was very reminiscent of what I saw while living in Malaysia, the lizard is an anole native to south Florida—and the photograph was taken at Fairchild Tropical Botanic Garden, in Miami.

plant in the understory and in rubber plantations as well, and that they had no idea why or how it produced such an electric blue color. So Brian and I began to study this little plant, and the results led me into further research on the nature and functional significance of colors in tropical plants.

We expect leaves to be green. The chlorophyll pigments in leaves absorb light at all visible wavelengths, but less in the green range. Light scattered from the leaf interior gives them their characteristic color (chapter 4). Any leaf color other than green is a consequence of wavelengths that are normally captured by chlorophylls instead of being reflected from the leaf. Thus less energy is captured for photosynthesis and growth. Why did these spectacular blue plants, and other species I later found growing in such deeply shaded environments, reflect light that would otherwise add to their photosynthetic efficiency? How did they produce this spectacular color? Developing the proper search image for these plants, and an intuition about where they grew, I discovered other blue plants in the shade of these forests: an iridescent blue "peacock" begonia, another blue begonia high in the mountains, some ferns, two blue understory plants of the melastome family, and even an understory sedge, a *Mapania*.

Later I looked for iridescent blue plants in the understory of rain forests in Central and South America—and found them. Many years later, I also found them in the shade of West African rain forests in Gabon.

It was apparent to Brian and me that the basis of color production in the *Selaginella* plants was not pigmentation, because the leaves lost their color when immersed in water. Later on, when the leaves dried out, they became blue again (fig. 10.3). Furthermore, only green chlorophyll and a few carotenoid pigments found in the chloroplasts could be extracted—and they were green and yellow. The blue colors in flowers (chapter 7) and fruits (chapter 8) are caused by modified anthocyanins, and anthocyanins were not present in these plants. Their leaves had to produce color by some physical means—by diffraction from a grating-like structure, by the selective scattering of small particles, or by the constructive interference produced by thin films (fig. 2.8).

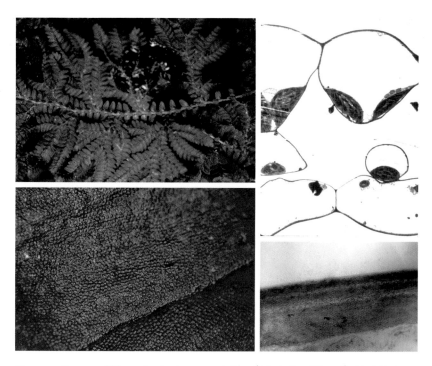

Figure 10.3 Structural blue color in the peacock fern (*Selaginella willdenowii*). *Top left*, the branch holds a water drop, which has cancelled much of the blue color. *Bottom left*, a detail of that surface color production. *Top right*, a transmission-electron-microscope photograph of a leaf section, revealing the two large chloroplasts in the outer layer, and the convexly curved outer walls. *Bottom right*, the two layers at the outer margin of the cell wall, producing the blue color.

In temperate regions and high elevations or desert areas, plants with waxy coverings are whitish blue. We previously saw the example of *Schima wallichii* (fig. 6.11), the tea relative found high up on Mount Kinabalu. The light blue color of the leaves of these plants is due to a waxy coating. More familiar to most of us are the waxy varieties of many conifer trees, particularly the blue spruce (fig. 10.4). The needles of this tree are coated with small wax particles. These particles are so small that they scatter radiation more effectively at short wavelengths. Given the distribution of sunlight and our visual sensitivity, we observe this scatter as a blue color. This selective scattering is a physical phenomenon, called Tyndall scattering, and it is also responsible for the blue color of the sky and the ocean.

For *Selaginella willdenowii*, it was easy to eliminate the first two explanations. Color production was of a single hue. Reflectance measurements gave a clear peak, not a continued increase at shorter wavelengths as would be expected from small-particle scattering. Finally, only a single color was produced, not the rainbow spectrum of colors that would result from diffraction.

Structural coloration is common in animals, particularly insects and birds, as in the peacock feather commented upon by Robert Hooke. Such colors are almost always caused by thin-film interference. The iridescent green wing chevrons of the majestic Rajah Brooke birdwing butterfly we occasionally encountered in Malaysian rain forests are produced in this way, by multilayered structures in the wing scales that interfere with visible

Figure 10.4 The powder blue color of the Colorado blue spruce (*Picea engelmannii*) is caused by small wax particles that scatter shorter wavelengths of light. *Right*, a scanning-electron-microscope photograph of wax deposits on the surface of the needles.

light (fig. 10.1). So are the more familiar brilliant blue wings of the morpho butterfly from South America.

The explanation for such color production was first provided by Thomas Young in 1801, and he referred to the earlier observations of Hooke. Imagine a material made up of transparent layers with different optical densities; in other words, where the index of refraction—the extent to which light waves are slowed down by the medium—is different for each layer. In this material, light reflects at any boundary. Which wavelengths of light pass through the layer (destructive interference) and which are reflected from the layer (constructive interference) depends on the layer's thickness, its refractive index, and the angle at which the light enters (fig. 2.8). When light passing through and back is retarded by half a wavelength, the interference is destructive, so that these wavelengths pass through the layer. When light is reinforced at a full wavelength, the interference is constructive, producing an intense reflected metallic color. If one layer has a lower refractive index than the layers above and beneath it, a phase shift occurs. Because the balance between light reflectance and absorption depends on the thickness of individual layers as well as their optical density, a filter of different thickness has the same effect as a thinner layer with a different refractive index. The layers in a filter can be extremely thin; at one-quarter of a wavelength, the interference effect is similar to that at a full wavelength of light. By measuring the reflectance perpendicular to the leaf surface and assuming a refractive index of 1.40, typical of the moistened cellulose of cell walls or the membranes of organelles, it is possible to predict what thickness of layers (or similar three-dimensional structures) in the leaves would produce an intense metallic coloration.

In selaginella leaves, the task of locating the site of interference was simplified by the fact that we were sure it was near the outer cell wall, where water could cancel its effect. Also, we could compare the cell structure in the iridescent blue leaves of some selaginella plants with that in the green leaves of plants of the same species developing in more sunny locations.

After four years in Malaysia, followed by a year of travel back to the United States, Carol and I moved to Montpellier, in southern France, where I worked in the tropical botany laboratory of Francis Hallé. While at the University of Montpellier, I teamed up with one of Francis's colleagues, Charles Hébant, who was a student of the functional anatomy of mosses—and an excellent electron microscopist. We examined the epidermal cells of the leaves by transmission electron microscopy. We discovered two layers of the predicted thickness, approximately 80 nanometers thick, and opaque in the electron microscope, in the cell walls of *Selaginella willdenowii* and *S. uncinata* (fig. 10.3). Such layers were absent from the green leaves of both species.

Later I worked on the structures of other iridescent blue plants with different students and colleagues. The simple explanation for iridescence in selaginella left me unprepared for the more elaborate structures that turned up in other plants of the rain-forest understory. In the New World tropical fern *Danaea nodosa*, repeated electron-opaque layers alternate with more transparent arcs of cellulose microfibrils, the long cylindrical fibers that make up plant cell walls (fig. 10.5). Similar patterns had been observed in the cuticles of arthropods by A. C. Neville and his colleagues at the University of Bristol. Neville's group showed that succeeding layers of fibrils are deposited at a regular angle in the formation of the cuticle, creating the patterns and the resulting iridescence.

The distance between the consecutive light and dark bands is a result of a step-by-step rotation, over a total of 180 degrees, in the orientation of the microfibrils. As a result, a slightly oblique cross section of a beetle cuticle or,

Figure 10.5 An irides-
cent blue fern (*Danaea
nodosa*), native to the
shade of tropical
rain forests in Latin
America. The leaves
are an iridescent blue
in young plants, but
become a normal
green in adult plants.
The basis for con-
structive interference
in this plant is the
multilayered outer cell
wall (*bottom*).

it turns out, a *Danaea nodosa* leaf, has a helicoidal appearance; it looks like a stack of coiled fibers (fig. 10.6). Neville and his co-workers demonstrated that this helicoidal structure is the cause of iridescent coloration in beetles.

Two mechanisms for color production may operate in these helicoidal layers. First, the periodicity of the layering provides the conditions for the

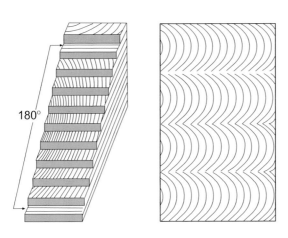

Figure 10.6 Layering by helicoidal thickening. The successive deposit of cellulose fibers at a uniformly different angle produces a slight difference in refractive index and polarization at repeats in the wall. This is the basis for constructive interference and the bright blue colors of some plants.

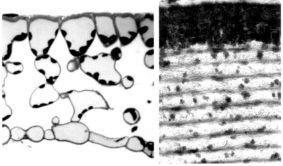

Figure 10.7 Two iridescent blue plants from the understory of tropical rain forests in Malaysia. On top is a fern (*Diplazium tomentosum*), and on the bottom is a flowering plant (*Phyllagathis rotundifolia*). *Bottom left*, a transverse section of the leaf of the fern. *Right*, a detail of the outer cell wall, showing the helicoidal thickening.

reinforcing iridescence of multiple layers. Second, the helicoidal structures cause the circular polarization of white light, which can produce a narrow spectral region of reflected color. In *Danaea nodosa* the layers were of the predicted thickness (about 160 nanometers) to produce blue colors through interference. These layers were either absent or of the wrong thickness in the green leaves of adult plants. Kevin Gould, now at the University of Otago, and I later figured out the helicoidal basis for blue leaf iridescence in two Malaysian understory ferns, *Diplazium tomentosum* (fig. 10.7) and *Lindsaea lucida*. We have not yet determined whether circular polarization also occurs.

Blue leaf iridescence is not limited to ferns and their allies, nor is blue iridescence limited to leaves. Some marine algae, particularly red algae, produce such color. Layers in the cuticle cause blue color by light interference in *Iridaea*, but structures within cells may produce this color in other algae. The iridescent blue algae pictured here, *Dictyota mertensii* and *Ochtodes secundiramea* (fig. 10.8), have not been analyzed. A variety of flowering plants produce leaf iridescence,

Figure 10.8 Iridescent algae from the Caribbean. *Top, Dictyota mertensii*, a brown alga; *bottom, Ochtodes secundiramea*, a red alga. Photograph courtesy of Diane and Mark Littler.

particularly in the Asian tropics. Kevin and I studied two distantly related flowering herbs from the Malaysian rain-forest understory that I had first observed in the 1970s. These studies helped us locate unusual structures that might produce the iridescence. In the peacock begonia (fig. 10.9) and in *Phyllagathis rotundifolia* (fig. 10.10), the ultrastructural basis is a remarkably modified chloroplast, which we have called an "iridoplast."

Since chloroplasts are the sites where the hard work of photosynthesis takes place, they are the first place to look for interactions involving light. In the iridoplasts, the normal aggregation of disk-shaped thylakoids, forming stacks called grana, was lost. Each iridoplast in the two plants we studied is shaped like a pancake and covers much of the bottom of the epidermal cell. Within, thylakoid stacks in close contact with one another (called appressed thylakoids) form the basis for the interference filter. Such

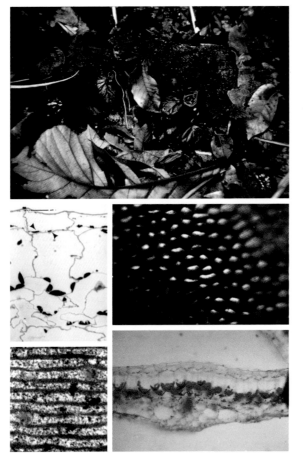

Figure 10.9 Iridescent blue color in the Malaysian peacock begonia (*Begonia pavonina*). *Top*, the plant with some peacock fern in the forest understory; *middle left*, a transverse section of the leaf, *arrows* pointing to iridoplasts in the top cell layer; *middle right*, the leaf surface, showing the convexly curved outer cells and the flecks of blue light coming from within those cells; *bottom left*, a detail of the iridoplast showing the layered membranes; *bottom right*, a fresh leaf section showing the chloroplasts in the palisade layer, and small plastids in the outer cells.

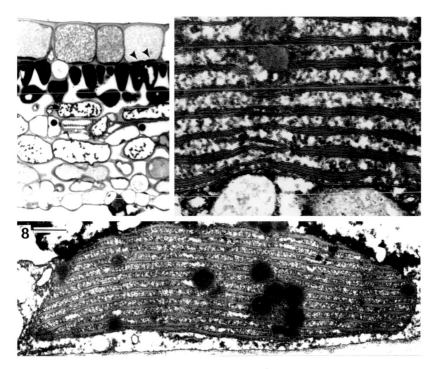

Figure 10.10 Details of iridescent structures, the iridoplasts, in the Malaysian flowering plant (*Phyllagathis rotundifolia*). *Top left*, the entire leaf section, *arrows* pointing to the locations of the iridoplasts; *bottom*, an iridoplast; *top right*, details of the iridoplasts, showing the appressed membranes responsible for the constructive interference of the blue color.

structures are not found in the green leaves of these plants, which contain normal chloroplasts in other parts of the leaves.

An iridescent blue-green filmy fern, *Trichomanes elegans*, grows in extremely shady and wet locations in New World tropical rain forests (fig. 10.11). Graduate students at the La Selva Research Station in Costa Rica, where Rita Graham and I studied this plant in detail, called it the dime store plant because it looks like a plastic copy of a real plant. The filmy fern's iridescent blue-green fronds contain modified chloroplasts in the epidermal cells (fig. 10.12). The grana stacks, which have five thylakoids each and are connected by extremely short stromal lamellae, form a repeating series of filters that produce the blue-green interference color.

There are many other remarkable blue plants to study. In Central and South America, the brilliant blue fronds of a strap fern, *Elaphoglossum herminieri*, frequently fall from this epiphytic plant that grows on branches in the crowns of large rain-forest trees (fig. 10.13). The graduate students at La

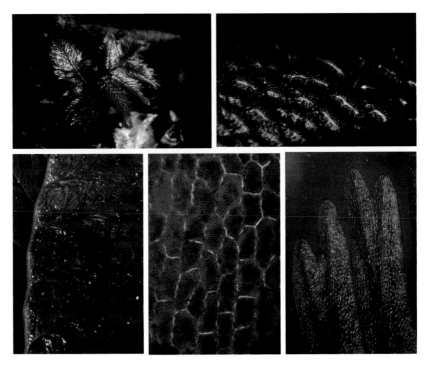

Figure 10.11 The "dime store" plant (*Trichomanes elegans*), growing in very moist and shady sites in tropical rain forests in the New World. These are various views of the plant, from the leaf down to the cells that produce the blue-green color.

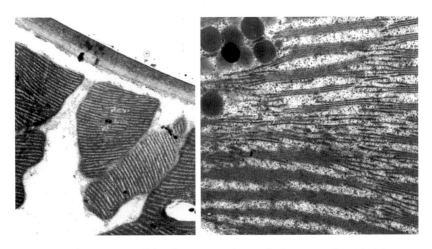

Figure 10.12 Ultrastructure of the "dime store" plant. *Left*, the outer cells with chloroplasts near the surface; *right*, details of the chloroplasts, showing the appressed thylakoids; they are the right thickness to produce the iridescent color.

Figure 10.13 Iridescent blue fronds of a strap fern (*Elaphoglossum herminieri*), native to tropical rain forests of the Neotropics. *Left*, the whole plant; *center*, a frond detail; *right*, a light-microscope view of a leaf section. Blue color could be produced by the cell wall (but color is not removed by wetting) or iridoplasts in the outer cell layer.

Selva may also call these "dime store" plants, because they look like they are made of plastic—maybe "Wal-Mart" plants would be better these days.

Iridescent Blue Fruits

About 10 percent of fleshy fruits are blue. As earlier discussed for flowers and fruits, such color is produced by modified anthocyanin pigments in the outer layers of the fruit wall. But constructive interference could produce blue color in some fruits, and it may be that we have not looked carefully enough for this phenomenon. Iridescent colors other than blue could also be produced. I fell into studying such fruits when washing the outer pulp of fruits of the rudraksha tree (fig. 10.14). This tree is sacred in India, where the stony inner fruit is highly ornamented and used for bracelets and necklaces (*malas*) for reciting prayers and sacred sounds. The sadhus (spiritual seekers) wandering the countryside frequently carry *malas* of rudraksha beads. I wanted to collect some seeds so we could grow this tree in Miami. In washing away the brilliant blue outer wall, I noticed that no blue or red pigments were extracted in the water. These fruits are a persistent electric blue, and I suspected that their brilliant color might be produced structurally. Subsequently I discovered that the interference color of the rudraksha fruits is indeed produced by a structure whose cellulose layers are of the predicted thickness to produce blue. This structure, which I have called an "iridosome," is different from those seen in leaves; it is secreted by the

Figure 10.14 Details of rudraksha (*Elaeocarpus angustifolius*), native to India and Southeast Asia. *Top left*, a flowering branch with the tiny fruit beginning to form; *top right*, a wandering sadhu, with rudraksha beads; *bottom left*, the brilliant blue fruits, each about 2 centimeters in diameter; *bottom right*, microscope view of the fruit surface, showing the intense reflection of blue color.

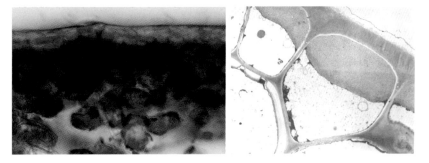

Figure 10.15 Details of the iridescent structure of rudraksha fruit. *Left*, a transverse section of the fresh fruit, showing a clear area between the chloroplasts and the outer wall, where the iridosomes are located. Note the extensive layer of photosynthetic tissue (chloroplasts) beneath the surface. *Right*, an electron-microscope photograph of the iridosome, revealing the individual layers that constructively interfere with light to produce the blue color.

epidermal cells of the fruit and is located outside the cell membrane but inside the cell wall (fig. 10.15). Most of the fruits of this tree genus are blue, and the trees are frequently found in forests throughout Southeast Asia, and as far south as New Zealand. Presumably, they share this mechanism of iridescent color production.

A few years later I received a message from Tony Irvine, now retired as a botanist for CSIRO, the Australian government research organization. Tony had noticed the brilliant blue fruits of a small tree in Queensland rain forests and had contacted me. The tree, *Delarbrea michieana*, is a member of the ginseng family (Araliaceae). We found a similar structure in the epider-

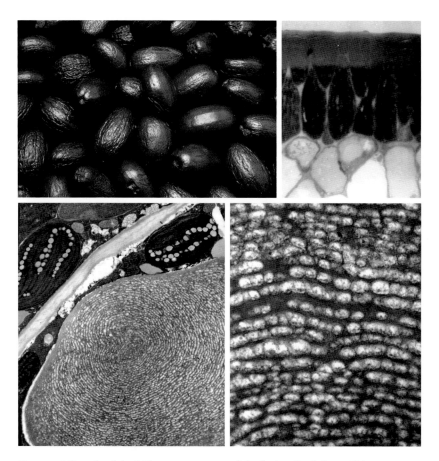

Figure 10.16 Details of the iridescent structure of the fruits of *Delarbrea michieana*, an understory tree in the Queensland tropical rain forest. *Top left*, iridescent blue fruits; *top right*, iridescent structures in the fruit section are immediately beneath the outer wall; *bottom left*, the iridosome; *bottom right*, details of the iridosome, showing the layers that account for the iridescence.

mal cells of the fruits, and the thickness of the layers accounted for the production of color at the measured wavelength (fig. 10.16). Certainly, careful search may reveal other fruits with structural color, perhaps even outside of the tropics. It is easy to test for this. A slightly acidic alcohol solution effectively extracts anthocyanin pigments from normal blue fruits, and a lack of extracted color indicates that iridescence may produce the blue color. I risk the danger of being called arrogant if I say that I am the world expert on this phenomenon, but what I'm really saying is that I am just about the only person who has studied it! After all, there is a great variety of flowering plants and ferns, and few scientists spread around to study them. I'm very unlikely to exhaust the mystery of this subject by myself, and there are plenty of other unstudied plants—especially in the tropics.

Mysteries of Function

What advantage would a plant gain in reducing the efficiency of absorption (and thereby presumably diminishing photosynthesis) in its leaves by producing blue iridescence? I have contemplated this puzzle for the past thirty years, and have not yet come up with a single explanation. In one habitat, the tropical rain-forest understory, we can find structural coloration among distantly related species of shade-tolerant plants, but not in other ecosystems. This observation leads me to conclude that there is a selective advantage to this leaf character in extreme shade. Why might leaves reduce their absorption (reflecting blue color) in environments with severely reduced light energy?

One day when I was first contemplating the blue selaginella of the Malaysian rain forests, I looked at the compound lens of my camera and had the intuition (aha!) that perhaps the interference of these leaf cell layers could also function as an antireflection coating, more efficiently capturing certain wavelengths for use by the plant. My camera lens had a blue sheen to it from the interference layers between each of the elements of its compound lens. A careful analysis of the optical properties of the blue leaves, compared to the green leaves of the same species, showed that the blue leaves reflected more of a blue wavelength than the green leaves, expected from the constructive interference. Theory predicts that destructive interference at longer wavelengths should decrease reflectance and increase absorption by the leaves. The iridescent blue leaves did absorb more radiation in the longer wavelengths of the visual range. In a sunny setting, chloroplasts have access to a wide spectrum of visible light. In the rain-forest shade, the longest visible wavelengths are slightly more available for photosynthesis (see fig. 12.6), and more efficiently absorbed by blue leaves. For selaginella, I also learned that the blue iridescence develops under experimental conditions that replicate the shift in spectral quality characteristic of forest shade conditions. In the laboratory, however, these conditions changed the leaves in other ways, so it was not possible to obtain direct physiological evidence for the advantage of iridescence in low-light conditions.

In the rest of the blue plants I studied, the interference layers are not at the surface, but somewhere within the leaf. Blue iridescence does not affect surface reflectance in them. Invariably when I give talks on this topic, colleagues ask, "Well, what is the function of blue iridescence in those plants?" and I haven't a very good answer. Iridescence must have some selective function in these extreme-shade plants. What may be important for them is the spectral environment within the iridescent leaves. Interference significantly reduces the amounts of blue wavelengths entering into the leaf interior, particularly at the wavelengths of 460 to 480 nanometers. Could there be an advantage in keeping these wavelengths out of the leaf interior? Kevin Gould and I, along with Steve Oberbauer and David Kuhn, two friends and colleagues at Florida International University where I work, have argued that this region may be more sensitive to damage from sudden flecks of high-intensity light (such as a sun fleck in the understory). During the autumn of 2005 (while finishing revisions to this book), I spent two months in Southeast Asia, particularly in Malaysia, to study the physiological responses of these iridescent blue plants to exposure to high light, hoping to find that the iridescent species are better defended. The protective effects observed in this study were not dramatic, and I am still puzzled. As is the rule in scientific research, I don't feel comfortable in saying anything unless the results are submitted for publication in a refereed scientific journal. Here I can respond to my critics and say that I have at least made a serious effort to answer the question.

The best evidence of the function of iridescence in plants comes from studies on the fruits of the rudraksha. The clue is the difference in reflectance and transmittance of light through the iridescent layer compared to a normal fruit with blue pigment (fig. 10.17). In a blue fruit, such as that of *Heliconia mariae* from Central America (fig. 8.11), the blue color is produced when light passes through epidermal cells with pigments concentrated in their large central vacuoles. The colorless, spongy tissue directly beneath scatters light back through the pigmented cells. The blue color we see is the small portion of the spectrum not absorbed by the pigments. In contrast, the interference filter in the epidermis of rudraksha fruits constructively

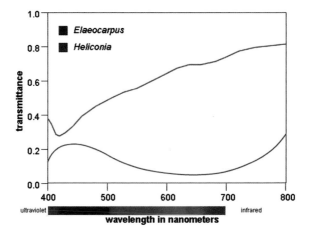

Figure 10.17 A comparison of transmittance through the outer wall of an iridescent fruit of rudraksha (*Elaeocarpus angustifolius*) with seeds of *Heliconia mariae*. In rudraksha the fruit-wall transmittance is lowest in the blue wavelengths because of constructive interference, yet the wall is relatively transparent at longer wavelengths. Its reflectance is low at longer wavelengths because of absorption by tissues beneath. Thus light can penetrate through the blue covering and be used in photosynthesis.

interferes in the blue range (best at wavelengths of about 430 nanometers), but the skin is relatively transparent at longer wavelengths. This means that the light can penetrate into the deeper tissues of the fruit.

Fruit pigeons, flightless cassowaries, and bowerbirds in Queensland and New Guinea find the fruits of these trees on the ground and avidly collect them, exposing and dispersing the tough rudraksha seeds within. The fruit color may be selectively advantageous because it is durable and lasts even during the breakdown of the pulp. The fruits of *Delarbrea michieana* are also found in the rain-forest understory in Queensland, along with the rudraksha. Both of these may be picked up by these birds.

In addition, this structural color may provide a physiological advantage. The ripe rudraksha fruits contain tissues rich in chlorophyll beneath their blue skin. Since the fruit skin is transparent, structural color may enable the fruits to reduce their costs of production by permitting light for photosynthesis to penetrate when the fruits are ripe, even after they have fallen from the parent tree. Because photosynthetic plants release carbon dioxide by respiration when they are in the dark, I looked at the amount of carbon dioxide lost by the rudraksha fruits under shady conditions compared to dark ones. I detected a reduction in the carbon dioxide loss — that is, an enhancement of the fruits' carbon budgets — a finding consistent with this hypothesis. The fruits of *Delarbrea michieana* also appear green and photo-synthetic beneath the iridescent layer, so the ripe fruits of these trees may also actively undergo photosynthesis.

The leaves and fruits of plants found in the deep shade of tropical rain forests have evolved especially sophisticated mechanisms for modify-ing light environments in their interiors. Their internal anatomy displays chloroplasts, or allows them to move, to optimize interception. Lenslike epidermal cells direct light toward these chloroplasts in shady conditions. Accessory pigments absorb particular wavelengths, providing protection from intense sunlight. Iridescent blue plants also produce structures that physically modify the interior light environments in another way, through constructive interference. We see this as a remarkable color display, color that rivals the brilliant feathers of birds and wings of butterflies. How it aids in the survival of these shade-tolerant plants, if at all, will require much more study. One thing is clear. The more we look, the more treasures we find in the world's tropical rain forests. Among them, iridescent blue plants are certainly the precious jewels.

Chapter Eleven
Why Leaves Turn Red

....

MARTHA: What is Autumn?
JAN: A second spring, where the leaves imitate the flowers.
Maybe it would be so too with human beings that you would
see bloom if only you helped them with your patience.

ALBERT CAMUS

A woodland in full color is awesome as a forest fire, in
magnitude at least, but a single tree is like a dancing tongue
of flame to warm the heart.

HAL BORLAND

....

There is no better time and place than New England
during the autumn. As I revised this chapter, I spent
the fall in central Massachusetts, studying the color changes
in the foliage of the surrounding woods. In late October
after a visit to Harvard's libraries, I visited Walden Pond
and walked to the site of Henry David Thoreau's cabin in
the woods near the pond's edge. I looked at the fall foliage
in the late afternoon sun and enjoyed the view across the
pond from beneath his cabin, as he must have.

> Here . . . is not merely the plain yellow of the grains, but
> nearly all the colors that we know, the brightest blue not
> excepted: the early blushing maple, the poison sumach
> blazing its sins as scarlet, the mulberry ash, the rich
> chrome yellow of the poplars, the brilliant red huckle-
> berry, with which the hills' backs are painted, like those
> of sheep. The frost touches them, and, with the slightest
> breath of returning day or jarring of earth's axle, see in

what showers they come floating down! The ground is all parti-colored with them.

Many of you have witnessed the spectacular autumn color changes in temperate deciduous forests, as in the Appalachian Mountains or throughout New England. Others may have observed flashes of red color in the developing foliage of tropical forests, or the undersurface coloration of leaves in the shade of tropical and temperate forests. Growing up in eastern Washington State, I took little notice of color changes in the autumn. There was no forest in the background, and the street trees in town mainly turned yellow. There was little red color in the autumn, and the principal red-maroon leaves belonged to the Japanese maples, which kept these colors during their entire growing season. So I didn't pay much attention to these things. Later, when I studied botany in college, I learned that the autumn colors of yellow and red were caused by the unmasking of the carotenoid and anthocyanin pigments, from the breakdown of chlorophyll in the senescing leaves, and left it at that.

We mostly think of anthocyanin pigments coloring flowers and fruits, but they are also produced in vegetative organs of flowering plants and in nonflowering plants, such as ferns and gymnosperms (fig. 11.1). Since animals generally did not pollinate or disperse seeds in these plants, anthocyanins probably had other functions in more ancient plants. The foliage of some mosses and liverworts also changes color, by producing pigments similar to anthocyanins (fig. 11.2). This ancient distribution of anthocya-

Figure 11.1 Left, anthocyanins in the young leaves of this maidenhair fern (*Adiantum* sp.), photographed in French Guiana. *Right*, the young leaves of this cycad (*Zamia skinneri*) are red due to the production of carotenoids, not of anthocyanins.

Figure 11.2 Color change in bryophytes. *Top*, a sphagnum moss near the summit of Mount Marcy, New York. *Bottom*, a liverwort (*Lepidolaena taylorii*) from New Zealand. Photograph courtesy of Cortwa Hooijmaijers.

nins among the land plants spurred my interest in their function in leaves, and I became really interested in the phenomenon of red leaf color when we lived in Malaysia.

During my sojourns in the forest, I learned that few leaves changed color during their senescence, yet many were red (sometimes even purple) during development. The expanding leaves of the jade vine, native to New Guinea but grown as an ornamental plant in Malaysia, are purple early in development (figs. 7.21, 11.3). Brian Lowry, my University of Malaya colleague and fellow forest sojourner, was an expert on the chemistry of plant flavonoids, and I learned much from him. Since different tree species lose leaves at different rates and times during the year, there was little color change due to leaf fall in these forests. The red color of aging leaves was a curious enough phenomenon to be an identifying character of the few trees with this characteristic, as for members of the genus *Elaeocarpus* (the same genus that has iridescent blue fruits; chapter 10). Along the coasts and in

Figure 11.3 A view through the surface of a young leaf of the jade vine (*Strongylodon macrobotrys*, see 7.21), whose veins are surrounded by cells with vacuoles full of anthocyanins, the young leaves appearing dark purple during development.

urban landscapes, the foliage of the Indian almond tree turns a brilliant red before shedding, and the trees then quickly recover a crop of new leaves (fig. 11.4).

On the other hand, it was a different story for the young developing leaves of many trees. They were often a brilliant red, sometimes even violet or blue. This was such a common phenomenon in these forests that tree crowns appeared to be in flower from a distance while producing these young leaves (as at Barro Colorado Island in Panama, fig. 6.27). The Indian almond leaves are also red while developing (fig. 11.5). This was common not only in the rain-forest trees, but also in many of the tree crops. Young leaves of cacao are spectacular in color (fig. 11.6), as are many varieties of mangoes (fig. 11.7). Brian Lowry informed me that this color was caused by anthocyanin pigments, and that their function in these young leaves was a mystery.

Once I began to pay attention to the appearance of anthocyanins in leaves, I began to find them in other plants in different environments. Many

Figure 11.4 Color during the leaf senescence of tropical trees. *Top*, rudraksha (*Elaeocarpus angustifolius*) and *bottom*, the Indian almond tree (*Terminalia catappa*).

of the plants of the dense understory of these forests produced leaves with red to purple undersurfaces. Some of the patterned leaves of understory plants also produced red splotches of anthocyanins (figs. 11.8, 11.9). I also noticed red colors on the undersurfaces of floating-leaved aquatic plants, such as water lilies (figs. 11.10, 11.11). Finally, I noticed that red discoloration was produced by leaves in response to a variety of stresses, such as lack of nutrients, physical injury, insect attack, and disease.

Anthocyanins are almost always responsible for the red color in rapidly expanding and aging leaves in woody plants. Occasionally, carotenoid pigments (often rhodoxanthin) produce red color in senescing leaves of some conifers, as well as bronze colors of the developing leaves of the cardboard zamia (fig. 11.1). Betalain pigments may color young leaves of plants within a single order of flowering plants, the Caryophyllales (such as bougainvilleas, 11.12). A few other miscellaneous pigments produce reds in very rare cases. Anthocyanin-like pigments produce color in the bryophytes, such as

Figure 11.5 Flushing young leaves of tropical trees, loaded with anthocyanins. *Top*, young leaves cover the crown of this Asian tropical forest tree (*Mesua ferrea*) growing in my yard in Miami. *Bottom*, young leaves of the Indian almond tree (*Terminalia catappa*), growing in Miami.

sphagnorubin in the sphagnum mosses and riccionidin in various liverworts (figs. 11.2, A.37, A.38).

Carol and I left Malaysia, traveled and lived in France, and then settled down to live in upstate New York for two years. There I experienced the color changes in the temperate deciduous forest, observed with a different perspective because of my experience in Malaysia. Not only was red a common color during the autumn, from the leaves of maples and oaks, and the purples from sumac and ash, but I noticed a little red color in the developing leaves of some plants during spring (fig. 11.13). This was the reverse of my tropical experience. There young red leaves were common, and dying red leaves were rare. When we moved to Miami, I again saw the color

Figure 11.6 Young leaf color in the cacao (or chocolate tree, *Theobroma cacao*). *Top left*, the entire plant, a small tree; *top right*, light passing through the young leaves reveals the pattern of veins; *bottom*, microscope view reveals that the anthocyanin-containing cells are associated only with the leaf veins, similar to the jade vine.

changes in the leaves of my old tropical tree friends, such as the mango and cacao, and I traveled to Central America and saw the same patterns in rain forests there as in Southeast Asia. I saw the young red flushing leaves, the red-purple leaf undersurfaces of shade plants (and red variegation too), and the purple undersurfaces of the floating leaves of aquatic plants. Then, as I began to visit and then do research in the Everglades, I noticed that many plants produced red leaves. Some were the young leaves of tropical trees, such as the paradise tree (fig. 11.14), which reaches the northern edge of its distribution in southern Florida. Others produced red leaves before leaf fall, such as the red maple at the southern end of its distribution. Finally,

Figure 11.7 Top, the bright red color of the expanding leaves is characteristic of certain varieties of the mango (*Mangifera indica*), such as this 'Glen' variety, growing in my yard in Miami. *Bottom*, leaf section shows the accumulation of anthocyanins in the middle (spongy mesophyll) of the leaf.

many aquatic plants produced red leaves and shoots. These observations reinforced my interest in understanding why leaves turn red.

Physiological Functions of Anthocyanins

Just after World War II, the eminent German plant physiologist Erwin Bünning, traveling in Southeast Asia, observed the anthocyanic reds of the young leaves of rain-forest trees. He hypothesized that this color could pro-tect against the strong ultraviolet radiation of tropical latitudes. Ultraviolet radiation could damage these vulnerable plant tissues by destroying DNA

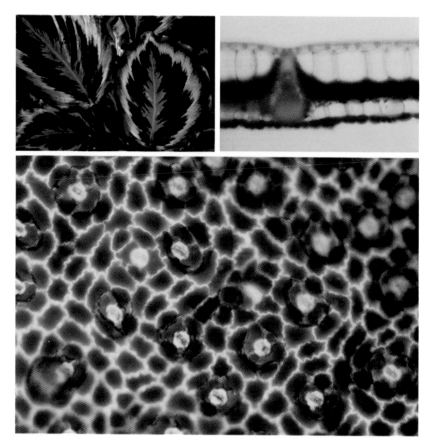

Figure 11.8 Red coloration in the undersurfaces of leaves of shade-loving plants. *Top left*, a cultivar of a prayer plant (*Calathea* hybrid), with variegated leaves; *top right*, leaf section showing anthocyanins on lower (epidermal) cell layer; *bottom*, the surface of that cell layer, with cell vacuoles full of anthocyanins.

and proteins rich in certain amino acids. I began to study leaf color in my laboratory at the University of Malaya, and obtained some results consistent with this hypothesis in studying some Malaysian species. Subjecting this hypothesis to a more critical test in research on mango and cacao after moving to Miami, our results with these trees showed clearly that anthocyanins could not have such a protective function against ultraviolet. They do not absorb radiation very strongly in the ultraviolet wavelengths that would be most damaging to plants, they are produced deep within the leaf tissue, and both species produce other flavonoid pigments located near the leaf surface that absorb ultraviolet wavelengths much more effectively. These results led me to consider alternative explanations.

Figure 11.9 Top, a wandering jew (*Tradescantia zebrina*), with silver variegation and purple undersurface. *Bottom*, view through lower surface, showing pigmentation and stomatal distribution.

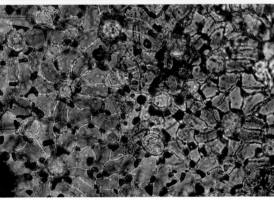

Figure 11.10 Leaf undersurface color in a floating-leaved aquatic plant, *Nymphaea odorata*. *Top right* reveals the undersurface color in detail; *bottom*, a microscopic view of the undersurface layer. Note that anthocyanins are found in bodies and not throughout the vacuole.

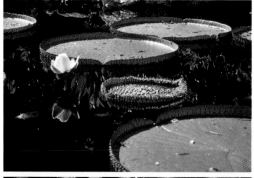

Figure 11.11 The undersurface of the giant floating leaf of the Amazonian lily (*Victoria amazonica*) is bright red contrasted with the yellow of veins.

Figure 11.12 Young leaves of bougainvillea (*Bougainvillea spectabilis*) often develop red coloration, from the betacyanin pigments characteristic of this and related plant families.

Figure 11.13 Young leaves develop anthocyanic color in temperate plants as well, how widely I don't know. *Top left, Photinia glabra; top right, Mahonia aquifolium* (the young shoots are brownish from a combination of anthocyanin and chlorophyll); *bottom,* a variety of hybrid tea rose ('Paradise').

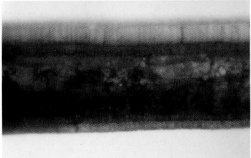

Figure 11.14 *Top,* red flushing young leaves in the paradise tree (*Simarouba glauca*) native to the Caribbean and south Florida. *Bottom,* a leaf section reveals the accumulation of anthocyanins in the upper cell layer only.

We assume, given the sophistication and spread of scientific research today, that such simple phenomena as plant color are well understood. There are several reasons explaining our relative ignorance about color, however, red leaves in particular. First, plants are not human diseases and don't benefit from the largesse of the National Institutes of Health. The few plants that are well-studied tend to be those that are economically the most important, such as maize and soybeans. Second, because anthocyanins are often responsible for the colors of fruits and flowers, most research has focused on pigmentation in these economically important organs for which the function of anthocyanin seems obvious—to attract animals for pollination and seed dispersal. Third, since Arthur Everest, an English chemist, and Willstätter published the structure of anthocyanin in 1913 (chapter 4), shortly after the rediscovery of Mendel's laws of inheritance, anthocyanins became early subjects of research in molecular genetics, rather than physiology. The inheritance of pea-flower colors studied by Mendel is an example of anthocyanin expression. Fourth, the discovery that light promoted the synthesis of anthocyanins led to the pigment's study in the rapidly expanding fields of photobiology and gene expression, again at the expense of research into anthocyanin function. Thus although there was much early speculation about the role of anthocyanins in leaves among late–nineteenth-century botanists, research has languished until the last decade or so.

During the past fifteen years, research on the functions of anthocyanins in leaves has been boosted by other discoveries in the physiological and ecological functions of plants and, for me, by establishing collaborations with other scientists. Kevin Gould, now at the University of Otago in New Zealand, contacted me about doing research in my lab during his sabbatical leave. I suggested to Kevin that he work with me on anthocyanin function in leaves, specifically in leaves of tropical rain-forest understory plants. Kevin was struck by the bizarre diversity of patterns of leaf colors in juvenile and adult forms of plants in New Zealand forests. From our collaboration, the hypothesis emerged that anthocyanins might alter the light environments within leaves to protect their chloroplasts from the inhibition of photosynthesis and, ultimately, damage to these organelles. During this time, the study of the suppression of photosynthesis by intense sunlight (photoinhibition) had emerged as an important factor explaining the functioning of many plants in their natural environments. Much of our ability to learn about photoinhibition was due to the discovery of the rapid changes in the fluorescence of green leaves after exposure to intense sunlight. Very elegant and portable instruments were developed to measure these fluorescence responses, which were correlated with photosynthesis and allowed us to learn about physiological changes in plants under natural conditions. These measurements can be made on plants in the field, very much like a physician uses a stethoscope to detect abnormalities in a patient's heartbeat. Kevin and I found, working with two of my university colleagues, that leaves of understory plants could be protected with a layer of anthocyanins observable from the undersurface. About the time we published these results, other researchers found a similar protective function of anthocyanins in temperate conifers and Antarctic mosses.

Nineteenth-century botanists studied the distribution of anthocyanins in plant organs and among different plant groups (fig. 11.15). Later, members of the school of physiological plant anatomy in Germany noticed that anthocyanin production was related to low temperatures and intense light conditions. This led to the widely accepted explanations that anthocyanins protect the photosynthetic structures against intense sunlight, and help to warm leaves and increase their rates of metabolism. Our hypothesis was really just a modern version of the protective hypothesis developed by botanists in the nineteenth century in Europe.

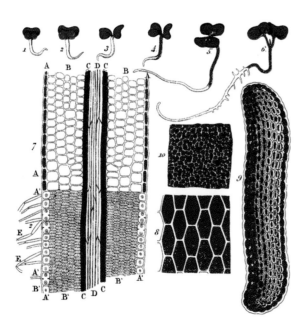

Figure 11.15 An early study of the distribution of anthocyanins in young red cabbage plants by the Belgian botanist Édouard Morren, published in 1858.

Anthocyanins in Autumn Leaves

Coming from a tropical perspective, research with Kevin inspired me to turn my attention to the phenomenon of temperate autumn leaf color (fig. 11.16). What were the details of the color changes, and what might the function of anthocyanins be during senescence? If anthocyanins protect the leaves from damage by excess light, why would this be important during the autumn, and why protect leaves about to fall from the plant? Producing anthocyanins in these leaves comes at some expense of energy, and why protect leaves whose photosynthetic life was about finished? In 1998, I began a collaboration with Harvard scientists Missy Holbrook and John O'Keefe, and Missy's Ph.D. student Taylor Feild, now at the University of Tennessee. We speculated that anthocyanins could protect the leaves to better enable them to take apart their chlorophyll pigments and photosynthetic enzymes and move the nitrogen present in these molecules back into the parent plants. Then they would be available for growth the following spring. Leaf nitrogen content is very strongly correlated with photosynthetic activity, and nitrogen is a valuable nutrient, which often limits plant growth at low concentrations. It is obtained from the soil, where it is principally fixed

Figure 11.16 A rural scene in northwest Massachusetts, the Deerfield River in the Berkshires during the autumn; the colors of foliage change are evident.

by the *Rhizobium* bacteria living in the root nodules of legume plants. The release of nitrogen, and its movement out of leaves and into the tree trunk (its recycling), is a sophisticated and highly regulated process. This process could easily be disrupted by photodamage to chloroplasts and the havoc in cells caused by the production of free radicals.

There was a lot for me to learn about autumn leaves. I had forgotten the little I had known about temperate plants after living in the tropics for twenty-five years. We had no idea of how commonly anthocyanins occurred among different plants during senescence, no knowledge of the timing and amounts of anthocyanin production, and no evidence for the photoprotective function in autumn leaves. Finally, we knew nothing about the levels of nitrogen left in senescing leaves with and without anthocyanin. If our hypothesis was correct, leaves with anthocyanins should have lower concentrations of nitrogen than those without anthocyanins, indicating that more nitrogen is resorbed by the trunks.

At the Harvard Forest in central Massachusetts, we observed eighty-nine woody species changing leaf color during the autumn (fig. 11.17). Some turned yellow, such as witch hazel, for which chlorophyll loss from degrading chloroplasts reveals carotenoid pigments in these organelles, producing the yellow color (fig. 11.18). However, 70 percent (62/89) contained anthocyanins, producing colors of brown to red, depending upon how much chlorophyll had been retained by the leaves.

We monitored the leaf-pigment contents in eighteen of those species (nine that turned yellow and nine that turned red) from late summer until leaf fall in early November. The leaves of witch hazel are typical of the species that turn yellow. As the aging chloroplasts lose chlorophyll, the ca-

Figure 11.17 A collage of leaves collected near the peak of autumn color change in mid-October, at the Harvard Forest in central Massachusetts. See notes for a guide to the identity of different leaves.

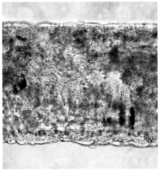

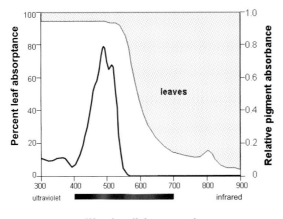

Figure 11.18 Top left, autumn yellow leaf color in witch hazel (*Hamamelis virginiana*); *top right*, leaf section showing aging plastids with carotenoids; *bottom*, degree of leaf absorption related to β-carotene absorbance.

rotenoid pigments become more visible. Eventually all chlorophyll breaks down, some carotenoid pigments remain, and the leaves are a brilliant yellow. The leaves of the red oak are fairly typical of the species that turn red. In red oak (fig. 11.19), anthocyanins begin to accumulate in the vacuoles of the palisade mesophyll layer of the leaf when part of the original chlorophyll concentration is lost. The final color is thus a blend between the anthocyanins produced in vacuoles, some carotenoids in the old chloroplasts, and the removal of chlorophyll. The anthocyanins are clearly being produced while leaves are in decline, when about half of the chlorophyll has already broken down. In virtually all cases, the anthocyanins accumulate in the vacuoles of photosynthetic cells, in palisade and spongy mesophyll layers, of the leaves. In a few evergreen species, anthocyanins are produced during the autumn and vanish the following spring. Both red and yellow leaves contain about the same amounts of carotenoid pigments, which de-

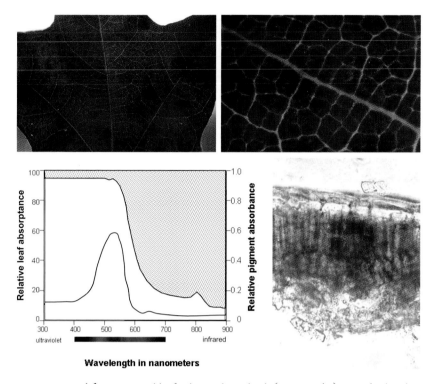

Relative leaf absorptance

100
80
60
40
20
0

1.0
0.8
0.6
0.4
0.2
0

Relative pigment absorbance

300   400   500   600   700   800   900

ultraviolet   infrared

Wavelength in nanometers

Figure 11.19 Top left, autumn red leaf color in the red oak (*Quercus rubra*); *top right*, detail of senescing leaves showing veins; *bottom right*, leaf section showing the concentration of anthocyanins in the vacuoles of the palisade cells; *bottom left*, degree of leaf absorption related to anthocyanin absorbance.

cline in concentration as chlorophyll degrades and the leaves prepare to detach from the trees.

Taylor Feild, who had excellent background in the technique of fluorescence measurement, took the lead in looking for a photoprotective function of anthocyanins in autumn leaves. We chose to study the red-osier dogwood (fig. 11.20). As in many plants, the timing of production of anthocyanins in the leaves of this species is influenced by exposure to sunlight. Thus we were able to compare the photosynthetic responses of senescing leaves that varied only in the presence or absence of anthocyanins in the palisade layer. We hypothesized that this pigment layer would protect intact photosynthetic tissues in those layers underneath. When exposed to intense simulated sunlight, the red leaves were photoinhibited less and recovered more quickly than the green leaves (fig. 11.21). The degree of photoprotection

Figure 11.20 Leaf color change and drop in the red-osier dogwood (*Cornus stolonifera*). *Top*, autumn plant with ripe red berries and foliage turning; *bottom right*, detail of leaf showing anthocyanins and veins; *bottom left*, leaf section showing anthocyanins accumulated in palisade cells; areas below are green.

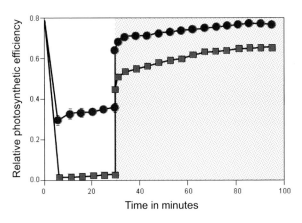

Figure 11.21 Relative photosynthetic efficiency based on fluorescence measurements in leaves of the red-osier dogwood, after exposure to high-intensity light (similar to sunlight). *Red circles* are leaves with anthocyanins, showing recovery. *Green squares* are leaves without anthocyanin, showing greater depression of photosynthetic efficiency and less recovery.

in the red leaves was even greater at low temperatures. When the red and green leaves were illuminated from their green undersurfaces, they both had the same degree of photoinhibition (where the anthocyanin layer could not protect the photosynthetic tissues). Thus we obtained direct evidence for the photoprotection of aging leaves by the anthocyanin pigments. However, we did not see any relationship between anthocyanin concentrations and amounts of nitrogen in the old leaves of the red-osier dogwood.

During the time that Kevin and I established our collaboration, Japanese scientists pioneered research in the remarkable antioxidant activity of anthocyanins. A variety of plant molecules are capable of neutralizing (or "scavenging") the free radicals that cause plant cell damage from exposure to intense light (or are associated with aging in animals, including humans). Protective molecules made by plants include vitamin E and various carotenoids. Some flavonoid compounds (the so-called bioflavonoids) are antioxidative, and anthocyanins—such as the cyanidin glucosides predominantly seen in leaves—are particularly effective. Cyanidin has about four times as much antioxidant capacity as carotene or vitamin E. Kevin launched a research program to test for the activity of anthocyanins as antioxidative compounds in leaves. He used one of his old New Zealand plant friends, the horopito tree (in the primitive family Winteraceae, fig. 11.22). Parts of the upper surfaces of its leaves are blotched red from accumulating anthocyanins in groups of palisade cells. When leaves were punctured with a fine needle, chloroplasts in the injured area produced a strong burst of the reactive oxygen-rich molecule hydrogen peroxide (H_2O_2). The concentration of hydrogen peroxide in cells could be observed under the microscope by adding fluorescent dye. Kevin and colleagues monitored changes in fluorescence levels over time, as a measure of hydrogen peroxide production. Hydrogen peroxide, a strong oxidant, was produced by chloroplasts in both the green and red regions of the wounded leaves (fig. 11.22). However, concentration differences between the two regions became apparent within minutes of injury. Hydrogen peroxide continued to accumulate in the green portions for 10 minutes longer, and then decreased only slowly. By contrast, levels of hydrogen peroxide in the red regions declined rapidly to background counts within the first 5 minutes, and maintained consistently low levels thereafter. This is evidence for the neutralizing of hydrogen peroxide in plant tissues by anthocyanins.

One problem in explaining antioxidant activity by anthocyanins is their accumulation in cell vacuoles; reactive oxygen molecules should damage chloroplasts the most, and then the cytoplasm. Although anthocyanins are made in the cytoplasm, they are rapidly pumped into the cell vacuoles. Most reactive oxygen structures do not move across the membrane into the vacuole, but hydrogen peroxide does. Thus this antioxidant activity could take place in the cytoplasm or against hydrogen peroxide that diffuses into the vacuole.

Anthocyanins can thus reduce photooxidative damage in autumn leaves

Figure 11.22 Leaves of New Zealand's horopito tree (*Pseudowintera colorata*), which make red splotches that persist through the year. *Bottom*, fluorescence stains and controls of leaf areas with varying amounts of anthocyanin. *From left to right*: lesion in green leaf; green leaf lacks fluorescence due to the accumulation of hydrogen peroxide; lesion in anthocyanic spot; increased fluorescence due to destruction of hydrogen peroxide.

via two mechanisms. First, by absorbing blue-green wavelengths of light, the anthocyanin pigments effectively shield chloroplasts beneath them from the higher-energy photons. Thus chloroplasts in red leaves received less light, and produced fewer free radicals as a consequence. Second, anthocyanins can chemically neutralize photooxidative chemicals, especially hydrogen peroxide.

For autumn leaves, the challenge remained to find evidence for a difference in the retention of nitrogen between species that turn red versus those that turn yellow. The physiologist William Hoch of the University of Wisconsin had published such a hypothesis in a long review in 2001, and he had predicted that the anthocyanin-containing red senescent leaves should retain less nitrogen than yellow or pale green ones.

In our survey at the Harvard Forest, we looked at the leaves of three trees of each of the nine species with aging red leaves and the nine species with aging yellow leaves. When we looked for a relationship between the amount of anthocyanin in the leaf tissues and the amount of nitrogen per unit leaf area, we did see a significant relationship: the more anthocyanins in the leaves, the less nitrogen in them. This was a confirmation of the hypothesis. Much better evidence was obtained by Hoch and his colleagues. They chose three plants with mutant forms, a dogwood, huckleberry, and viburnum, each of which lacked anthocyanin production during the autumn. These could be compared with normal anthocyanin-producing forms. In addition they examined a common species, paper birch, that does not produce anthocyanins and turns a brilliant yellow during the autumn. As we found with the red-osier dogwood, the varieties without anthocyanins had the greatest change in fluorescence, indicating photoinhibition. This was particularly true for plants grown in conditions of bright light and low temperature. Furthermore, they showed that the plants with anthocyanins in autumn leaves had less nitrogen in the leaves before they fell, and presumably had resorbed more nitrogen into the trunk. These results are a strong confirmation of the nutrient-resorption hypothesis, that anthocyanins protect the senescing leaves to allow more nitrogen to be stored by trees during the winter.

Both photoprotection and antioxidant activity could function simultaneously in protecting plant tissues from intense sunlight. The absorption of light by anthocyanins would reduce the rate of photoinhibition and photooxidative damage, and those free radicals that are produced could be neutralized by the anthocyanins. Both mechanisms could serve to protect vegetative tissues that are especially vulnerable to such damage, such as those under environmental stress, or leaves during development when the photosynthetic structures are being assembled in the cells of the leaves. So these may be explanations for the production of anthocyanins in leaves in a variety of circumstances, although we need to do much more research on plants in these different environments.

Ecological Functions of Anthocyanins

Anthocyanins may defend plant cells by absorbing excess radiation or by neutralizing free radicals, and they may be important in our diets, but there is little evidence that they are biologically active, either against diseases or herbivorous animals. They are not poisonous, as anyone who eats blueberries or strawberries knows. However, the red appearance of leaves may warn animals of their poor palatability and therefore reduce herbivory, or they may camouflage leaves from being seen by animals that would otherwise eat them. In many tropical trees, young red-purple leaves hang from branch tips prior to their rapid greening at the end of their development. Lissy Coley of the University of Utah has shown that these anthocyanic leaves are little damaged by herbivores and have very low nitrogen content. She believes that the coloration could turn animals away from eating nonnutritious foliage and help keep tender leaves from being eaten during their development. Although red color often appears as the leaves develop in tropical woody plants (and possibly in temperate plants as well), most have more slowly expanding leaves that contain some chlorophyll and protein such as cacao or mango. Paul Lucas and colleagues, of the University of Hong Kong, have argued to the contrary, that young red leaves in tropical forests are tender and palatable, important in the diets of leaf-eating primates. They have shown that primates of the Kibale Forest, in Uganda, preferred young red leaves. These young leaves were highly palatable because of high protein levels and tenderness. Field tests of feeding activities suggested to them that red to green shifts in leaf color are important cues that led to the evolution of three-color vision in certain primates (including humans), unlike all other mammals.

In rare cases, anthocyanin color in leaves may attract animals to consume the persistent fruits and disperse the seeds. Leaves could serve as flags, advertising the availability of a tasty reward. However, few of the red-senescing plants produce fruits that persist so late in the growing season.

Autumn coloration could also protect foliage from being eaten by animals, but only if the feeding occurred in subsequent years. The late William Hamilton of the University of Oxford and his associate Sam Brown argued that autumn coloration could warn aphids against laying eggs on trees with greater defenses of plant compounds. Thus the autumn coloration of the previous year could prevent herbivory in the following year after the eggs had hatched. So far this research is primarily based on literature surveys and modeling, with few direct field observations. Their article has provoked a series of responses. Simcha Lev-Yadun of the University of Haifa has proposed that anthocyanins could repel insects because reducing the effectiveness of camouflage makes them more visible to their predators, and odd variants on the original ecological hypothesis keep appearing. The physiological and ecological hypotheses are not mutually exclusive. In the Mediterranean region, the kermes oak produces anthocyanins during leaf development. This accumulation provides a little photoprotection and is associated with a reduction in herbivory. Clearly this remains a controversial area of research.

Many Unsolved Problems Remain

The research on red leaves I've described in this chapter is a step toward the solution of an old mystery, but is just a beginning. If anthocyanins protect plants physiologically or ecologically, then we need to understand how plants that do not produce such pigmentation coexist alongside the red-leaved species. For instance, Hoch and colleagues found that paper birches were about as efficient as the anthocyanin-producing plants in their resorption of nitrogen. The tendency to produce anthocyanins in leaves appears to be influenced by evolutionary history, and there may be other mechanisms of protection that have evolved in certain plants. Some plants produce red betacyanin pigments in their young leaves, such as bougainvilleas. Perhaps these compounds protect similarly to anthocyanins, and they have lots of antioxidant activity. Finally, sometimes the production of anthocyanins in leaves is correlated with production elsewhere, as in flowers or fruits. After all, both of these organs develop from leaflike structures. So perhaps what goes on in leaves is merely the by-product of production elsewhere, or vice versa. Refinements of experiments like those done by Hoch are an important strategy in teasing out these entangling factors.

Many odd things happen during the autumn. In boxwood, the reddish leaves are produced by a specialized carotenoid, anhydroeschscholzxanthin (what a name!), like the coloring of the California poppy (fig. 7.15). In fiddlewood, a tropical hardwood hammock tree native to south Florida, the leaves turn orange with carotenoids before falling off in April (fig. 11.23). In the famous ginkgo tree of China, the color of the autumn yellow leaves (fig. 11.23) is enhanced by a molecule somewhat akin to a laundry whitener. The molecule, 6-hydroxykynurenic acid, is derived from the amino acid tryptophan. It is produced in the leaves before they drop, captures ultraviolet radiation, and fluoresces it in the yellow wavelengths. It thus increases color

Figure 11.23 Two unusual color displays during senescence. *Top*, leaves of ginkgo (*Ginkgo biloba*) produce a fluorescent pigment that absorbs ultraviolet radiation and emits it as yellow, producing more intense color. *Bottom*, leaves of fiddlewood (*Citharexylum fruticosum*) produce a carotenoid pigment that makes the leaves orange before they fall in the spring.

Figure 11.24 *Top*, autumn foliage color change in the Chinese larch (*Larix potaninii*). Carotenoid pigments are retained before leaf drop. *Bottom*, autumn foliage color of the trembling aspen (*Populus tremuloides*) in southern Colorado. Note the well-defined area where the trees turn red rather than yellow.

intensity in these leaves. Not only broadleaf flowering trees turn color in the autumn (along with the ginkgo!), but needles of the larch, a conifer, turn to gold before they drop (fig. 11.24). Some broadleaf trees vary dramatically in the colors they produce during autumn. Sugar maples range from gold to bright red. The trembling aspen mainly turns yellow, but turns red in a few isolated localities (fig. 11.24). In the maples, red is associated with a reduction in nitrogen content of the leaves, and geology may play a role in determining the color of autumn aspen foliage.

Perhaps these ideas will give you pause in contemplating the patterns of leaf color in nature, especially during the autumn. I'm not sure that Walt Whitman would be pleased by such scientific revelations (see his words quoted in the preface). To me it adds to the beauty of the pageant. Knowing a little about the science of plant color makes the phenomenon even more complex, more elegant. During the autumn of 2004 in Massachusetts, I looked at the patterns of coloration in the herbs along roadsides, near streams, and in the forests of this beautiful region. I'm predicting that perennials that live in more exposed areas should have the most anthocyanins, perennials in more shady and protected areas should have less, and the annuals (which die and have nowhere to resorb their nitrogen) should not produce anthocyanins at the end of their lives. It would be nice to have a complete answer to the mystery of anthocyanin function, but at least now we some partial answers. Such results will refine those questions and lead to more experimentation and, ultimately, an unraveling of this puzzle.

Chapter Twelve Chlorophilia

....

The force that through the green fuse drives the flower
Drives my green age; that blasts the roots of trees
Is my destroyer.
And I am dumb to tell the crooked rose
My youth is bent by the same wintry fever.

DYLAN THOMAS

I thank you God for this most amazing day,
for the leaping greenly spirits of trees,
and for the blue dream of sky and for everything
which is natural, which is infinite, which is yes.

E. E. CUMMINGS

....

Green is my favorite color, by far. Perhaps such a preference is expected in a botanist. We associate green
with plants, and the color evokes the symbolic associations of plants: fertility and productivity, roots reaching
deep into the unconscious, seeds of enormous potential,
branches reaching toward the celestial sky, and fruits of
forbidden knowledge. So the title of this chapter is derived from the root *chloro-* (Greek for "green") and the
suffix *-philia* (Greek for "love of").

When I was a young child, our yard in the desert of
the Columbia plateau in Washington boasted bright floral colors, yellows, blues, and reds. I favored the colors
of my mother's garden over the green of the lawn and
foliage. My tastes shifted during childhood, influenced by
summers spent in the forests of the Cascade Range. Lake
Wenatchee YMCA Camp was located in that thick forest,

about 150 kilometers west of my hometown of Ephrata. The camp had a curriculum, probably shared by other YMCA camps, that was a mixture of Christianity, love for nature, and Native American spirituality. My principal memory of the camp, now over half a century later, is of the forest (fig. 12.1). The camp was nestled among the dark green conifers that covered the slopes of the Cascades. Later, those camp experiences inspired my friends and me to explore the mountains in the summer months, hiking through the forest into the high country.

In the 1950s, my hometown was isolated, but I learned about other countries, and the tropics, by reading voraciously and collecting stamps. I learned about the possible ecologies of imagined worlds by reading science fiction, about the tropics by reading authors who had traveled there, such as Joseph Conrad, W. H. Hudson, and H. M. Tomlinson, and from articles in *National Geographic*. Travel to New Zealand and the South Pacific helped me decide to become a botanist (fig. 12.2). During my college years near Puget Sound, and in graduate school in New Jersey, I continued to explore the nearby landscapes, to satisfy my aesthetic and emotional needs, and to study their plants. In a sense, my life became a progression of living near and visiting forests; this book, with its photographs and descriptions of plants and their colors, documents this forest journey (fig. 12.3).

We eventually settled in south Florida, the United States' continental

Figure 12.1 Trail to the swimming beach, Lake Wenatchee YMCA Camp, in the Cascade Range, Washington, 1954.

Figure 12.2 Beech forest, Te Anau Lake and the Southern Alps of the South Island of New Zealand, March 1964.

Figure 12.3 Tropical rain forest in the central highlands of French Guiana, near Saül, 1978.

foot in the tropics. Miami is a place where plants and landscapes take on particular cultural importance, odd Edenic visions inspired by a multitude of gardening clubs and plant societies; thousands belong to the orchid and palm clubs alone. Miami-Dade county supports cultivation of tropical fruits and foliage plants. The fruits (mango, avocado, star fruit, lychee, mamey, and jackfruit) are avidly eaten by local "tropicalists" and dress up salads and deserts in upscale table settings in the north. The foliage plants fill the lobbies of professional offices and the atria of malls and airports everywhere (and remind me of my times in the shade of tropical rain forests). They provide a reassurance of green nature in these otherwise sterile environments, and they grow verdantly in the urban setting of Miami.

In Miami, Carol and I created our own little tropical paradise, with mangoes and palms, and the windows of our house are surrounded by tropical foliage filtering the sunlight. Fairchild Tropical Botanic Garden became my favorite place for experiencing the greenness of nature, particularly in its Tropical Rainforest (fig. 12.4). There were also remnants of tropical forests to be explored, hammocks, with their understory light filtered by

Figure 12.4 The Rainforest, Fairchild Tropical Botanic Garden, Miami, October 2005.

the foliage of trees and branches festooned with epiphytes of orchids and air plants. The open Everglades also beckoned, a subtle landscape and soft color palette.

Miami is much more than a green landscape; it is also full of bright tropical flowers that bloom throughout the year. However, as much as we enjoy the bright colors of flowers and fruits, our principal color experience in nature is green. And I find comfort and tranquillity in its shade. My preference for these forested environments fits nicely with the experience of the Ituri Pygmies of the Congo, described by Colin Turnbull.

> If you *are* of the forest it is a very different place. What seems to other people to be eternal and depressing gloom becomes a cool, restful, shady world with light filtering lazily through the tree tops that meet high overhead and shut out the direct sunlight—the sunlight that dries up the nonforest world of the outsiders and makes it hot and dusty and dirty.

The Pygmies live in some tension with the outside tribes at the edge of the forest. The villagers, Bantu and Sudanese people of the savanna, fear the forest. When Turnbull took his Pygmy friend Kenge away from the forest to the edge of the savanna, he shrank from the openness and could not estimate the distances properly, thinking that the buffalo grazing several miles away were insects. In Malaysia, the Temoin we knew in the Ulu Langat were at home in the forest. Clearly, the Pygmies and Temoin have made peace with the environments in which they live, reflected in their cultures and attitudes. *Jungle* is a very inadequate word for such places, derived from a Sanskrit word to describe a dry wasteland, full of spiny shrubs and short trees. *Forest* is a better word (fig. 12.5).

As I was working on this book, I took some time off to help run the annual conference of the Association of Tropical Biology and Conservation, ATBC as we call it, held in Miami in July 2004. It is a society promoting the study of life in the tropics. We share a love for green tropical nature . . . but we don't talk about it. One of our evenings was a reception at the Fairchild garden, and for many the rain forest evoked the beauty of the tropics.

Is this affinity for the green natural world universal? Might the solace and tranquillity experienced in the company of plants be shared among all people? We certainly care about clean air, and the preservation of natural environments. Maintaining parks is high on the public lists of priorities, as politicians continually relearn. Gardening is our most important leisure-time activity, to which the inventories of the Home Depot and Wal-Mart stores attest, perhaps even more popular than watching stock car races.

Figure 12.5 Pasoh Forest, Negeri Sembilan, Malaysia, 1992.

Gardening and growing plants has become an important therapeutic "modality" helping a variety of clients. Travel to our domestic landscape shrines, the national parks, increases in popularity each year, so much so that intensity of use threatens some of them. Visits to the public parks in our communities increase as well. This all suggests, at the very least, a strong affinity to these green environments—chlorophilia. Where does it originate?

An Ancient Relationship with Green Landscapes

Our experiences with green originate in landscapes, in nature. Thus it is difficult to separate our responses to color from our experience of nature. Green is symbolic of nature; the Green Party focuses on political environmental issues. What are the origins of these preferences shared by people from all sorts of cultures and backgrounds? It is difficult to determine the depths of our responses, at what level they operate: social, cultural, psychological, and/or biological.

At a most basic level, we may respond to color and natural environments biologically, perhaps the result of our evolutionary past. All humans likely share predispositions toward certain experiences in nature as a consequence of this history. We have also evolved in a cultural sense, and those experiences may become embodied in our oral and written histories, in myths and in language. These ancient influences then may become embedded in the influences and mannerisms of contemporary cultures, as well as in our languages and in the values transmitted by our families and by our educational institutions. Finally, contemporary styles and fashion (politics included) may influence our attitudes, particularly given the influence of mass media.

How far in our evolutionary history might we go back to find these influences? Would it be the origin of modern humans, *Homo sapiens*, less than 150,000 years ago, or the origin of the genus *Homo* around 4 million years ago? Could it be the origin of hominids (walking around on two legs) some 6 million years ago? Perhaps it could even be in the origins of our arboreal primate lifestyle, some 55 million years ago. During most of this history, our ancestors lived in trees, and we've only been on the ground for the past 6 million years.

Paul Shepard has written about the ecology and landscapes of human origins more eloquently than anyone else, I believe. Clues about origins are seen in our eyes, which are those of an arboreal creature. Our densest packing of cones forms a circular area, not biased to scan the horizon like a savanna creature, but to evenly sweep the dimensions of the forest canopy. We are sensitive to a range of light conditions, from intense sunlight to deep shade. Our ancestors evolved the capacity to perceive the three primary colors, adding red to the blues and greens of savanna animals. This enabled us to select the ripest fruits and the most tender young leaves in the canopy. The single-color vision of our rods peaks in sensitivity in the green wavelengths, more abundant in the forest understory. Only after this long evolutionary history in the forest, our ancestors moved onto the East African savanna, into the sunlight, as Shepard wrote.

> The bush apes ancestral to man gradually worked farther from the forest. They took the sea eye out of the gloaming and into the radiance of the open day, though not without some enduring nostalgia. The awareness of glare in the open air and a sense of vulnerability when in it have not been completely dispelled. An affinity for shade, trees, the nebulous glimmering of the forest interior, the tracery of branches against homogeneous surfaces, climbing,

the dizzy childlike joy of looking down from a height, looking through windows and into holes, hiding, the mystery of the obscure, the bright reward of discovered fruits are all part of the woody past. Restfulness to the eyes and temperament, unspoken mythological and psychic attachments, remain part of the forest's contributions to the human personality.

It was not possible to see the forest while in it. Once out, we acknowledge the bond by remaining near its edge, cutting it back the right distance or planting forest-edge growth near us.

Color in Natural Environments

Vegetation, as in the understory of a forest or in open grassland, is not actually very green. The distribution of wavelengths is dominated by the principal sources of radiation, either the sun, blue sky, or white clouds. Thus the different colors visible to us (the spectral distribution, as we say) are not very different than the source of radiation. There is a slight increase in the green wavelengths, but nothing as dramatic as being in a totally green environment. John Endler has identified four major light environments that animals experience moving from the forest understory to a gap in the forest: yellow-green, blue-gray, reddish, and "white" ambient light spectra. Animals vary in their sensitivities to different portions of this spectrum. Humans are sensitive to the entire range, but some of our primate ancestors were less sensitive to the red region, and unable to distinguish it from green.

The greatest shift in the distribution of wavelengths in the shade compared to sun is barely beyond the limits of human vision, in the far-red region of the spectrum (fig. 12.6). This is particularly well documented by the ratio of the red and far-red regions. In sunlight the ratio is around 1.2, and in forest shade it is around 0.3. Plants have evolved an elegant system of perception of the shifts in these two bands with the phytochrome pigment (fig. A.4). This pigment detects the ratio of these two wavelengths as an equilibrium between the chemically distinct red (R) and far-red (FR) forms of the molecule. Since this ratio is strongly influenced by green foliage, either transmitted through it or reflected from the foliage surface, plants have thus evolved an elegant means of determining the presence of nearby plants. Many of the plant growth responses to the red:far-red ratio improve their ability to compete against other plants, as in growing taller by stretching more and allocating fewer resources to make roots.

My own research on plant responses to the light environments in tropi-

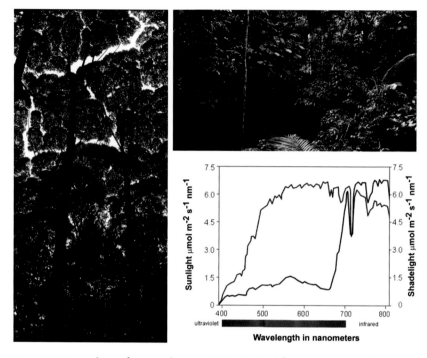

Figure 12.6 Tropical rain-forest understory, Malaysia. *Top left*, crown shyness, where individual branches are separated; *top right*, understory vegetation in Bukit Lanjang Forest Reserve; *bottom right*, comparison of the spectral distribution of sunlight and understory (note different scales).

cal forests has led me to speculate whether humans might possess an analogous system that could function to detect subtle changes in the shade environment (such as detected by plants), and which might affect us psychologically. The advantage for the phytochrome system in plants is that it detects shifts in the ratios of wavelengths. In contrast, single pigments detect the increase in intensity at a single narrow wavelength. Plants use blue-absorbing pigments, as well as the chlorophyll pigments (which absorb much of the visible spectrum; fig. 3.5), to detect changes in intensity. The phytochrome pigments may be better at sensing light changes related to the presence of neighboring plants because detection of shifts in ratios may be more accurate than detecting the amount of radiation at a specific wavelength. The situation may be analogous to the greater fidelity of the FM radio (frequency modulation) compared to AM (amplitude modulation).

Animals detect light through vision and through nonvisual light-detecting systems. In the vertebrate animals, the pineal gland (part of the endocrine system) is also an optical organ. It sits between the brain and the skull, and in all vertebrates except mammals can detect filtered light. Light reception controls the secretion of melatonin. It in turn promotes the secretion of other hormones, which collectively influence various physiological responses, including the sleep cycle. We know much about the pineal function in poultry, and their responses to light have been exhaustively studied. We alter poultry-shed environments to increase egg-laying activity and the size and muscle mass of chickens and turkeys. The chicken pineal gland is sensitive to red wavelengths, but I don't think that the analogy with plant red and far-red ratios applies.

Humans have a pineal gland; it produces melatonin; and it is affected by light. It is not a light-sensitive organ by itself, but receives information about light from the eye via the optic nerve. We are now quite sure that certain cells in the eye (but not the rods or cones) contain a rhodopsin-like pigment whose peak absorbance is in the blue part of the spectrum. These sensory cells send information about the light/dark cycle to the pineal gland. Melatonin, which affects the body in a multitude of ways, is thus tied into the sensory activity of those cells. We may also be sensitive to light through our skin (perhaps our blood, but the evidence is controversial), and have no idea of how that might affect our physiology, let alone our behavior.

If we could perceive the subtle wavelength changes produced by vegetation, it would not be the color green, but would more likely be the color blue—which is reduced in the understory where there is a slight increase in green. So light effects might be in a wavelength range that is a consequence of the absorption of sunlight by foliage, and not green directly. I have sometimes fantasized about doing experiments on human light responses, extrapolating from my work with plants, *way* beyond my professional expertise. Such research would establish a human physiological basis for the beneficial effects of filtered forest shade on our behavior, related to melatonin production. Perhaps the nonvisual and visual perception of light could team up in a manner analogous to the phytochrome system of plants. If I were to discover such a subtle influence on human behavior, however, it would certainly be a mixed blessing. Such knowledge would help us to understand our responses to natural environments. It would also provide others with tools to manipulate our behavior in an economically profitable direction. Already the discovery of our nonvisual sensitivity to blue light is being put to good use to help cure sleep disorders. Perhaps in the future it might be used to affect behavior, whether for individuals' health or for private economic benefit.

Green Psychology

Maybe our response to nature, particularly to vegetation, is hardwired to the level of our visual sense and our sensitivity to color. Certain colors, and thus certain wavelengths of radiation, might evoke psychological responses in us. Those responses could be detected through careful exposure to colors and the objective measurements of response. They might even be detected physiologically, through brain activity, blood pressure, pulse, or some other measurement. The practical motivation for this research has been to find colors or color combinations that affect human behavior, and thus help create environments that reduce "negative behaviors," such as aggression and anger, and promote "positive behaviors," such as tranquillity or (unfortunately) passivity or a tendency to buy things. Such research goes back a century, and then some. It is not very definitive but has been used by color consultants in designing human environments, even in dressing for success. One of the problems is defining color, and taking into account brightness and saturation in addition to hue (chapter 2). Earlier research neglected this, and even the area of the color used in a survey (or whether it is produced on a flat surface or produced by passing light through a film) can affect the outcome. Nonetheless, there is evidence that humans are attracted to colors, that bright colors (such as red) promote activity, and that darker colors (such as green and blue) are associated with tranquillity. Babies are attracted to bright colors. School children prefer bright colors (including red, yellow, green, and blue) over dark (like gray and black).

Certainly green is an attractive color for many people, but not uniquely so. It is probably dangerous to read too much into surveys of these color choices, and color tests may not tell us very much. Nonetheless, such color preferences and responses have been used to design building interiors and architectural projects. We have moved far beyond the antiseptic green institutional spaces (supposedly designed to decrease aggressive behavior) to sunny and pleasant color combinations. My preference for the color green is shared by others, but not by everyone. What is calming and cool to me may be passive or boring to someone else.

If there are fundamental human responses to colors, we might also expect to find them in various languages. However, what we find among languages is an enormous variation in the names of colors used to describe the visual spectrum. This spectrum (400–700 nanometers, our blue through red) is detectable by all humans. However, there appears to be a relatively small number (eleven) of basic color terms used to partition the color spectrum. These are the "fundamental neural response categories" (FNRs) used in the World Color Survey (chapter 2). What this survey of over 100 languages demonstrates is a dramatic variation in the color terms of different languages. Some do not have a color term for blue, others for yellow, and so forth. Berlin and Kay, who initially developed these concepts, postulated a sequence in the evolution of languages, starting with those having only two FNRs, black and white, then adding red, and then other colors, including green. Thus green is not a fundamental color term in all languages. How color terms are added to languages should be related to the environments where these cultures were established, but these relationships are not very clear. Languages established by peoples living in rain-forest environments (where green should be an important color) vary in the use of color terms along with those developed in other environmental settings. For groups that use green as a basic color term, however, it is often associated with growth, productivity, and fertility. This connotation occurs in the languages of the Siona-Secoya as well as the Hanunóo (chapter 1). Green may be seen as moist, in contrast with red or brown, seen as dry. It is also likely that people sense and describe colors by analogy with objects collectively experienced in their environments, such as flowers or animals, but may not have a basic color word for that category. Colors may also be associated with ripening, and it may be difficult to distinguish between color words and words for development. Thus the Xhosa people of Africa have twenty-six color descriptors for cattle, partly based on the colors of different breeds. Eskimos supposedly have many words for snow (as we've all heard), and thus descriptors for the color white.

To make a long story short, and to summarize a large body of research (much of it useless), there is not much evidence supporting the intrinsic psychological attractiveness of green color.

Influences of Landscape

Our attitudes toward the colors of nature may be revealed by our preferences toward different landscapes, and these preferences may have been influenced by our evolutionary history. Gordon Orians, an ecologist, and Judith Heerwagen, an architect (both at the University of Washington), hypothesized that our preferences toward nature, and our attraction to certain types of landscapes, originate in our evolutionary experiences as we became human in the savanna landscapes of East Africa. Indeed, there is an extensive literature surveying landscape preferences within cultural groups, either by showing people photographs of different landscapes or by giving them cameras to photograph landscapes that they like or do not like. Much of this research was conducted by Rachel and Steven Kaplan and their students at the University of Michigan. By and large, people favor landscapes that are green and parklike, and feature trees (fig. 12.7). However, people also prefer landscapes that are somewhat open (not too many trees) with a water feature. People do not like crowded landscapes of dense foliage. In evolutionary psychological terms, the landscapes of dense foliage provide cover for enemies and predatory animals, and could be threatening. More open landscapes combine the reassurance of a vegetated environment with the safety of having a view of any threats and an escape route, or a clear destination. Such preferences have been documented among groups in the United States (but too often the groups are students in university psychology courses), and for people living in quite distinct cultures, for example Bali.

Similar relationships appear in myths created in traditional cultures.

Figure 12.7 Forest and cultivation, near Mukkali in the Western Ghats of Kerala, India, January 1989.

I am particularly partial to Indian mythology, the *Ramayana* in particular, because I have spent a lot of time in India. In this story, both history and myth, Lord Rama, with his wife Sita and brother Laksmana, were banished to the southern forest of Dandaka for seventeen years, because of a promise Rama's kingly father made to one of his wives, jealous of the king's favoritism toward Rama. The three moved into the forest, Sita was kidnapped by the clever and evil Ravana, Rama got help from the monkey king Hanuman and his army, they defeated Ravana in his island kingdom of Lanka, they rescued Sita, and they all lived happily ever after—well, not exactly. The epic was set down in verse by the poet Valmiki about 600 BCE from earlier versions, so this may be a story that is 3,000 or more years old. It is a drama performed in shadow-puppet plays in Malaysia and southern Thailand, and in the dance dramas and music of Java, Bali, and Benares. It has also been serialized in Indian television. Valmiki's poetry evokes the attitudes toward nature in those early times, and certain passages convey these two views of nature, threatening and bucolic:

> But soon they came to a trackless, dreadful-looking forest, and Rāma Aiksvāka, son of the best of kings, asked the bull among sages: "What a forbidding forest this is! Echoing with swarms of crickets, it swarms with fearsome beasts of prey and harsh-voiced vultures. It is filled with all sorts of birds, screeching fearsome cries, as well as lions, tigers, boars and elephants. It is full of *dhava, asvakarna, Kakubha, bilva, tinduka, patala,* and *badari* trees. What dreadful forest is this?"

In this dangerous forest, even the trees, listed by their Sanskrit names, are full of thorns and have dark and somber foliage. Perhaps this is the landscape interpreted by environmental psychologists as universally disliked. Other passages in the book of the *Ramayana* that describe the period of exile in the forest give a more positive sense of the forest landscape.

> As they traveled on with Sītā, they saw varied mountain landscapes, forests, lovely rivers with cranes and sheldrakes upon the sandbanks, ponds covered with lotuses and thronged with water birds, dappled antelopes massed in herds, rutting horned buffaloes and boars, and elephants butting at trees.

The forest landscape described in this passage has been opened up, probably modified by human use of fire, to provide distant vistas and routes through the landscape. The references to water are important in this culture because of the seasonality of the monsoon rains; their failure has lead to famine throughout South Asian history.

Reflecting on my own preferences for natural environments, I rank the savanna pretty low. This grassland environment has clumps of trees here and there. A seasonal tropical climate creates the savanna. When we are not the primary modifiers through our use of fire, savannas are the result of climates of short rainy seasons and long dry seasons. Early humans would have experienced the savanna as a lush and parklike environment during the brief wet season and shortly after, and then struggled to obtain clean water during the extremely hot dry season while avoiding attack by predators congregating in the same places. Abundant and more evenly distributed rainfall creates a more closed forest environment, my personal preference. The forests of the *Ramayana* were influenced by a strong seasonality of rainfall. Two years in the semiforested countryside north of Mumbai, not far from the location of Rama's exile hermitage of Dandaka, taught me the extremes of such a climate. This forest is conditioned by an annual rainfall of almost 3 meters, all of if falling in about three and a half months, from June into September. The months leading up to the monsoon were exceedingly dry and uncomfortable, and we rejoiced in the abrupt and dramatic arrival of the monsoon rains and the greening of the forests and agricultural landscapes. The culture of the indigenous people of this region, the Warli, enabled them to survive in this environment (but barely so during drought years), as have indigenous people throughout the world, enjoying the abundance of the seasonally dry forests and savannas during good times and fighting for survival during bad times.

Many of the remnant forests had connections with myths and tribal history, were considered sacred, and are protected (fig. 12.8). Throughout India, many places are associated with the mythic poems of its culture, such as the *Ramayana* and *Mahabharata*. The wanderings of their heroes sanctified the landscape, and these have become sacred places, often sacred forests. Today in India, they are probably the greatest collective preserve of biodiversity, although now burdened by hordes of spiritual tourists.

Forests have been deemed sacred and preserved in other parts of the world as well. Early Europeans worshiped nature and trees, and celebrated their forests as sacred. In a comedy sketch in the NPR radio show *Prairie Home Companion*, Jim and Barb, representing the American Ketchup Council, discussed where to go on vacation. Barb suggested that they go to the beach, and Jim replied, "We don't do beaches. We're Caucasians, people of the forest." The origins of this European connection to forests can be found in the roots of the proto-Indo-European (PIE) language and culture. The early language was established about 10,000 years ago in central Asia, in forests comprising oaks, ash, beech, linden, . . . and apples. These

Figure 12.8 Temple and sacred forest adjacent to Manali, a village in the Kulu valley in the Indian Himalayas of Himachal Pradesh, 1976.

trees were sacred to those early people and became sacred in the cultures that followed. The Celts had a language of trees, derived from their forests. The Norse had a religion and mythology in which trees played an important part. The ancient Germans lived in a great forest of oaks and beeches when confronted by the Romans. Their resistance to conquest, described by Tacitus in the first century CE, depended upon their sylvan camouflage. This history fueled a mythology of a noble race from the primeval forest, and the tension between natural-forest living versus civilized urban life has been an important influence in European history. Such "woodland ethnicity" was used by the National Socialists (Nazis) in their rise to power. So Jim's opinion about where to go on vacation has some hidden baggage. Rather than exploiting the ancient tension between trees and masonry (Germans versus Romans), it is better to interpret this as a conflict between oaks (a hunting and gathering existence) versus olives (an agrarian but green and settled existence). Even in the ancient forests of Dandaka, Rama and his companions frequently came across ashrams, spiritual settlements in the forest, always described paradisiacally.

> Rama, the self-disciplined and invincible prince, saw a circle of ashrams where ascetics dwelt. Kúsa grass and bark garments were strewn about it,

and, flooded with brahmanical splendor, it was as luminous and blinding to the eye as the sun's circle in heaven. It was a place of refuge for all creatures; its grounds were always kept immaculate, and troupes of apsarases ever paid homage there and danced.

Today, the elderly Indian gentleman enters retirement seeking enlightenment. He "goes to the forest." The destination is usually not a wilderness, but the company of other seekers in a forest park, perhaps some contemporary equivalent of the ashrams described by Valmiki. Such a tradition is also found in China (fig. 12.9). We also enter such landscapes for the purposes of recreation (re-creation), and venerate our public nature temples: our national parks. The preservation of wilderness in our landscapes, areas where human influences are almost entirely absent, has provided an opportunity for many to experience nature with few distractions from our contemporary culture.

Several writers of science fiction have dealt with the role of colors in nature and environmental response. Some of these books were attempts to create worlds with different ecologies, and some imagined vegetated

Figure 12.9 Huatingsi Buddhist Monastery, Western Hills, Kunming, Yunnan, China, November 2005.

worlds with colors other than green. These are myths of a different sort. Sometimes this artifice was just part of making a new world exotic or different, but in other cases it was a means of exploring our attraction to the natural landscape. In *Snare*, Katherine Kerr created a world inhabited by colonists, who brought visions of an Eden-like paradise with them from their home, the distant planet Earth. Yet the natural vegetation of the planet was quite different.

> All around (Ammadin) the lavender grasslands stretched out to an endless horizon. As she rode, the grass crackled under her horse's hooves. . . . Along a violet stream bank, the red spears leapt from the earth and towered, far taller than a rider on horseback. Close up they appeared to have grown as a single leaf, wound around and around on itself to the thickness of a child's waist, but down at the base, hidden by a clutter of mosses and ferns, were the traces of old leaves that had died back and withered. The spears grew in clumps from long tuberous roots, spiraling out from a mother plant. How the mother plants got their start no one knew.

We also use our imaginations and resources to create artificial environments that look green and vegetative. Thus we hide cell-phone towers within artificial palms and giant conifers, and create groves of plastic trees in shopping malls and airports (fig. 12.10). This is not to say that plastic

Figure 12.10 Plastic date palms, Dubai International Airport, United Arab Emirates. Photograph courtesy of Preserved Treescapes International (www. Treescapes.com).

plants are bad (although they certainly don't remove carbon dioxide and refresh our air with moisture and oxygen). This green and plastic vegetation illustrates the lengths we go to create some substitute for nature in our built environments, when the real thing will not survive.

Green Therapy

Natural environments, with green as an important element, affect our behavior, particularly in reducing stress. Such behavioral responses have been studied in hospitals, prisons, dentist offices, office work spaces, and urban housing projects. Roger Ulrich, who led much of this research in the 1980s, conducted an analysis of the rates of recovery of gall-bladder surgery patients in a suburban Pennsylvania hospital. The patients recovered in identical hospital rooms that varied in whether their single window opened to a brick wall or to a small stand of deciduous trees. He found that the twenty-three patients with the view of trees recovered significantly more quickly and had fewer postoperation complaints than the twenty-three patients who looked at the wall. Wall-viewing patients also needed more pain-reducing medications. Sick calls of inmates at a Michigan maximum-security prison were reduced when they were assigned cells that had some view of trees and grass. In inner-city housing developments in Chicago, landscaped areas with trees near the buildings promoted more informal social contact among neighbors. These landscaped areas also reduced the incidence of aggression and violent crime around the buildings. Heerwagen and Orians analyzed windowless offices at the University of Washington; they found a much higher frequency of nature pictures compared to the art-work displayed in offices with a view. Heerwagen also found that landscape photographs reduced stress among patients in a dental office. Similarly, astronauts preferred landscape pictures during extensive periods in orbit.

Ulrich went further in demonstrating the effects of landscape scenes on stress levels and well-being by comparing the brain activities of students preparing for examinations who were shown various photographs, includ-ing nature and urban scenes. Observing landscape photographs reduced pulse rate, increased alpha waves as measured by an electroencephalograph, and improved their sense of well-being. Not only do such green experiences seem to improve behavior, but they may also aid in learning, particularly among those with deficits. In Chicago, disadvantaged children from public housing projects improved in school performance when moved to housing in a more sylvan setting. In children with attention-deficit/hyperactivity disorder (ADHD), symptoms improved with regular recreation in a land-scaped setting. These results are consistent with the Kaplans' theory (bor-rowed from the psychologist William James) that environments demand-ing our constant attention exhaust us, and the particular advantage of green landscapes is their provision of experience of passive nondirected and non-focused attention, which is highly restorative. It is pretty clear that experi-ences in nature, particularly those that are "green" (include vegetation) and have a water element, reduce stress and improve our moods (fig. 12.11).

Long before Ulrich conducted his research on the therapeutic value of landscapes and the Kaplans developed a research program on landscape preferences, the Menninger Clinic, in Topeka, Kansas, began pioneering work on the use of plants in psychotherapy. Psychiatrists there reported on the success of plant-related activities, such as gardening and landscaping, in relieving psychological problems. A new discipline, horticultural therapy, came into being, and graduates were trained at nearby Kansas State Univer-sity. Today, horticultural therapists work in a variety of settings throughout

Figure 12.11 Walkway through the woods, the Bloedel Reserve, Bainbridge Island, Washington, October 1994.

the world. In the United States such work is performed in hospitals (both general and psychiatric), prisons, facilities for the mentally handicapped, retirement homes, and many other places. Ann Parsons, who was trained in horticultural therapy and was education director at Fairchild Tropical Botanic Garden, participated in course on the "meaning of gardens" that I taught. She recounted a dramatic story from another therapist concerning a patient in a mental hospital who was incapacitated because she always kept her fists clenched to keep her energy from escaping from her body. Her clenched fists made it impossible for her to feed and care for herself. She began to hang around the greenhouse, the location of the horticultural therapy program. For weeks she observed but did not participate. Then one day, she was present when an urgent need arose to move plants that were on hangers to another location. She stuck out a single finger, and plants were hung on it so that she could carry them. That was the beginning of the transformation of opening up her hands, opening up to other people, and beginning to care for herself. Horticultural therapists share these stories when they come together for meetings. They must ask themselves, what is it about plants that make them such effective vehicles of therapeutic transformation? Perhaps it is not their "plantness" but their "livingness," as animals are also effective vehicles of therapy, and dogs can be trained to serve as therapeutic animals. Still, there may be something special about plants, and their greenness, that makes them so therapeutic. Millions of gardeners worldwide can certainly attest to the therapeutic benefits of living with plants.

Thus this book ends with an intuition about the vital importance of plants, elements in green landscapes, in our lives. There may be no mechanism for how they benefit us psychologically and emotionally, no actual scientific proof that this is so, but their soothing and calming presence has been important in my life. This is an experience shared by others, but less by scientists and more by therapists, poets, gardeners, and the like. Paul Shepard used the term "phyto-resonance of the true self" for this affinity, and I use the word "chlorophilia." Shepard's ideas are consistent with those of David Abram in writing about traditional people throughout the world, living close to the natural landscape, with heightened perceptions of their environments and a more authentic existence in their world. In his book *In Praise of Plants*, Francis Hallé wrote of two botanies. One with the capital *B* is the science—of DNA sequences and measurements of photosynthesis. The other, with a lowercase *b*, is the less formal and intuitive appreciation of plants by poets and artists, gardeners and orchid growers, and some farmers and foresters. I seem to belong to both botanies, and have found that my scientific research illuminates my love for the amazing complexity and beauty of plants, and my deep appreciation has helped to guide my scientific study.

Today we are witnessing a profound transformation of the green landscapes of our planet. The crush of expanding populations is reducing the amount of remaining forest, sometimes replacing it with uncultivable wastelands. India, now with 1.1 billion people, has reduced its forest cover to less than 15 percent. That amount is impressive, given the pressure on its forest resources—especially since 70 percent of the population still uses firewood for cooking. In the United States, about 20 percent of our land surface is covered by, or directly impacted by, roads and paving, and built-up urban areas now take up an area of the size of Indiana. The human and economic forces that drive the de-greening of our planet are extremely powerful, and it is hard to envision how we might deflect them.

Scientifically (as in Botany with a B), we attempt to come up with logical reasons, backed by research, that assign economic value to vegetation and nature. Our scientific hope is that logic and reason will win out, and will guide economic forces to more beneficial results. Such approaches have

Figure 12.12 Mixed broadleaf and coniferous forest in the Swift River Reservation, Petersham, Massachusetts, preserved for its beauty, September 2005.

led to limited success, as in the debt for nature swaps and carbon trading to preserve tropical rain forests in the tropics. The danger in playing this game is that the economic forces can enlist the numbers to a short-sighted economic advantage, and we end up playing along with them. Part of the attractiveness of plants is the way that they add color to our lives. If this attractiveness had a quantifiable value, the people who count up what things are worth could include it in their calculations. As a result, nature might more often be appropriated for commercial and monetary interests, rather than being available for long-term benefit to humanity (fig. 12.12).

Unscientifically (botany with a *b*), the strongest reasons for preserving green landscapes, whether as wilderness, a patch of wildflowers by a road, or small gardens in a housing development, are really emotional. Our motivation is derived from our personal and collective responses to plants. An important part of our connection to plants is the color with which plants illuminate our lives, green foremost, but also the flecks of bright colors with which we sanctify the important moments of our existence. I don't know why or how we respond to plants the way we do, but I know that our visual perception of them is important. Ultimately the reasons that we preserve what is valuable to us, and I am speaking particularly to the amazing diversity of plants and other organism that inhabit our planet with us, is that we are emotionally connected to them. Perhaps the best that I can do as a scientist is to reveal this beauty to those unaware, or in ways not yet appreciated, to increase knowledge, appreciation, and—ultimately—a feeling for plants. Despite the political and environmental problems we face, we still live in an awesomely complex and beautiful world, mirrored by the colors with which plants illuminate our lives.

APPENDIX A: PLANT PIGMENTS AND RELATED MOLECULES

....

The plant pigments and related molecules in this appendix are described in the text. They are listed here in the order their synthesis is described in chapter 3, and by the pathway they are part of or by general types of structures. Information on these structures was obtained from references cited in the notes for chapter 3. The structure names are sometimes simplified, so that cyanidin-3-glucoside is used instead of cyanidin-3-O-glucose.

Porphyrins and Tetrapyrroles

A.1 chlorophyll *a*

A.2 chlorophyll *b*

A.3 phycoerythrin

A.4 phytochrome

The Isoprene Pathway

A.5 latex

Carotenoids

A.6 phytoene

A.7 lycopene

A.8 β-carotene

A.9 fucoxanthin

A.10 eschscholzxanthin

A.11 bixin

A.12 crocetin

A.13 capsanthin

A.14 retinol

Pigments and Related Compounds of the Phenylpropanoid Pathway

A.15 alizarin

QUINONES

A.16 plastoquinone-*n*

A.17 phylloquinone

NAPTHOQUINONES

A.18 naphthoquinone

A.19 lawsone

A.20 juglone

A.21 lapachol

A.22 4-methoxydalbergione

LIGNINS

A.23 R = H p-coumaryl alchohol
R = OCH₃ coniferyl alcohol

A.24 lignin

TANNINS

A.25 gallic acid

A.26 condensed tannin $(n = 1–30)$
proanthocyanidin

A.27 virolin

A.28 mopanol

catechol

A.29 phytomelanin

Pigments of the Flavonoid Pathway

CHALCONES

A.30 naringenin chalcone

A.31 butein

AURONE

A.32 aureusidin

DIHYDROFLAVONOL

A.33 dihydrokaempferol

FLAVONE

A.34 luteolin

FLAVONOL

A.35 kaempferol

ANTHOCYANIN-LIKE FLAVONOIDS

A.36 apigeninidin-5-glucoside

A.37 riccionidin-a

A.38 sphagnorubin

ANTHOCYANINS

A.39 cyanidin

A.40 cyanidin-3-glucoside

A.41 pelargonidin

A.42 malvidin

A.43 peonidin

A.44 delphinidin

Nitrogen-containing Pigments

A.45 indican

A.46 indigotin

A.47 betanidin

A.48 indicaxanthin

Miscellaneous Pigments

A.49 coleone E

A.50 dracorubin

A.51 santalin

A.52 hematoxylin

A.53 thiarubrine A

A.54 curcumin

A.55 mangostin

A.56 urushiol

APPENDIX B: SEPARATING PLANT PIGMENTS BY PAPER CHROMATOGRAPHY

....

Here are very simple techniques for separating most of the plant pigments described in this book. By using these techniques (appropriate for middle and high school students and teachers), you can separate simple mixtures of pigments in plant tissues. By varying the elution solvent (with which the pigment spots are separated), you can determine the basic classes of pigments occurring in plants. I have devised these techniques to be used with inexpensive materials available in local hardware and grocery stores.

The theory is very simple. The molecules have varying affinity for the paper's cellulose fibers and the solvent. The strength of those two affinities determines the rate of movement up the paper as the solvent is absorbed by the fibers and moves up (fig. B.1). A water-soluble pigment such as an anthocyanin will move up the paper in a solvent in which it is soluble (and those with some affinity with it), but not up the paper in a solvent in which it is not soluble, such as the mineral spirits.

Materials

1 small box of large coffee filters (Melitta or cheaper substitute)

1 very small ceramic cup or dish

1 small eyedropper

1 small spoon

1 or more small narrow jars with tops

The following solvents in small volumes

1 bottle of 91% isopropyl alcohol (rubbing alcohol)

1 bottle of mineral spirits

1 bottle of clear vinegar (3% acetic acid)

1 bottle of ammonia cleaning solution (ammonium hydroxide)

Procedure

1. Make the extraction and elution solvents.
   - For carotenoids and chlorophylls, use the mineral spirits without any dilution or mixing.
   - For flavonoids, use the rubbing alcohol and mix with clear vinegar in a ratio of 9:1. This will give an approximately 80% aqueous solution of isopropanol, mildly acidified.
   - For betalains, mix rubbing alcohol with ammonia cleaning solution in a ratio of 9:1. This will give an approximately 80% aqueous solution of isopropanol, strongly basic.
2. Add about 1 cm depth of the elution solvent to a small narrow jar. Let it equilibrate for a few hours, keeping it covered.
3. Cut coffee filters into long strips approximately 3 cm wide.
4. Use a very small ceramic cup or dish. Add a few drops of the appropriate solvent, such as the mineral spirits for extracting leaf pigments. Add a small square of fresh tissue (1 cm²), and mash the tissue with the flat of a small spoon. Continue mashing and grinding tissue and solvent until you obtain a small volume of darkly colored extract.
5. Using the eyedropper, suck up a little of the dark slurry and carefully let one small drop soak into a designated place on the filter paper, 2 cm from the end. Let the spot dry, and repeat this procedure several times. Try to keep the spot as small as possible, which is helped by allowing the previous drop to dry.
6. Carefully place the spotted paper strips into the jar, making certain not to immerse the spot in the solvent.
7. Let the solvent soak into, and move up, the paper. Let the solvent run until it is near the top of the paper. Mark its position with a pencil. Dry the paper and keep it in a dark place. Depending upon the pigment and the solvent, the spot may or may not move up the paper. Expect the following results.
   - Carotenoids (yellow spot) do not move in the rubbing-alcohol solution. They move in the mineral-spirit solution. See figures B.1A and B.

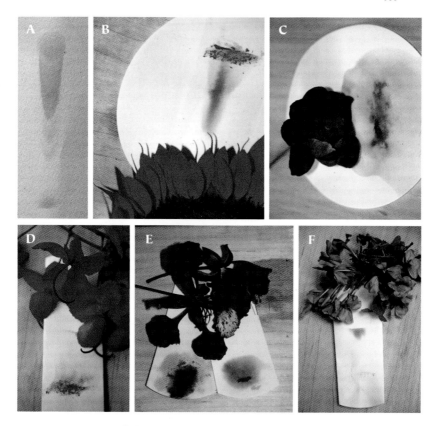

Figure B.1 Separation of plant pigments, using different solvents. A, leaf mineral-spirit extract from umbrella tree, spotted for elution by mineral spirits. This procedure separates the two chlorophylls, *a* and *b*. A yellow spot is carotenoid, mainly β-carotene. B, ray-flower extract from sunflower in mineral spirits, spotted for elution by mineral spirits. The yellow area moves up the paper with the solvent. C, rose petal ('Woodbury') extracted in acidified rubbing alcohol. Spot moves up with the same solvent on the filter paper. D, yellow shower petals extracted in acidified rubbing alcohol. The spot, a flavonol, moves up with the same solvent on the filter paper. Note that this spot would not move with the mineral spirits, as for B. E, poinciana flower extracts, in acidified rubbing alcohol on the right and in mineral spirits on the left. In the left strip, the red color does not move, but the yellow color (carotenoid) does. In the right strip, the red color moves with the solvent, but the yellow color does not. The orange-red flowers in this spectacular tree are produced by a combination of these quite different pigments. F, lead plant flower extract. The spot moves up in acidic rubbing alcohol, but its color changes from blue to pink. This indicates loose copigmentation with another phenylpropanoid pigment.

- Anthocyanin (red spot) does not move in the mineral-spirit solution, and loses its color in the basic rubbing-alcohol solution. It moves nicely in the acidified rubbing-alcohol solution. See figure B.1C and F.
- Flavonols (yellow spot) do not move in the mineral-spirit solution. They move in the rubbing-alcohol solutions. See figure B.1D.
- Betalains (yellow or red spot) do not move in the mineral-spirit solution. They move in the rubbing-alcohol solutions, and retain their color in the basic solution.

CHAPTER NOTES

....

Preface and Acknowledgments

The quotation at the beginning of the preface is from W. Whitman, *Specimen Days* (1880; reprint, Boston: D. R. Godine, 1971). Throughout the book I have benefited from frequent reference to D. J. Mabberley, *The Plant Book*, 2nd ed. (Cambridge: Cambridge University Press, 1997). I have also been influenced by many ideas put forward by Francis Hallé in his *Éloge de la Plante*, particularly so because of the time I spent translating his book into English (*In Praise of Plants* [Portland, OR: Timber Press, 2002]).

Chapter One: Coloring Our Bodies with Plants

The beginning quotation is from G. Snyder, *Turtle Island* (New York: New Directions Books, 1974). The second quotation is from F. Delamare and B. Guineau, *Colors: The Story of Dyes and Pigments* (New York: Harry N. Abrams, 2000). This book is a concise and useful introduction to colors and dyes, from plants and other sources.

The burial of Shanidar IV was described in two papers (R. S. Solecki, "Shanidar IV: A Neanderthal flower burial in northern Iraq," *Science* 190 [1975]: 880–81; and A. Leroi-Gourhan, "The flowers found with Shanidar IV: A Neanderthal burial in Iraq," *Science* 190 [1975]: 562–64) and in a personal account (R. S. Solecki, *Shanidar: The First Flower People* [New York: Alfred A. Knopf, 1971]). A general description of the stratigraphy of the cave was provided in R. S. Solecki, R. L. Solecki, and A. P. Agelarakis, *The Proto-Neolithic Cemetery in Shanidar Cave* (College Station: Texas A&M Press, 2004). For a more general account of the Neanderthal, I consulted C. Stringer and C. Gamble, *In Search of the Neanderthals: Solving the Puzzle of Human Origins* (New York: Thames and Hudson, 1993); E. Trinkhaus, *The Shanidar Neandertals* (New York: Academic Press, 1983); and I. Tattersal, *The Last Neanderthal?* (New York: Macmillan, 1995). Remarkable data on DNA from Neanderthal bones (M. Krings, A. Stone, R. W. Schmitz, H. Krainitzki, M. Stoneking, and S. Pääbo, "Neanderthal DNA sequences and the origin of modern humans," *Cell* 90 [1997]: 19–30; and I. V. Ovchinnikov, A. Götherström, G. P. Romanova, V. M. Kharitonov, K. Lidén, and W. Goodwin, "Molecular analysis of Neanderthal DNA from the northern Caucasus," *Nature* 404 [2000]: 490–93) has led to a reassessment of the relation between modern humans and the Neanderthal (S. Pääbo, "The mosaic that is our genome," *Nature* 421 [2003]: 409–12), as has the discovery of Neanderthal child burial in Portugal (C. Duarte, J. Mauricio, P. B. Pettitt, P. Souto, E. Trinkaus, H. van der Plicht, and J. Zilhao, "The early upper Paleolithic human skeleton from the Abrigo do Lagar Velho (Portugal) and modern human emergence in Iberia," *Proceedings of the National Academy of Sciences U.S.* 96 [1999]: 7604–9; also see I. Tattersall and J. H. Schwartz, "Hominids and hybrids: The place of Neanderthals in human evolution," *Proceedings of the National Academy of Sciences U.S.* 96 [1999]: 7117–19); and E. Daynes's Web site, http://daynes .com. A recent novel, W. M. Gear and K. O'Neal Gear, *Raising Abel* (New York: Warner Books, 2002), has also influenced my views.

Information on the Secoya was provided by discussions with my colleague William Vickers, who has studied and helped these people for the past thirty years. Some of this extensive research was published in W. T. Vickers, "The territorial dimensions of Siona-Secoya and Encabellado adaptation," in *Adaptive Responses of Native Amazonians*, ed. R. B. Hames and W. T. Vickers, Studies in Anthropology (New York: Academic Press, 1983), 451–78; and W. T. Vickers and T. Plowman, "Useful plants of the Siona and Secoya Indians of eastern Ecuador," *Fieldiana Botany*, n.s., no. 15 (1984).

The following books on Tutankhamen and his times provided some general background for the discussion on the use of plant colors in his tomb: C. Desroches-Noblecourt, *Tutankhamen: Life and Death of a Pharaoh* (New York: New York Graphic Society, 1963); and E. I. S. Edwards, *Tutankhamun, His Tomb, and Its Treasures* (New York: Metropolitan Museum of Art, 1976). Several other books provided information about the contents of the tomb, particularly from a botanical standpoint: Howard Carter, *The Tomb of Tutankhamen* (London: Barrie and Jenkins, 1972); F. N. Hepper, *Pharaoh's Flowers: The Botanical Treasures of Tutankhamun* (London: Her Majesty's Stationery Office, 1990); R. Germer, *Die Pflanzenmaterialien aus dem Grab des Tutanchamun* (Hildesheim: Gerstenberg Verlag, 1989); P. E. Newbury, "Report on the floral wreaths found in the coffins of Tutankhamen," in *The Tomb of Tutankhamen*, H. Carter (London: Barrie and Jenkins, 1972), 232–34. The quotation from the Egyptian Book of the Dead is from O. Goelet, *The Egyptian Book of the Dead* (San Francisco: Chronicle Books, 1998). Information on medicinal and psychoactive plants used by the Egyptians and elsewhere in the Middle East was from D. S. Flattery and M. Schwartz, *Haoma and Harmaline: The Botanical Identity of the Indo-Iranian Sacred Hallucinogen "Soma" and Its Legacy in Religion, Language, and Middle Eastern Folklore* (Berkeley and Los Angeles: University of California Press, 1989); J. F. Nunn, *Ancient Egyp-*

tian Medicine (Norman: University of Oklahoma Press, 1996); R. E. Schultes and
A. Hofmann, *Plants of the Gods* (Rochester, VT: Healing Arts Press, 1992); and C. J. S.
Thompson, *The Mystic Mandrake* (Hyde Park, NY: University Books, 1968).

Evidence for the early use of plants for color in human culture, particularly in
dyes and the practice of tattooing, comes from the following sources: D. Brothwell,
The Bog Man and the Archaeology of People (Cambridge, MA: Harvard University Press,
1987); J. H. Dickson, K. Oeggl, and L. L. Handley, "The iceman reconsidered," *Sci-
entific American* 288, no. 5 (2003): 70–79; F. B. Pyatt, E. H. Beaumont, D. Lacy, J. R.
Magilton, and P. C. Buckland, "*Non isatis sed vitrum*; or, The colour of Lindow man,"
Oxford Journal of Archaeology 10 (1991): 61–73; R. Rolle, *The World of the Scythians*, trans.
F. G. Walls (Berkeley and Los Angeles: University of California Press, 1989);
T. Schick, "Nahal Hemar cave: Cordage, basketry, and fabrics," *Atiquot*, English ser.,
18 (1988): 31–43; and W. Van Zeist and W. Waterbolk van-Rooijen, "Two interest-
ing floral finds from third millennium B.C. Tell Hamman et-Turkman, northern
Syria," *Vegetation History and Archaeobotany* 1 (1992): 157–61. The quotation by Caesar
is from J. Caesar, *The Gallic Wars*, trans. H. J. Edwards (Cambridge, MA: Harvard
University Press, 1979). Use of textiles and dyes in Egypt was described in G. M.
Eastwood, *Egyptian Dyes and Colours* (Edinburgh: National Museum of Antiquities of
Scotland, 1984); R. Germer, *Die Textilfarberei und die Verwendung gefärbter Textilien im alten
Aegypten* (Wiesbaden: Otto Harrassowitz, 1992); L. Keimer, *Die Gartenpflanzen im alten
Agypten* (Hildesheim: Georg Olms Verlagsbuchhandlung, 1967); and C. G. Thomp-
son and E. W. Gardner, *The Desert Fayum* (London: Royal Anthropological Institute
of Great Britain and Ireland, 1934). General information on makeup and tattooing
was obtained from M. Angeloglou, *A History of Make-up* (London: Macmillan, 1970).
More detailed information on the tradition of tattooing in Japanese culture was
obtained from T. Yoshida and R. Yuki, *Japanese Print-making: A Handbook of Traditional
and Modern Techniques* (Rutland, VT: Charles E. Tuttle, 1966); and W. R. Van Gulik,
Irezumi: The Pattern of Dermatography in Japan (Leiden: E. J. Brill, 1982).

The importance of dyes in the progress of chemistry in the eighteenth and nineteenth centuries can be seen in virtually any book on the history of chemistry, and I
referred to B. Bensaude-Vincent and I. Stengers, *A History of Chemistry*, trans. D. van
Dam (Cambridge, MA: Harvard University Press, 1996); W. H. Brock, *The Norton
History of Chemistry* (New York: W. W. Norton, 1992); and M. J. Nye, *Before Big Science:
The Pursuit of Modern Chemistry and Physics, 1800–1940* (New York: Twayne, 1996). Two
dye plants were particularly well covered in recent monographs: J. Balfour-Paul,
Indigo in the Arab World (Richmond, Surrey: Curzon Press, 1997); and R. Chenciner,
Madder Red: A History of Luxury and Trade (Richmond, Surrey: Curzon Press, 2000).
For a description (and quotation) of the practices of dyeing fabric in the Italian
city-states, see G. Rosetti, *The Plictho. Instructions in the Art of the Dyers Which Teaches the
Dyeing of Woolen Cloths, Linens, Cottons, and Silk by the Great Art as Well as by the Common*,
translated from the first edition of 1548 by S. M. Edelstein and H. C. Borghetty
(Cambridge, MA: MIT Press, 1969). The commercial roles of natural plant dyes

in the Industrial Revolution are documented by F. Braudel, *Civilization and Capitalism, 15th–18th Centuries*, vol. 2, *The Wheels of Commerce* (Berkeley and Los Angeles: University of California Press, 1992); A. Nieto-Galan, *Colouring Textiles: A History of Natural Dyestuffs in Industrial Europe* (Dordrecht: Kluwer Academic, 2001); K. Pomeranz and S. Topik, *The World That Trade Created: Society, Culture, and the World Economy, 1400–the Present* (Armonk, NY: M. E. Sharpe, 1999); and R. S. Lopez and I. W. Raymond, *Medieval Trade in the Mediterranean World* (New York: Columbia University Press, 1955), for the list of "spices." The importance of synthetic dyes in replacing cultivated dye plants is documented in A. S. Travis, *The Rainbow Makers: The Origins of the Synthetic Dyestuffs Industry in Western Europe* (Bethlehem, PA: Lehigh University Press, 1993), as well as the works on the history of chemistry mentioned previously.

The following plants, listed in the chapter by common name and not mentioned in the figures are given here with scientific names, authorities and family affiliations: ayahuasca = *Banisteriopsis caapi* (Griseb.) Morton (Malpighiaceae); genip = *Genipa americana* L. (Rubiaceae); tree cotton = *Gossypium hirsutum* L. (Malvaceae); joint sedge = *Cyperus articulatus* L. (Cyperaceae); papyrus = *Cyperus papyrus* L. (Cyperaceae); white lotus lily = *Nymphaea alba* L. (Nymphaeaceae); oxtongue = *Picris radicata* Less. (Asteraceae); cornflower = *Centaurea cyanus* L. (Asteraceae); chamomile = *Anthemis pseudocotula* Boiss. (Asteraceae); date palm = *Phoenix dactylifera* L. (Arecaceae); pomegranate = *Punica granatum* L. (Punicaceae); ashwagandha = *Withania somnifera* (L.) Dunal (Solanaceae); mandrake = *Mandragora officinarum* L. (Solanaceae); wild celery = *Apium graveolens* L. (Apiaceae); persea = *Mimusops laurifolia* (Forssk.) Friis (Sapotaceae); flax = *Linum usitatissimum* L. (Linaceae); safflower = *Carthamus tinctorius* L. (Asteraceae); white acacia = *Acacia albida* Del. (Fabaceae); gamboge = *Garcinia xanthochymus* Hook. f. ex T. Anderson (Clusiaceae); hemp = *Cannabis sativa* L. (Cannabaceae); jute = *Corchorus capsularis* L. (Malvaceae); ramie = *Boehmeria nivea* (L.) Gaudich. (Urticaceae); woad = *Isatis tinctoria* L. (Brassicaceae); indigo = *Indigofera tinctoria* L. (Fabaceae); kermes oak = *Quercus coccifera* L. (Fagaceae); red sanders = *Pterocarpus santalinus* L. f. (Fabaceae); sappan wood = *Caesalpinia sappan* L. (Fabaceae).

Chapter Two: Light, Vision, and Color

The quotations at the beginning of the chapter are from F. Birren, *Color and Human Response: Aspects of Light and Color Bearing on the Reactions of Animals and the Welfare of Human Beings* (New York: Van Nostrand Reinhold, 1978); and P. Teilhard de Chardin, *The Phenomenon of Man* (New York: Harper Colophon Books, 1975). I was assisted on the general principles outlined in this chapter by consulting several general books on the subject: C. J. Mueller and M. Rudolph, *Light and Vision* (New York: Time-Life Books, 1966); M. I. Sobel, *Light* (Chicago: University of Chicago Press, 1987); N. J. Wade, *A Natural History of Vision* (Cambridge, MA: MIT Press, 1998); and B. A. Wandell, *Foundations of Vision* (Sunderland, MA: Sinauer Associates, 1995).

For the structure and function of eyes, I consulted the works above plus H. Kolb, "How the retina works," *American Scientist* 91 (2003): 28–35; and H. Kolb, E. Fernandez, and R. Nelson, "Webvision: The organization of the retina and visual system," at http://www.webvision.med.utah.edu (2003). For the origin of the vertebrate eye, I consulted J. R. Cronly-Dillon, "Origin of invertebrate and vertebrate eyes," in *The Evolution of the Eye and Visual System*, ed. J. R. Cronly-Dillon and R. L. Gregory (Boca Raton, FL: CRC Press, 1991), 15–51. A number of articles in the series Perception of Color were particularly helpful: J. D. Dodge, "Photosensory systems in eukaryotic algae," in *The Evolution of the Eye and Visual System*, ed. J. R. Cronly-Dillon and R. L. Gregory (Boca Raton, FL: CRC Press, 1991), 323–40; T. H. Goldsmith, "The evolution of visual pigments and colour vision," in *Vision and Visual Dysfunction*, ed. P. Gouras (Boca Raton, FL: CRC Press, 1991), 62–89; J. K. Bowmaker, "Visual pigments, oil droplets, and photoreceptors," in *Vision and Visual Dysfunction*, 108–27; J. N. Lythgoe, "The evolution of visual behaviour," in *The Evolution of the Eye and Visual System*, 3–14; and C. Neumeyer, "Evolution of colour vision," in *The Evolution of the Eye and Visual System*, 284–305. For the spectral sensitivity of vision in different animals, I also consulted R. Menzel, "Spectral sensitivity and colour vision in invertebrates," in *Comparative Physiology and Evolution of Vision in Invertebrates*, vol. A, *Invertebrate Photoreceptors*, ed. H. Autrum (Berlin: Springer-Verlag, 1979), 503–80. The general books on vision and color, cited above, described the history of systems of color perception; these articles were also important: G. Derefeldt, "Colour appearance systems," in *Vision and Visual Dysfunction*, 218–61; P. Gouras, "The history of colour vision," in *Vision and Visual Dysfunction*, 1–9; and L. Chittka and N. E. Raine, "Recognition of flowers by pollinators," *Current Opinion in Plant Biology* 9 (2006): 428–35.

For the cultural anthropological description of color naming in different cultures, several classical works were pointed out to me by William Vickers: H. C. Conklin, "Hanunóo color categories," *Southwestern Journal of Anthropology* 11 (1955): 339–44; H. H. Barlett, "Color nomenclature in Batak and Malay," *Papers of the Michigan Academy of Science, Arts, and Letters* 10 (1929): 1–52; and B. Berlin and P. Kay, *Basic Color Terms: Their Universality and Evolution* (Berkeley and Los Angeles: University of California Press, 1969). The different points of view on color naming were provided in a 1991 issue of the *Journal of Linguistic Anthropology*, including P. Kay, B. Berlin, and W. Merrifield, "Biocultural implications of systems of color naming," 1:12–25; and B. A. C. Saunders and J. van Brakel, "Are there nontrivial constraints on colour categorization?" *Behavior and Brain Sciences* 20 (1997): 167–228. I consulted these articles on synesthesia: V. S. Ramachandran and E. M. Hubbard, "Hearing colors, tasting shapes," *Scientific American* 288, no. 5 (2003): 53–59; F. Beeli, M. Esslen, and L. Jäncke, "When coloured sounds taste sweet," *Nature* 434 (2005): 38; and S. Hornik, "For some, pain is orange," *Smithsonian Magazine* 31, no. 11 (2003): 48–56. For general discussions on the psychology and philosophy of color perception, I consulted C. W. Riley II, *Color Codes* (Hanover, NH: University Press of New England, 1995); M. Merleau-Ponty, *The Phenomenology of Perception* (London: Routledge, 1958); M. Merleau-Ponty, *The Primacy of Perception* (Evanston, IL: Northwestern University Press, 1964); quotation from M. Merleau-Ponty, *Basic Writings* (London: Routledge,

2004), 36; J. Gage, "Colour and culture," in *Colour Art and Science*, ed. T. Lamb and J. Bourriau, 175–93 (Cambridge: Cambridge University Press, 1995); R. Kuschel and T. Monberg, "'We don't talk much about colour here': A study of colour semantics on Bellona Island," *Man* 9 (1974): 213–42; U. Ecco, "How culture conditions the colours we see," in *On Signs*, ed. M. Blonsky (Baltimore: Johns Hopkins University Press, 1985), 157–75. For the affective value of red, see R. A. Hill and R. A. Barton, "Red enhances human performance in contests," *Nature* 435 (2005): 293.

John Moorfield of the University of Otago in New Zealand provided information on the Maori use of color, adding to information first revealed by W. W. Colenso, "On the fine perception of colours possessed by the ancient Maori," *Transactions of the New Zealand Institute* 14 (1882): art. 3, 49–76.

The common name of this plant was mentioned in chapter: Madagascar periwinkle = *Catharanthus roseus* (L.) G. Don f. (Apocynaceae).

Chapter Three: The Palette

The quotation at the beginning of the chapter is attributed to Monet, but I cannot find the original source. That by Edwin Arlington Robinson comes from his early work *Captain Craig and Other Poems*, included in *The Poetry of E. A. Robinson* (New York: Modern Library, 1999).

Several general books on plant biochemistry and natural-products chemistry were particularly important in this chapter. For a modern account of the major pathways of pigment synthesis, I used B. B. Buchanan, W. Guissem, and R. L. Jones, *Biochemistry and Molecular Biology of Plants* (Rockville, MD: American Society of Plant Physiologists, 2000). For chapters on individual pigments, including a wide variety of the less commonly described ones, T. W. Goodwin, ed., *Chemistry and Biochemistry of Plant Pigments*, 2nd ed., vol. 1 (London: Academic Press, 1976), is still unsurpassed. I used chapters on the individual pigment classes for this book. Another old but extremely useful book on plant natural-products chemistry is T. Robinson, *The Organic Constituents of Higher Plants*, 4th ed. (North Amherst, MA: Cordus Press, 1980). Toward the end of preparing the book, I was able to use a new book: K. Davies, ed., *Plant Pigments and Their Manipulation* (Boca Raton, FL: CRC Press, 2004). This is the most useful book on plant pigments to be published in the past twenty-five years.

I used many other articles for details on specific classes of plant pigments. For carotenoid pigments, see R. Vetter, G. Englert, N. Rigassi, and U. Schwieter, "Spectroscopic methods," in *Carotenoids*, ed. O. Isler (Basel: Birkhäuser Verlag, 1971), 189–266. For the use of carotenoid pigments by crustaceans, see G. Britton, R. J. Weesie, D. Askin, J. D. Warburton, L. Gallardo-Guerrero, F. J. Jansen, H. J. M. de Groot, J. Lutenburg, J.-P. Cornard, and J.-C. Merlin, "Carotenoid blues: Structural studies on carotenoproteins," *Pure and Applied Chemistry* 69 (1997): 2075–84; and V. A. Olson and I. P. F. Owens, "Costly sexual signals: Are carotenoids rare, risky, or required?" *Trends in Ecology and Evolution* 13 (1998): 510–14. For flavonoid pigments, see K. R. Markham and T. J. Mabry, "Ultraviolet-visible and proton magnetic resonance spectroscopy of flavonoids," in *The Flavonoids*, part 1, ed. J. B. Harborne, T. J. Mabry, and H. Mabry (New York: Academic Press, 1975), 45–77; and K. R. Markham, *Techniques of Flavonoid Identification* (London: Academic Press, 1982). Helen Stafford's review of betalain synthesis was also quite helpful: H. A. Stafford, "Anthocyanins and betalains: Evolution of the mutually exclusive pathways," *Plant Science* 101 (1994): 91–98; and more recently, D. Streck, T. Vogt, and W. Schliemann, "Recent advances in betalain research," *Phytochemistry* 62 (2003): 247–69.

I also consulted U. Steiner, W. Schliemann, H. Böhm, and D. Strack, "Tyrosinase involved in betalain biosynthesis of higher plants," *Planta* 208 (1999): 114–24, for information on the pathways of synthesis. For floral fluorescence by betaxanthins, see F. Gandia-Herrero, F. Garcia-Carmona, and J. Escribano, "Floral fluorescence effect," *Nature* 437 (2005): 334. Anthocyanin synthesis and function was reviewed by K. S. Gould and D. W. Lee, eds., *Anthocyanins in Leaves*, Advances in Botanical Research 37 (London: Academic Press, 2002). The historical descriptions (and quotations) of Richard Willstätter's discoveries of anthocyanin and chlorophyll are from his autobiography, *From My Life: The Memoirs of Richard Willstätter*, trans. Lilli S. Hornig (New York: W. A. Benjamin, 1965).

Common names of these plants were mentioned in the chapter: Carolina redroot = *Lachnanthes tinctoria* (Lam.) Dandy (Haemodoraceae); mangosteen = *Garcinia mangostana* L. (Clusiaceae); ragweed = *Ambrosia artemisiifolia* L. (Asteraceae).

Chapter Four: The Canvas

Rauschenberg's words on painting and the canvas are from C. Tomkins, *Off the Wall: Robert Rauschenberg and the Art World of Our Time* (Garden City, NY: Doubleday, 1980); those attributed to Kandinsky can be found in K. C. Lindsay and P. Vergo, *Wassily Kandinsky: Complete Writings on Art* (New York: De Capo Press, 1994). Cage's comments were quoted in Tomkins's book.

For general reviews on the optics of plant tissue, I used the following articles: L. Fukshansky, "Optical properties of plants," in *Plants and the Daylight Spectrum*, ed. H. Smith, 21–40 (London: Academic Press, 1981); and T. C. Vogelmann, "Plant tissue optics," *Annual Review of Plant Physiology and Plant Molecular Biology* 44 (1993): 231–51. For interaction of light with the leaf surface, see J. H. McClendon, "The microoptics of leaves: 1, Patterns of reflection from the epidermis," *American Journal of Botany* 71 (1984): 1391–97; and D. W. Lee, "Unusual strategies of light absorption in rainforest herbs," in *The Economics of Plant Form and Function*, ed. T. Givnish, 105–31 (New York: Cambridge University Press, 1986). For focusing effects of light, see G. Haberlandt, *Physiological Plant Anatomy*, 4th ed. (London: Macmillan, 1914), for a historical perspective; and B. E. Juniper and C. E. Jeffree, *Plant Surfaces* (London: Edward Arnold, 1983). For the effects of scales and pubescence, see J. Ehleringer, "Leaf absorptances in Mohave and Sonoran desert plants," *Oecologia* 49 (1981): 366–70; and B. M. Eller and P. Willi, "The significance of leaf pubescence for the absorption of global radiation by *Tussilago farfara* L.," *Oecologia* 49 (1981): 179–87.

Because of its relevance to remote sensing, there is a large body of research on the penetration of light into plant tissues, and models of light scattering in tissues. The general reviews above were helpful, as were the following articles, used to support specific points in the chapter: J. H. McClendon and L. Fukshansky, "On the interpretation of absorption spectra of leaves: 2, The non-absorbed ray of the sieve effect and the mean optical pathlength in the remainder of the leaf," *Photochemistry and Photobiology* 51 (1990): 211–16; J. Ramus, "Seaweed anatomy and photosynthetic performance: The ecological significance of light guides, heterogenous absorption, and multiple scatter," *Journal of Phycology* 14 (1978): 352–62; and M. Seyfried and L. Fukshansky, "Light gradients in plant tissue," *Applied Optics* 22 (1983): 1402–8. For models of light distribution in plant tissues, see W. A. Allen, H. W. Gausman, and A. J. Richardson, "Willstätter-Stoll theory of leaf reflectance evaluated by ray tracing," *Applied Optics* 12 (1973): 2448–53; and R. Willstätter and A. Stoll, *Untersuchungen über die Assimilation der Kohlensäure* (Berlin: Springer-Verlag, 1918), for a historical perspective.

Chapter Five: Patterns

For the quotations at the beginning of the chapter, see G. C. Lichtenberg, *Aphorisms* (New York: Penguin Books, 1990); and D. Michals, *The Vanishing Act*, reprinted in *The Essential Duane Michals* (London: Thames and Hudson, 1997). Documentation for the discussion on inheritance and gene expression can be found in any textbook in biology or botany, such as W. K. Purves, D. Sadava, G. H. Orians, and H. C. Heller, *Life: The Science of Biology* (Sunderland, MA: Sinauer and Associates, 2003). For a general description of meristem structure and plant development, I consulted T. A. Steeves and I. M. Sussex, *Patterns in Plant Development*, 2nd ed. (Cambridge: Cambridge University Press, 1989); and K. Esau, *The Anatomy of Seed Plants*, 2nd ed. (New York: John Wiley, 1977), still the best general text on plant anatomy. For background on shoot organization and chimeras, see R. A. E. Tilney-Bassett, *Plant Chimeras* (London: Edward Arnold, 1986). For examples, see R. N. Stewart and H. Dermen, "Ontogeny in monocotyledons as revealed by studies of the developmental anatomy of periclinal chloroplast chimeras," *American Journal of Botany* 66 (1979): 47–58; R. N. Stewart, P. Semeniuk, and H. Dermen, "Competition and accommodation between apical layers and their derivatives in the ontogeny of chimeral shoots of *Pelargonium × hortorum*," *American Journal of Botany* 61 (1974): 54–67; R. W. Korn, "Analysis of shoot apical organization in six species of the Cupressaceae based on chimeric behavior," *American Journal of Botany* 88 (2001): 1945–52; and M. Marcotrigiano, "Genetic mosaics and the analysis of leaf development," *International Journal of Plant Science* 162 (2001): 513–25. For a survey of viral diseases expressed through meristems, see K. M. Smith, *A Textbook of Plant Virus Diseases* (New York: Academic Press, 1972).

Because we know about the genetics of pattern formation in petunias and snapdragons better than for other flowers, I have referred to some of the extensive literature on these two plants: J. Almeida, R. Carpenter, T. P. Robbins, C. Martin, and E. S. Coen, "Genetic interactions underlying flower color patterns in *Antirrhinum majus*," *Genes and Development* 3 (1989): 1758–67; H. Wiering and P. de Vlaming, "Inheritance and biochemistry of pigments," in *Petunia*, ed. K. C. Sink (Berlin: Springer-Verlag, 1984), 49–76; A. Cornu, "Genetics," in *Petunia*, ed. Sink, 34–48; and A. G. M. Gerats, W. Veerman, P. de Vlaming, A. Cornu, E. Farcy, and D. Maizonnier, "*Petunia hybrida* linkage map," in *Genetic Maps*, ed. S. J. O'Brien, vol. E (Woodbury, NY: Cold Spring Harbor Laboratory Press, 1987), 746–52. Since patterns in these flowers involve the expression of transposable elements, I have referred to some of the general literature and papers on these two plants: P. Nevers, N. S. Shepherd, and H. Saedler, "Plant transposable elements," *Advances in Botanical Research* 12 (1986): 104–203; E. S. Coen, R. Carpenter, and C. Martin, "Transposable elements generate novel spatial patterns of gene expresssion in *Antirrhinum majus*," *Cell* 47 (1986): 285–96; M. Doodeman and F. Bianchi, "Genetic analysis of instability in *Petunia hybrida*: 3, Periclinal chimeras resulting from frequent mutations of unstable alleles," *Theoretical and Applied Genetics* 69 (1985): 297–304; J. N. M. Mol, A. R. Stuitje, A. G. M. Gerats, A. R. van der Krol, and R. Jorgensen, "Saying it with genes: Molecular flower breeding," *Trends in Biotechnology* 7 (1989): 148–53; and H. N. M. Mol, A. R. Stuitje, and A. van der Krol, "Genetic manipulation of floral pigmentation genes," *Plant Molecular Biology* 13 (1989): 287–94.

For information on the cultural value of pigmented maize among Native Americans, I consulted A. Ortiz, "Some cultural meanings of corn in aboriginal North America," in *Corn and Culture in the Prehistoric New World*, ed. S. Johannessen and C. A. Hastorf (Boulder, CO: Westview Press, 1994), 527–44. Finally, I consulted A.-G. Rolland-Lagan, J. A. Bangham, and E. Coen, "Growth dynamics underlying petal shape and asymmetry," *Nature* 422 (2003): 161–65, for a more detailed discussion of the role of a transposable element in the patterning of snapdragon petals.

The common name of this plant was mentioned in the chapter: malanga = *Xanthosoma sagittifolium* (L.) Schott (Araceae).

Chapter Six: Leaves

The quotations at the beginning of the chapter are from W. Whitman, *The Illustrated Leaves of Grass* (New York: Madison Square Press, 1971); and K. Raine, "Vegetation," in *The Collected Poems of Kathleen Raine* (London: H. Hamilton, 1968). For the discussion of leaf gas exchange in the umbrella tree, I used W. C. Fonteno and E. L. McWilliams, "Light compensation points and acclimation of four tropical foliage plants," *Journal of the American Society of Horticultural Science* 103 (1978): 52–56; S. Oberbauer, D. Clark, D. A. Clark, P. Rich, and G. Vega, "Light environment, gas exchange, and annual growth of saplings of three species of rain forest trees in Costa Rica," *Journal of Tropical Ecology* 9 (1993): 511–23; and A. M. Hetherington and F. I. Woodward, "The role of stomata in sensing and driving environmental change," *Nature* 424 (2003): 901–7. For estimates of human gas exchange, I used

W. D. McCardle, F. I. Katch, and V. L. Katch, *Exercise Physiology*, 4th ed. (Baltimore: Williams and Wilkins, 1996).

For a review of the variation in leaf structure and form, and their influences on optical properties, see I. Roth, *Stratification of Tropical Forests as Seen in Leaf Structure* (The Hague: W. Junk, 1984); T. J. Givnish, "Leaf and canopy adaptations in tropical forests," in *The Physiological Ecology of Plants of the Wet Tropics*, ed. E. Medina, H. A. Mooney, and C. Vásquez-Yanes (The Hague: W. Junk, 1984), 51–84; T. J. Givnish, "Comparative studies of leaf form: Assessing the relative roles of selection pressures and phylogenetic constraints," *New Phytologist* 106, suppl. (1987): 131–60; and the reviews by Vogelmann, Lee, and Ehleringer cited for chapter 4. I have reviewed results reported in J. Ehleringer, "Leaf absorptances in Mohave and Sonoran desert plants," *Oecologia* 49 (1981): 366–70; D. W. Lee and R. Graham, "Leaf optical properties of rainforest sun and extreme shade plants," *American Journal of Botany* 73 (1986): 1100–1108. For the relationships between surface structures, absorptance, and plant function, see J. Ehleringer, O. Björkman, and H. A. Mooney, "Leaf pubescence: Effects on absorptance and photosynthesis in a desert shrub," *Science* 192 (1976): 316–77; I. Terashima and T. Saeki, "Light environment within a leaf: 1, Optical properties of paradermal sections of camellia leaves with special reference to differences in the optical properties of palisade and spongy tissues," *Plant and Cell Physiology* 24 (1983): 1493–1501; D. W. Lee, R. A. Bone, S. L. Tarsis, and D. Storch, "Correlates of leaf optical properties in tropical forest sun and extreme-shade plants," *American Journal of Botany* 77 (1990): 370–80; D. W. Lee, S. F. Oberbauer, P. Johnson, B. Krishnapilay, M. Mansor, H. Mohamad, and S. K. Yap, "Effects of irradiance and spectral quality on leaf structure and function in seedlings of two Southeast Asian *Hopea* (Dipterocarpaceae) species," *American Journal of Botany* 87 (2000): 447–55; J. B. Clark and G. R. Lister, "Photosynthetic action spectra of trees: 2, The relationship of cuticle structure to the visible and ultraviolet spectral properties of needles from four coniferous species," *Plant Physiology* 55 (1975): 407–13; K. Kitayama, R. Pattison, S. Cordell, D. Webb, and D. Mueller-Dombois, "Ecological and genetic implications of foliar polymorphism in *Metrosideros polymorpha* Guad. (Myrtaceae) in a habitat matrix on Mauna Loa, Hawaii," *Annals of Botany* 80 (1997): 491–97; D. W. Lee and J. B. Lowry, "Plant speciation on tropical mountains: *Leptospermum* (Myrtaceae) on Mount Kinabalu, Borneo," *Botanical Journal of the Linnaean Society* 80 (1980): 223–42; W. K. Smith, T. C. Vogelmann, E. H. Delucia, D. T. Bell, and K. A. Shepherd, "Leaf form and photosynthesis," *BioScience* 47 (1997): 785–93; and D. G. Rhoades, "Integrated antiherbivore, antidesiccant, and ultraviolet screening properties of creosote bush resin," *Biochemical Systematics and Ecology* 5 (1979): 281–90. Other articles cited in chapter 4 are also relevant.

For information on leaf movements and appearance, see C. R. Darwin, *The Power of Movement in Plants* (New York: Appleton, 1881); J. Ehleringer and I. N. Forseth, "Solar tracking by plants," *Science* 210 (1980): 1094–98; T. J. Herbert and P. B. Larsen, "Leaf movement in *Calathea lutea* (Marantaceae)," *Oecologia* 67 (1985): 238–

43; T. C. Vogelmann, "Site of light perception and motor cells in a sun-tracking lupine (*Lupinus succulentus*)," *Physiologia Plantarum* 62 (1984): 335–40; and R. L. Satter, S. E. Guggino, T. A. Lonergan, and A. W. Galston, "The effects of blue and far-red light on rhythmic leaflet movements in *Samanea* and *Albizia*," *Plant Physiology* 67 (1981): 965–68.

For the phenomenon of variegation, see T. J. Givnish, "Leaf mottling: Relation to growth form and leaf phenology and possible role as camouflage," *Functional Ecology* 4 (1990): 463–74; and A. P. Smith, "Ecology of leaf color polymorphism in a tropical forest species: Habitat segregation and herbivory," *Oecologia* 69 (1986): 283–87. For variegation in crotons, I consulted B. F. Brown, *Crotons of the World* (Valkaria, FL: Valkaria Tropical Garden, 1995).

In my discussion on the biological significance of leaf color I relied on the research of the following: B. C. Stone, "Protective coloration of young leaves in certain Malaysian palms," *Biotropica* 11 (1979): 126; S. Lev-Yadun, A. Dafni, M. A. Flaishman, M. Inbar, I. Izhaki, G. Katzir, and G. Ne'eman, "Plant coloration undermines herbivorous insect camouflage," *BioEssays* 26 (2004): 1126–30; H. M. Schaefer and D. M. Wilkinson, "Red leaves, insects, and coevolution: A red herring?" *Trends in Ecology and Evolution* 19 (2004): 616–18; and S. Lev-Yadun, "Aposematic (warning) coloration associated with thorns in higher plants," *Journal of Theoretical Biology* 210 (2001): 385–88.

The following articles provided information on chloroplast movements and their effects on leaf absorptance and function: J. Augustynowicz and H. Gabrys, "Chloroplast movements in fern leaves: Correlation of movement dynamics and environmental flexibility of the species," *Plant, Cell, and Environment* 22 (1999): 1239–48; M. Kasayara, T. Kagawa, K. Oikawa, N. Suetsugu, M. Miyao, and M. Wada, "Chloroplast avoidance movement reduces photodamage in plants," *Nature* 420 (2002): 829–32; M. Wada, T. Kagawa, and Y. Sato, "Chloroplast movement," *Annual Reviews of Plant Biology* 54 (2003): 455–68; and D. C. McCain, "Chloroplast movement can be impeded by crowding," *Plant Science* 135 (1998): 219–25.

For leaves with functions other than photosynthesis, see information on poinsettias in R. N. Stewart, S. Asen, D. R. Massie, and K. H.Norris, "The identification of poinsettia cultivars by HPLC analysis of their anthocyanin content," *Biochemical Systematics and Ecology* 7 (1979): 281–87; and for pitcher plants, P. M. Sheridan and R. J. Griesbach, "Anthocyanidins of *Sarracenia* L. flowers and leaves," *HortScience* 36 (2001): 384; and P. M. Sheridan and R. R. Mills, "Presence of proanthocyanidins in mutant green *Sarracenia* indicate blockage in late anthocyanin biosynthesis between leucocyanidin and pseudobase," *Plant Science* 135 (1998): 11–16.

Common names of these plants were mentioned in the chapter: castillo = *Kalanchoe beharensis* Drake (Agavaceae); Spanish moss = *Tillandsia usneoides* L. (Bromeliaceae); greasewood = *Sarcobatus vermiculatus* (Hook.) Torr. (Chenopodiaceae); *Codiaeum moluccanum* Decne (Euphorbiaceae); poinsettia = *Euphorbia pulcherrima* Willd. ex Klotsch (Euphorbiaceae); mussaenda = *Mussaenda frondosa* L. (Rubiaceae)

Chapter Seven: Flowers

For quotations at the beginning of the chapter, see J. Giraudoux, *The Enchanted: A Comedy in Three Acts* (New York: Random House, 1933); and R. Verey, *The Scented Garden* (New York: Van Nostrand Reinhold, 1981). For the general structure of flowers, and orchids in particular, I used L. Van der Pijl and C. Dodson, *Orchid Flowers: Their Pollination and Evolution* (Coral Gables, FL: Fairchild Tropical Botanic Garden / University of Miami Press, 1966). A number of references on orchids provided the background for the story of the hybrid Blc 'Hillary Rodham Clinton' early in the chapter, along with much advice from Bill Peters, an American Orchid Society judge: J. Arditti and M. H. Fisch, "Anthocyanins of the Orchidaceae: Distribution, heredity, functions, synthesis, and localization," in *Orchid Biology: Reviews and Perspectives*, vol. 1, ed. J. Arditti (Ithaca, NY: Cornell University Press, 1977), 118–55; C. van den Berg, W. E. Higgins, R. L. Dressler, W. M. Whitten, M. A. Soto Arenas, A. Culham, and M. W. Chase, "A phylogenetic analysis of Laeliinae (Orchidaceae) based on sequence data from internal transcribed spacers (ITS) of nuclear ribosomal DNA," *Lindleyana* 15 (2000): 96–114; J. T. Curtis and R. E. Duncan, "The inheritance of flower color in cattleyas," *American Orchid Society Bulletin* 10 (1942): 283–307; F. Fordyce, "Hybridizer's notebook, 1–4," *American Orchid Society Bulletin* 59 (1990): 794–96; K. Jones, "Cytology and the study of orchids," in *The Orchids: Scientific Studies*, ed. C. Withner (New York: John Wiley and Sons, 1974), 383–92; S. Matsui and M. Nakamura, "Distribution of flower pigments in perianth of *Cattleya* and allied genera: 1, Species," *Journal of the Japanese Horticultural Society* 57 (1988): 222–32; G. A. L. Mehlquist, "Some aspects of polyploidy in orchids, with particular reference to *Cymbidium*, *Paphiopedilum*, and the *Cattleya* alliance," in *The Orchids: Scientific Studies*, 393–409; R. T. Northern, *Home Orchid Growing*, 3rd ed. (New York: Van Nostrand Reinhold, 1970); M. N. S. Stort, "Phylogenetic relationship between species of the genus *Cattleya* as a function of crossing compatibility," *Revista de Biología Tropical* 32 (1984): 223–26; M. Suichino, "Carotenoids in *Cattleya* flowers," *Lindleyana* 9 (1994): 33–38; and C. L. Withner, *The Cattleyas and Their Relatives*, vol. 1, *The Cattleyas* (Portland, OR: Timber Press, 1988). For information on orchids, I consulted the extensive Web site of the American Orchid Society (orchidweb.org). The classification and naming of orchids is in the process of being radically revised, based on the sequencing and comparing of chloroplast and nuclear genes. Orchid systematists are finding that some of the traditional groups are polyphyletic, and seek to employ new nomenclature to reflect natural and evolutionary-based groups. Orchid fanatics are bewildered by the activities of these "taxonomic lawyers," and grouse about the name changes at the society meetings. Reflecting these changes, note the use of *Rhyncholaelia* for *Brassavola* in *R. digbyana*.

For the origins and pigmentation of roses, I used the Web site of the American Rose Society (www.ars.org). I also used S. Asen, "Identification of flavonoid chemical markers in roses and their high-pressure liquid chromatographic resolution and quantitation for cultivar identification," *Journal of the American Society of Horticultural*

Science 107 (1982): 744–50; J. P. Biolley and M. Jay, "Anthocyanins in modern roses: Chemical and colorimetric features in relation to the colour range," *Journal of Experimental Botany* 44 (1993): 1725–34; H. H. Marshall, C. G. Campbell, and L. M. Collicutt, "Breeding for anthocyanin colors in *Rosa*," *Euphytica* 32 (1983): 205–16; J. Ogata, Y. Kanno, Y. Itoh, H. Tsugawa, and M. Suzuki, "Anthocyanin biosynthesis in roses," *Nature* 435 (2005): 757–58. For orange-red pigments in the Gesneriaceae, I referred to C. S. Winefield, D. H. Lewis, E. E. Swinny, H. Zhang, H. S. Arathoon, T. C. Fischer, H. Halbwirth, et al., "Investigation of the biosynthesis of 3-deoxyanthocyanins in *Sinningia cardinalis*," *Physiologia Plantarum* 124 (2005): 419–30.

I used the following references for documenting the pigments producing flower color in the examples of the chapter.

Alstroemeria species: N. Saito, M. Yokoi, M. Ogawa, M. Kamijo, and T. Honda, "6-Hydroxyanthocyanidin glycosides in the flowers of *Alstroemeria*," *Phytochemistry* 27 (1988): 1399–1401; and F. Tatsuzawa, N. Murata, K. Shinoda, R. Suzuki, and N. Saito, "Flower colors and anthocyanin pigments in 45 cultivars of *Alstroemeria* L.," *Journal of the Japanese Society for Horticultural Science* 72 (2003): 243–51.

Bougainvillea species: M. Piatelli and F. Imperato, "Pigments of Centrospermae: 11, Betacyanins from *Bougainvillea*," *Phytochemistry* 9 (1970): 455–58; and M. Piattelli and F. Imperato, "Pigments of Centrospermae: 13, Pigments of *Bougainvillea glabra*," *Phytochemistry* 9 (1970): 2557–60.

Helianthus species: K. Egger, "Xanthophyll esters of sunflowers," *Zeitschrift für Naturforschung B* 23 (1968): 731–33; L. R. G. Valadon and R. S. Mummery, "Carotenoids of certain Compositae flowers," *Phytochemistry* 6 (1967): 983–88.

Eschscholzia californica: M. A. Beck and H. Haberlein, "Flavonol glycosides from *Eschscholzia californica*," *Phytochemistry* 50 (1999): 329–32; L. Jain, M. Tripathy, V. B. Pandey, and G. Ruker, "Flavonoids from *Eschscholzia californica*," *Phytochemistry* 41 (1996): 661–62; and A. M. Wakelin, C. E. Lister, and A. J. Conner, "Inheritance and biochemistry of pollen pigmentation in California poppy (*Eschscholzia californica* Cham.)," *International Journal of Plant Science* 164 (2003): 867–75.

Tulipa species: M. Nieuwhof, J. P. Van Eijk, and W. Eikelboom, "Relation between flower colour and pigment composition of tulip (*Tulipa* L.)," *Netherlands Journal of Agricultural Science* 37 (1989): 365–70.

Viola species: T. Endo, "Biochemical and genetical investigations of flower color in *Viola tricolor*: 1, Interrelations of pigment constituents occurring in ten varieties," *Japanese Journal of Botany* 14 (1954): 187–93; and M. Farzad, R. Griesbach, and M. R. Weiss, "Floral color change in *Viola cornuta* L. (Violaceae): A model system to study regulation of anthocyanin production," *Plant Science* (Amsterdam) 162 (2002): 225–31.

Antirrhinum species: S. Asen, K. H. Norris, and R. N. Stewart, "Copigmentation of aurone and flavone from petals of *Antirrhinum majus*," *Phytochemistry* 11 (1972): 2739–41.

Delonix regia: F. B. Jungalwala and H. R. Cama, "Carotenoids in *Delonix regia* flower," *Biochemical Journal* 85 (1962): 1–8.

Petunia hybrida: R. J. Griesbach, "The inheritance of flower color in *Petunia hybrida* Vilm.," *Journal of Heredity* 87 (1996): 241–45.

Fuchsia hybrida: Y. Yazaki, "Co-pigmentation and color change with age in petals of *Fuchsia hybrida*," *Botanical Magazine* (Tokyo) 89 (1976): 45–57.

Bauhinia purpurea: R. Ramachandran and B. C. Joshi, "Chemical examination of *Bauhinia purpurea* flowers," *Current Science* 36 (1967): 574–75; K. P. Tiwari, M. Masood, Y. K. S. Rathore, and P. K. Mihocha, "Anthocyanins from the flower of some medicinal plants," *Vijnana Parishad Anusandhan Patrika* 21 (1978): 177–78.

Butea monosperma: M. Mishra, Y. Shukla, and S. Kumar, "Chemical constituents of *Butea monosperma* flowers," *Journal of Medicinal and Aromatic Plant Sciences* 24 (2002): 19–22; and A. P. Sinha and R. Kumar, "Extraction of colouring material from *Butea monosperma* (*frondosa*) (Tesu, Palash) flowers," *Asian Journal of Chemistry* 11 (1999): 1536–48.

Eustoma species: A. F. M. Jamal Uddin, F. Hashimoto, S. Nishimoto, K. Shimizu, and Y. Sakata, "Flower growth, coloration, and petal pigmentation in four lisianthus cultivars," *Japanese Journal of Horticultural Science* 71 (2002): 40–47; and K. R. Markham and D. J. Ofman, "Lisianthus flavonoid pigments and factors influencing their expression in flower color," *Phytochemistry* 34 (1993): 679–85.

For the complex interactions among pigments in producing flower colors, I consulted the following articles: S. Asen, R. N. Stewart, K. H. Norris, and D. R. Massie, "A stable blue non-metallic co-pigment complex of delphanin and C-glycosylflavones in Prof. Blaauw iris," *Phytochemistry* 9 (1970): 619–27; S. Asen, R. N. Stewart, and K. H. Norris, "Co-pigmentation of anthocyanins in plant tissues and its effect on color," *Phytochemistry* 11 (1972): 1139–44; R. Brouillard, "Flavonoids and flower colour," in *The Flavonoids: Advances in Research since 1980*, ed. J. B. Harborne (London: Chapman and Hall, 1988), 525–38; O. Dangles, N. Saito, and R. Brouillard, "Anthocyanin intramolecular copigment effect," *Phytochemistry* 34 (1993): 119–24; M. M. Giusti, L. E. Rodríguez-Saona, and R. W. Wrolstad, "Molar absorptivity and color characteristics of acylated and non-acylated pelargonidin-based anthocyanins," *Journal of Agricultural and Food Chemistry* 47 (1999): 4631–37; R. J. Griesbach, "Effects of carotenoid-anthocyanin combinations on flower color," *Journal of Heredity* 75 (1984): 145–47; and T. Toto and T. Kondo, "Structure and molecular stacking of anthocyanins: Flower color variation," *Angewandte Chemie* 30 (1991): 17–33 (particularly helpful). For changes in floral pigmentation over time, see individual species and the general review by M. R. Weiss, "Floral color change: A widespread functional convergence," *American Journal of Botany* 82 (1995): 167–85.

I have discussed pigment interactions in producing blue colors based on these articles: T. Kondo, K. Yoshida, A. Nagakawa, T. Kawai, H. Tamura, and T. Goto, "Structural basis of blue-colour development in flower petals from *Commelina com-*

munis," *Nature* 358 (1992): 515–18; R. N. Stewart, K. H. Norris, and S. Asen, "Micro-spectrophotometric measurement of pH and pH effect on color of petal epidermal cells," *Phytochemistry* 14 (1975): 937–42; Y. Toyama-Kato, K. Yoshida, E. Fujimori, H. Haraguchi, Y. Shimizu, and T. Kondo, "Analysis of metal elements of hydrangea sepals at various growing stages by ICP-AES," *Biochemical Engineering Journal* 14 (2003): 237–41; K. Yoshida, T. Kondo, Y. Okazaki, and K. Katou, "Cause of blue petal colour," *Nature* 373 (1995): 291; and M. Shiono, N. Matsugaki, and K. Takeda, "Structure of the blue cornflower pigment," *Nature* 436 (2005): 791. For biotechnological attempts to produce blue flower colors, see the general review Y. Tanaka, Y. Katsumoto, F. Brugliera, and J. Mason, "Genetic engineering in floriculture," *Plant Cell Tissue and Organ Culture* 80 (2005): 1–24; and these two Web sites: http://www.florigene.com and http://www.physorg.com/printnews.php?neewsid=3581.

These articles are the background for the discussion on color and function in the reproductive parts (anthers and stigmas): S. A. Mehlenbacher and D. C. Smith, "Inheritance of pollen color in hazelnut," *Euphytica* 127 (2002): 303–7; Y. Mo, C. Nagel, and L. P. Taylor, "Biochemical complementation of chalcone synthase mutants defines a role for flavonols in functional pollen," *Proceedings of the National Academy of Sciences U.S.* 89 (1992): 7213–17; C. A. Napoli, D. Fahy, H. Wang, and L. P. Taylor, "White anther: A petunia mutant that abolishes pollen flavonol accumulation, induces male sterility, and is complemented by a chalcone synthase transgene," *Plant Physiology* 120 (1999): 615–22; V. V. Roschina, E. V. Mel'nikova, N. A. Spiridonov, and L. V. Kovaleva, "Azulenes: Blue pigments of pollen," *Doklady Akademii Nauk* 340 (1995): 715–18; and B. Ylstra, B. Busscher, J. Franken, P. C. H. Hollman, J. N. M. Mol, and A. J. van Tunen, "Flavonols and fertilization in *Petunia hybrida*: Localization and mode of action during pollen tube growth," *Plant Journal* 6 (1994): 201–12.

Structural influences on color production are discussed in these articles: B. J. Glover and C. Martin, "Evolution of adaptive petal cell morphology," in *Developmental Genetics and Plant Evolution*, ed. Q. C. B. Cronk, R. M. Bateman, and J. A. Hawkins (London: Taylor and Francis, 2002), 160–72; Q. O. N. Kay and H. S. Daoud, "Pigment distribution, light reflection, and cell structure in petals," *Botanical Journal of the Linnaean Society* 83 (1981): 57–84; K. Noda, B. J. Glover, P. Linstead, and C. Martin, "Flower colour intensity depends on specialized cell shape controlled by a Myb-related transcription factor," *Nature* 369 (1994): 661–64; and R. G. Strickland, "The nature of the white colour of petals," *Annals of Botany* 38 (1974): 1033–37.

For the effects of pigmentation and flower color, these general articles and books provide good background: N. Biedinger and W. Barthlott, *Untersuchungen zur Ultraviolettreflexion von Angiospermenblüten: 1, Monocotyledoneae* (Stuttgart: Franz Steiner Verlag, 1993); M. T. Clegg and M. L. Durbin, "Tracing floral adaptations from ecology to molecules," *Nature Reviews Genetics* 4 (2003): 206–15; V. Grant, "Floral isolation between ornithophilous and sphingophilous species of *Ipomopsis* and *Aquilegia*," *Proceedings of the National Academy of Sciences U.S.* 89 (1992): 11828–31; V. Grant, "Origin of floral isolation between ornithophilous and sphingophilous plant species,"

Proceedings of the National Academy of Sciences U.S. 90 (1992): 7729–33; J. B. Harborne, "Functions of flavonoids in plants," in *Chemistry and Biochemistry of Plant Pigments*, ed. T. Goodwin (London: Academic Press, 1976), 736–78; J. B. Harborne, "Correlations between anthocyanin chemistry and pollination ecology in the Polemoniaceae," *Biochemical Systematics and Ecology* 6 (1978): 127–30; M. Proctor, P. Yeo, and A. Lack, *The Natural History of Pollination* (Portland, OR: Timber Press, 1996); N. Saito and J. B. Harborne, "Correlations between anthocyanin type, pollinator, and flower colour in the Labiatae," *Phytochemistry* 31 (1992): 3009–15; R. Tandon, K. R. Shivanna, and H. Y. Mohan Ram, "Reproductive biology of *Butea monosperma* (Fabaceae)," *Annals of Botany* 92 (2003): 715–23; and L. Chittka and J. Walker, "Do bees like Van Gogh's sunflowers?" *Optics and Laser Technology* 38 (2006): 323–28.

I have discussed the relationships between color and pollinator in several wildflowers, and I list the important articles under each plant, below. For general reviews see chapters in K. M. Davies, ed., *Plant Pigments and Their Manipulation* (Boca Raton, FL: CRC Press, 2004).

Mimulus species: H. D. Bradshaw, S. M. Wilbert, K. G. Otto, and D. W. Schemske, "Genetic mapping of floral traits associated with reproductive isolation in monkeyflowers (*Mimulus*)," *Nature* 376 (1995): 762–65; H. D. Bradshaw, K. G. Otto, B. E. Frewen, J. K. McKay, and D. W. Schemske, "Quantitative trait loci affecting differences in floral morphology between two species of monkey flower (*Mimulus*)," *Genetics* 149 (1998): 367–92; R. Medel, C. Botto-Mahan, and M. Kalin-Arroyo, "Pollinator-mediated selection on the nectar guide phenotype in the Andean monkey flower, *Mimulus luteus*," *Ecology* 84 (2003): 1721–32; D. W. Schemske, "Pollinator preference and the evolution of floral traits in monkeyflowers (*Mimulus*)," *Proceedings of the National Academy of Sciences U.S.* 96 (1999): 11910–15; and H. G. Pollock, R. K. Vickery, and K. G. Wilson, "Flavonoid pigments in *Mimulus cardinalis* and its related species: 1, Anthocyanins," *American Journal of Botany* 54 (1967): 695–701.

Antirrhinum species: K. N. Jones and J. S. Reithel, "Pollinator mediated selection on a flower color polymorphism in experimental populations of *Antirrhinum* (Scrophulariaceae)," *American Journal of Botany* 88 (2001): 447–54; E. Odell, R. A. Raguso, and K. N. Jones, "Bumblebee foraging responses to variation in floral scent and color in snapdragons (*Antirrhinum*: Scrophulariaceae)," *American Midland Naturalist* 142 (1999): 256–65; and B. J. Glover and C. R. Martin, "The role of petal cell shape and pigmentation in pollination success in *Antirrhinum majus*," *Heredity* 80 (1998): 778–84.

Raphanus sativus: R. E. Irwin, S. Y. Strauss, S. Storz, A. Emerson, and G. Guibert, "The role of herbivores in the maintenance of a flower color polymorphism in wild radish," *Ecology* 84 (2003): 1733–43; M. L. Stanton, "Reproductive biology of petal color variants in wild populations of *Raphanus sativus* L.: 1, Pollinator response to color morphs," *American Journal of Botany* 74 (1987): 176–85; and S. Y. Strauss, R. E. Irwin, and V. M. Lambrix, "Optimal defence theory and flower petal colour predict variation in the secondary chemistry of wild radish," *Journal of Ecology* 92 (2004): 132–41.

Ipomopsis species: E. Meléndez-Ackerman, D. R. Campbell, and N. M. Waser, "Hummingbird behavior and mechanisms of selection on flower color in *Ipomopsis*," *Ecology* 78 (1997): 2532–41; D. M. Smith, C. W. Glennie, J. B. Harborne, and C. A. Williams, "Flavonoid diversification in the Polemoniaceae," *Biochemical Systematics and Ecology* 5 (1977): 107–15; and articles by Grant above.

Costus species: D. W. Schemske, "Floral convergence and pollinator sharing in two bee-pollinated tropical herbs," *Ecology* 62 (1981): 946–54; and K. M. Kay and D. W. Schemske, "Pollinator assemblages and visitation rates for 11 species of neo-tropical *Costus* (Costaceae)," *Biotropica* 35 (2003): 198–207.

Stapelia species: L. E. Knaak, R. P. Porter, and N.-C. Chang, "Anthocyanins of *Stapelia*," *Lloydia* 32 (1969): 148–52; and A. Jürgens, S. Dötterl, and U. Meve, "The chemical nature of fetid floral odours in stapeliads (Apocynaceae—Asclepiadoi-deae—Ceropegieae)," *New Phytologist* 172 (2006): 452–68.

Finally, for comments on the interactions between pigments in flowers and other plant parts, and the complications of learning about the functions of pigments, see E. L. Simms and M. A. Bucher, "Pleiotropic effect of flower-color intensity on herbivore performance on *Ipomoea purpurea*," *Evolution* 50 (1996): 957–63; and J. Warren and S. Mackenzie, "Why are all colour combinations not equally represented as flower-colour polymorphisms?" *New Phytologist* 151 (2001): 237–41.

The common name of this plant was mentioned in the chapter: lady's tresses = *Spiranthes romanzoffiana* Cham. (Orchidaceae).

Chapter Eight: Fruits and Seeds

The citations for the quotations at the beginning of the chapter are L. Hall, *Abe Ajay* (Seattle: University of Washington Press, 1990); and J. Raban, *For Love and Money: Writing, Reading, Traveling, 1969–1987* (London: Collins Harril, 1987).

The best up-to-date text on economic botany and domestication is B. B. Simpson and M. C. Ogorzaly, *Economic Botany*, 3rd ed. (New York: McGraw-Hill, 2001); I used this book extensively for discussions of crops and domestication. For the lengthy discussion of apples and domestication, I used the following sources: S. K. Brown, "Genetics of apple," *Plant Breeding Reviews* 9 (1992): 333–66; Frank Browning, *Apples* (New York: North Point Press, 1998); F. S. Cheng, N. F. Weeden, and S. K. Brown, "Identification of co-dominant RAPD markers tightly linked to fruit skin color in apple," *Theoretical and Applied Genetics* 93 (1996): 222–27; S. Kondo, K. Hiraoka, S. Kobayashi, C. Honda, and N. Terahara, "Changes in the expression of anthocyanin biosynthetic genes during apple development," *Journal of the American Society of Horticultural Science* 127 (2002): 971–76; J. E. Lancaster, "Regulation of skin color in apples," *Critical Reviews in Plant Sciences* 10 (1992): 487–502; J. E. Lancaster, J. E. Grant, and C. E. Lister, "Skin color in apples: Influence of co-pigmentation and plastid pigments on shade and darkness of red color in five genotypes," *Journal of the American Society of Horticultural Science* 119 (1994): 63–69; M. Pollan, *The Botany of Desire* (New York: Random House, 2001); S. Singha, T. A. Baugher, E. C. Townsend, and M. C. D'Souza, "Anthocyanin distribution in 'Delicious' apples and the relationship between anthocyanin concentration and chromaticity values," *Journal of the American Society of Horticultural Science* 116 (1991): 497–99; and D. Zohary and M. Hopf, *Domestication of Plants in the Old World* (Oxford: Oxford University Press, 2000). The probable origin of the apple has been reviewed in S. A. Harris, J. P. Robinson, and B. E. Juniper, "Genetic clues to the origin of the apple," *Trends in Genetics* 18 (2002): 426–30; and V. G. M. Bus, F. N. D. Laurens, W. E. van de Weg, R. L. Rusholme, E. H. A. Rikkerink, S. E. Gardiner, H. C. M. Bassett, L. P. Kodde, and K. M. Plummer, "The *Vh8* locus of a new gene-for-gene interaction between *Venturia inaequalis* and the wild apple *Malus sieversii* is closely linked to the *Vh2* locus in *Malus pumila* R12740-7A," *New Phytologist* 166 (2005): 1035–49. The correct scientific name for the wild apple, *Malus pumila* Mill., was determined by David Mabberley and recently pointed out in an excellent popular article on the domestication of the apple: B. E. Juniper, "The mysterious origin of the sweet apple," *American Scientist* 95 (2007): 44–51.

Interest in fruits and vegetables for their health values, particularly in anti-oxidation and aging, has increased dramatically in the last five years. The proper mixture of colors in diets is now the vogue. I used M. A. Lila's review for this topic: "Plant pigments and human health," in *Plant Pigments and Their Manipulation*, ed. K. Davies, Annual Plant Reviews 14 (Boca Raton, FL: CRC Press, 2004), 248–74. For a good example of potential health benefits of fruits, see J. A. Joseph, B. Schukitt-Hale, N. A. Denisova, D. Bielinski, A. Martin, J. J. McEwen, and P. C. Bickford, "Reversals of age-related declines in neuronal signal transduction, cognitive, and motor behavioral deficits with blueberry, spinach, or strawberry dietary supplementation," *Journal of Neuroscience* 19 (1999): 8114–21.

For the discussions on the pigments producing colors in various common fruits, as well as the history of their domestication, I list the fruits and the sources of information.

Citrus: J. B. Harborne, "The flavonoids of the Rutales," in *Chemistry and Chemical Taxonomy of the Rutales*, ed. P. G. Waterman and M. F. Grundon (London: Academic Press, 1983), 147–73; H. S. Lee, "Characterization of carotenoids in juice of red navel orange (Cara Cara)," *Journal of Agricultural and Food Chemistry* 49 (2001): 2563–68; H. S. Lee, "Characterization of major anthocyanins and the color of red-fleshed Budd blood orange (*Citrus sinensis*)," *Journal of Agricultural and Food Chemistry* 50 (2002): 1243–46; and C. W. Wilson III and P. E. Shaw, "Separation of pigments, flavonoids, and flavor fractions from citrus oils by gel permeation chromatography," *Journal of Agricultural and Food Chemistry* 25 (1977): 221–24.

Eggplant: S. Doganier, A. Frary, M.-C. Daunay, R. N. Lester, and S. D. Tanksley, "Conservation of gene function in the Solanaceae as revealed by comparative mapping of domestication traits in eggplant," *Genetics* 161 (2002): 1713–26; N. Matsuzoe, M. Yamaguchi, S. Kawanobu, Y. Watanabe, H. Higashi, and Y. Sakata, "Effect of dark treatment of the eggplant on fruit skin color and its anthocyanin component," *Journal of the Japanese Society for Horticultural Science* 68 (1999): 138–45; and J. R. Stommel and B. D. Whitaker, "Phenolic acid content and composition of eggplant fruit in a germplasm core subset," *Journal of the American Society of Horticultural Science* 128 (2003): 704–10.

Peach: K. R. Tourjee, D. M. Varrett, M. V. Romero, and T. M. Gradziel, "Measuring flesh color variability among processing clingstone peach genotypes differing in carotenoid composition," *Journal of the American Society for Horticultural Science* 123 (1998): 433–37; and L. O. Van Blaricom and T. L. Senn, "Anthocyanin pigments in freestone peaches grown in the Southeast," *Proceedings of the American Society for Horticultural Science* 90 (1967): 541–45.

Pepper: V. M. Russo and L. R. Howard, "Carotenoids in pungent and non-pungent peppers at various developmental stages grown in the field and glass-house," *Journal of the Science of Food and Agriculture* 82 (2002): 615–24. Recent reviews on the domestication of peppers are S. Yamamoto and E. Nawata, "*Capsicum frutescens* L. in Southeast and East Asia, and its dispersal routes into Japan," *Economic Botany* 59 (2005): 18–28; B. M. Walsh and S. B. Hoot, "Phylogenetic relationships of *Capsicum* (Solanaceae) using DNA sequences from two noncoding regions: The chloroplast *atpB-rbcL* spacer region and nuclear waxy introns," *International Journal of Plant Science* 162 (2001): 1409–18; and E. J. Votava, J. B. Baral, and P. W. Bosland, "Genetic diversity of Chile (*Capsicum annuum* var. *annuum* L.) landraces from northern New Mexico, Colorado, and Mexico," *Economic Botany* 59 (2005): 8–17.

Mango: I learned much about domestication and variation from Ray Schnell, a geneticist at Chapman Field, the USDA research station in Miami, and my own personal observations. Mangoes are a big deal in Miami, with an annual festival and about 160 varieties locally grown.

Squash: H. S. Parris, "Gene for broad, contiguous dark strips in cocozelle squash (*Cucurbita pepo*)," *Euphytica* 115 (2000): 191–96; H. S. Parris, "Multiple allelism at a major locus affecting fruit coloration in *Cucurbita pepo*," *Euphytica* 125 (2002): 149–53; O. I. Sanjur, D. R. Piperno, T. C. Andres, and L. Wessel-Beaver, "Phylogenetic relationships among domesticated and wild species of *Cucurbita* (Cucurbitaceae) inferred from a mitochondrial gene: Implications for crop plant evolution and areas of origin," *Proceedings of the National Academy of Science U.S.* 99 (2002): 535–40; A. A. Schaffer, C. D. Boyer, and T. Gianfagna, "Genetic control of plastid carotenoids and transformation in the skin of *Cucurbita pepo* L. fruit," *Theoretical and Applied Genetics* 68 (1984): 493–501; O. Shifriss, "A developmental approach to the genetics of fruit color," *Journal of Heredity* 40 (1949): 233–41; O. Shifriss, "Origin, expression, and significance of gene B in *Cucurbita pepo* L.," *Journal of the American Society for Horticultural Science* 106 (1981): 220–32; and B. D. Smith, "The initial domestication of *Curcubita pepo* in the Americas 10,000 years ago," *Science* 276 (1997): 932–34.

Strawberry: J. B. Adams and M. H. Ongley, "Degradation of anthocyanins in canned strawberries: 1, Effect of various processing parameters on the retention of pelargonidin-3-glucoside," *Journal of Food Technology* 8 (1973): 139–45; and C. A. Lundergan and J. N. Moore, "Inheritance of ascorbic acid content and color intensity in fruits of strawberry (*Fragaria* ×*ananassa* Duch.)," *Journal of the American Society for Horticultural Science* 100 (1975): 633–35.

Tomato: Y.-S. Liu, A. Gur, M. Causse, R. Damidaux, M. Buret, J. Hirschberg, and D. Zamir, "There is more to tomato fruit colour than candidate carotenoid genes," *Plant Biotechnology Journal* 1 (2003): 195–207.

The attractive seeds illustrated in this chapter are generally toxic, documented by the following: Y. Endo, K. Mitsui, M. Motizuki, and K. Tsurugi, "The mechanism of action of ricin and related toxic lectins on eukaryotic ribosomes: The site and the characteristics of the modification in 28S ribosomal RNA caused by the toxins," *Journal of Biological Chemistry* 262 (1987): 5908–12; K. Folkers and F. Koniuszy, "*Erythrina* alkaloids: 9, Isolation and characterization of erysodin, erysopine, erysocine, and erysovine," *Journal of the American Chemical Society* 62 (1940): 1677–83; J. N. Hughes, C. D. Lindsay, and G. D. Griffiths, "Morphology of ricin and abrin exposed endothelial cells is consistent with apoptotic cell death," *Human and Experimental Toxicology* 15 (1996): 443–51; S. Olsnes, K. Refsnes, and A. Pihl, "Mechanism of action of the toxic lectins abrin and ricin," *Nature* 249 (1974): 627–31; S. C. Pando, M. L. R. Macedo, M. G. M. Freire, M. H. Toyama, J. C. Novello, and S. Marangoni, "Biochemical characterization of a lectin from *Delonix regia* seeds," *Journal of Protein Chemistry* 21 (2002): 279–85; and M. Richardson, F. A. P. Campos, J. Xavier-Filho, M. L. R. Macedo, G. M. C. Maia, and A. Yarwood, "The amino acid sequence and reactive (inhibitory) site of the major trypsin isoinhibitor (DE5) isolated from seeds of the Brazilian Carolina tree (*Adenanthera pavonina* L.)," *Biochimica et Biophysica Acta* 872 (1986): 134–40.

For my discussion on the domestication and seed-coat color of the common bean, I consulted M. J. Bassett, K. Hartel, and P. Mclean, "Inheritance of the Anasazi pattern of partly colored seed coats in common bean," *Journal of the American Society of Horticultural Science* 125 (2000): 340–43; C. W. Beninger, G. L. Hosfield, and M. J. Bassett, "Flavonoid composition of three genotypes of dry bean (*Phaseolus vulgaris*) differing in seedcoat color," *Journal of the American Society of Horticultural Science* 124 (1999): 514–18; C. W. Beninger and G. L. Hosfield, "Chemical and morphological expression of the *B* and *Asp* seedcoat genes in *Phaseolus vulgaris*," *Journal of the American Society of Horticultural Science* 125 (2000): 52–58; R. A. Emerson, "Inheritance of color in the seeds of the common bean, *Phaseolus vulgaris*," *Annual Report of the Nebraska Agricultural Experiment Station* 22 (1909): 65–101; R. Freyre, R. Ríos, L. Guzmán, D. G. Debouck, and P. Gepts, "Ecogeographic distribution of *Phaseolus* spp. (Fabaceae) in Bolivia," *Economic Botany* 50 (1996): 195–215; P. Gepts, "Origin and evolution of common bean: Past events and recent trends," *HortScience* 33 (1998): 1124–30; G. L.

Hosfield, "Seed coat color in *Phaseolus vulgaris* L., its chemistry, and associated health related benefits," *Annual Report of the Bean Improvement Cooperative* 44 (2001): 1–6; L. Kaplan and L. N. Kaplan, "*Phaseolus* in archaeology," in *Genetic Resources of Phaseolus Beans*, ed. P. Gepts (Boston: Kluwer, 1988), 125–42; P. E. McClean, R. K. Lee, C. Otto, P. Gepts, and M. J. Bassett, "Molecular and phenotypic mapping of genes controlling seed coat pattern and color in common bean (*Phaseolus vulgaris* L.)," *Journal of Heredity* 93 (2002): 148–52; S. P. Sing, P. Gepts, and D. G. Debouck, "Races of common bean (*Phaseolus vulgaris*, Fabaceae)," *Economic Botany* 45 (1991): 379–96; and L. Pallottini, E. Garcia, J. Kami, G. Barcaccia, and P. Gepts, "The genetic anatomy of a patented yellow bean," *Crop Science* 44 (2004): 968–77.

For general background on fruit display and seed dispersal, I consulted the following: E. J. H. Corner, "The durian theory or the origin of the modern tree," *Annals of Botany*, n.s., 13 (1949): 367–414; A. Estrada and T. H. Fleming, eds., *Frugivores and Seed Dispersal* (Dordrecht, The Netherlands: W. Junk, 1986); T. H. Fleming, "Fruiting plant–frugivore mutualism: The evolutionary theater and the ecological play," in *Plant-Animal Interactions: Evolutionary Ecology in Tropical and Temperate Regions*, ed. P. W. Price, T. M. Lewinsohn, G. W. Fernandes, and W. W. Benson (New York: John Wiley and Sons, 1991), 119–44; H. N. Ridley, *The Dispersal of Plants throughout the World* (Ashford, Kent: L. Reeve, 1930); L. van der Pijl, *Principles of Dispersal in Higher Plants*, 3rd ed. (New York: Springer-Verlag, 1982), particularly useful on arils and sarcotestae; M. F. Willson and J. N. Thompson, "Phenology and ecology of color in bird-dispersed fruits, or why some fruits are red when they are 'green,'" *Canadian Journal of Botany* 60 (1982): 701–13; M. F. Willson and C. J. Whelan, "The evolution of fruit color in fleshy-fruited plants," *American Naturalist* 116 (1990): 790–809; and D. McKey, "The ecology of coevolved seed dispersal systems," in *Coevolution of Animals and Plants*, ed. L. E. Gilbert and P. H. Raven (Austin: University of Texas Press, 1975), 159–91, a particularly good discussion of the possible significance of contrast-colored seeds. The possibility of an extended visual range for fruit perception by birds was discussed by D. Burkhardt, "Birds, berries, and UV: A note on some consequences of UV vision in birds," *Die Naturwissenschaften* 69 (1982): 153–57.

The following articles document the attraction of fruits to bird and other dispersers in different ecosystems: A. Gautier-Hion, J.-M. Duplantier, R. Quiris, F. Feer, C. Sourd, J.-P. Decoux, G. Subost, et al., "Fruit characters as a basis of fruit choice and seed dispersal in a tropical forest vertebrate community," *Oecologia* 65 (1985): 324–37; M. Gopalokrishnan, K. Rajaraman, and A. G. Matthew, "Identification of the mace pigments," *Journal of Food Science and Technology* 16 (1979): 261–62; C. M. Herrera, "Are tropical fruits more rewarding to dispersers than temperate ones?" *American Naturalist* 118 (1981): 896–907; C. M. Herrera, "Seasonal variation in the quality of fruits and diffuse coevolution between plants and avian dispersers," *Ecology* 63 (1982): 773–85; H. F. Howe and G. A. Vande Kerckhove, "Removal of wild nutmeg (*Virola surinamensis*) crops by birds," *Ecology* 62 (1981): 1092–1106;

R. S. Knight and W. R. Siegfried, "Inter-relationships between type, size, and colour of fruits and dispersal in southern African trees," *Oecologia* 56 (1983): 405–12; C. H. Janson, "Adaptation of fruit morphology to dispersal agents in a Neotropical forest," *Science* 219 (1983): 187–89; T. K. Pratt, "Examples of tropical frugivores defending fruit-bearing plants," *Condor* 86 (1984): 123–29; T. K. Pratt and E. W. Stiles, "How long fruit-eating birds stay in the plants where they feed: Implications for seed dispersal," *American Naturalist* 122 (1983): 797–805; S. T. Skeate, "Interactions between birds and fruits in a northern Florida hammock community," *Ecology* 68 (1987): 297–309; D. W. Snow, "Tropical frugivorous birds and their food plants: A world survey," *Biotropica* 13 (1981): 1–14; E. W. Stiles, "Patterns of fruit presentation and seed dispersal in bird-disseminated woody plants in the eastern deciduous forest," *American Naturalist* 116 (1980): 670–88; E. W. Stiles, "Fruits, seeds, and dispersal agents," in *Plant-Animal Interactions*, ed. W. G. Abrahamson (New York: McGraw-Hill, 1986), 87–122; A. Traveset, M. F. Willson, and M. Verdú, "Characteristics of fleshy fruits in southeast Alaska: Phylogenetic comparison with fruits from Illinois," *Ecography* 27 (2004): 41–48; F. J. Turcek, "Color preference in fruit and seed-eating birds," *Proceedings of the International Ornithological Congress* 13 (1963): 285–92; N. T. Wheelwright, "Fruit size, gape width, and the diets of fruit-eating birds," *Ecology* 66 (1985): 808–18; N. T. Wheelwright, W. A. Haber, K. G. Murray, and C. Guindon, "Tropical fruit eating birds and their food plants: A survey of a Costa Rican lower montane forest," *Biotropica* 16 (1984): 173–92; N. T. Wheelwright and C. H. Janson, "Colors of fruit displays of bird-dispersed plants in two tropical forests," *American Naturalist* 126 (1985): 777–99.

The common name of this plant was mentioned in the chapter: peach = *Prunus persica* (L.) Batsch (Rosaceae).

Chapter Nine: Stems and Roots

Ketzel Levine's quotation was taken from her Web site: Talking Plants, http://www .npr.org. The other quotation is from A. de Saint-Exupéry, *Wind, Sand, and Stars* (New York: Harcourt Brace Jovanovich, 1967).

For the general example of the complex chemistry and color of ponderosa pine bark, I used J. B. Harborne, *Introduction to Ecological Biochemistry*, 4th ed. (London: Academic Press, 1993); and V. R. Franceschi, P. Krokene, E. Christiansen, and T. Krekling, "Anatomical and chemical defenses of conifer bark against bark beetles and other pests," *New Phytologist* 167 (2005): 353–76. For the general structure of bark, I used K. Esau, *The Anatomy of Seed Plants*, 2nd ed. (New York: John Wiley, 1977); R. F. Evert, *Esau's Plant Anatomy*, 3rd ed. (Hoboken, NJ: Wiley Interscience, 2006) (discussion on bark is particularly up-to-date and useful); I. Roth, *Structural Patterns of Tropical Barks*, vol. 9-3 of *Encyclopedia of Plant Anatomy* (Berlin: Gebrüder Borntraeger, 1991); H. Vaucher, *Tree Bark: A Color Guide* (Portland, OR: Timber Press, 2003); and T. T. Kozlowsky, P. J. Kramer, and S. G. Pallardy, *The Physiological Ecology of Woody Plants* (San Diego: Academic Press, 1991). Bark structure varies dramatically among species within large families, well-documented by M. M. Chattaway, "The anatomy of bark: 1, The genus *Eucalyptus*," *Australian Journal of Botany* 1 (1953): 402–33; and T. C. Whitmore, "Studies in systematic bark morphology: 3, Bark taxonomy in Dipterocarpaceae," *Gardens Bulletin* (Singapore) 24 (1962): 321–71. The importance of stems in photosynthesis was shown by K. Muthuchelian, "Photosynthetic characteristics of bark tissues of the tropical tree of *Bombax ceiba* L.," *Photosynthetica* 26 (1992): 633–36. The presence of carotenoids producing the red color of the leaf sheaths of the sealing wax palm was observed by Franz Hoffman, relayed to me by Barry Tomlinson.

For the complex chemistry of bark and stems, consult the general reviews for chapter 3, and Harborne, *Introduction to Ecological Biochemistry*. For the chemistry of latex and resins, I used specific examples based on the following research: A. Arnone, G. Nasini, O. U. de Pava, and L. Merlini, "Constituents of dragon's blood: 5, Dracoflavans Bs, B2, C1, C2, D1, D2, new A-type deoxyproanthocyanidins," *Journal of Natural Products* 60 (1997): 971–75; H. Baer, "The poisonous Anacardiaceae," in *Toxic Plants*, ed. A. D. Kinghorn (New York: Columbia University Press, 1979), 161–70; M. D. Corbett and S. Billets, "Characterization of poison oak urushiol," *Journal of Pharmaceutical Sciences* 64 (1975): 1715–18; D. M. A. De Oliveira, A. M. Porto, V. Bittrich, I. Vencato, A. J. Marsaioli, and A. J. Tidsskrift, "Floral resins of *Clusia* spp.: Chemical composition and biological function," *Tetrahedron Letters* 37 (1996): 6427–30; A. G. González, F. León, L. Sánchez-Pinto, J. L. Padrón, and J. Bermejo, "Phenolic compounds of dragon's blood from *Dracaena draco*," *Journal of Natural Products* 63 (2000): 1297–99; J. S. Mills and R. White, "Natural resins and lacquers," in *The Organic Chemistry of Museum Objects*, ed. J. S. Mills (Oxford: Butterworth-Heinemann, 1994), 95–128; and P. B. Oeirichs, J. K. Macleod, A. A. Seawright, and J. C. Ng, "Isolation and characterization of urushiol components from the Australian native cashew (*Semecarpus australiensis*)," *Natural Toxins* 5 (1977): 96–98. I learned of the story by Richard Howard from his long-time friend, Larry Schokman.

I used the example of root color in ragweed from J. E. Page and G. H. N. Towers, "Anthocyanins protect light-sensitive thiarubrine phototoxins," *Planta* 215 (2002): 478–84. For the benefits of purple sweet potatoes, I used M. Philpott, K. S. Gould, C. Lim, and L. R. Ferguson, "*In situ* and *in vitro* antioxidant activity of sweetpotato anthocyanins," *Journal of Agricultural and Food Chemistry* 52 (2004): 1511–13; and received information directly from Kevin Gould.

For the discussion on stem chemistry and plant defenses, see I. T. Baldwin and C. A. Preston, "The eco-physiological complexity of plant responses to insect herbivores," *Planta* 208 (1999): 137–45; J. X. Becerra, "Squirt-gun defense in *Bursera* and the chrysomelid counterploy," *Ecology* 75 (1994): 1991–96; J. X. Becerra, "Insects on plants: Macroevolutionary chemical trends in host use," *Science* 276 (1997): 253–56; D. E. Dussourd and T. Eisner, "Vein-cutting behavior: Insect counterploy to the latex defense of plants," *Science* 237 (1987): 898–901; D. E. Dussourd and R. F. Denno, "Deactivation of plant defense: Correspondence between insect behavior and secretory canal architecture," *Ecology* 72 (1991): 1383–96; B. D. Farrell, D. E. Dussourd, and C. Mitter, "Escalation of plant defense: Do latex and resin canals spur plant diversification?" *American Naturalist* 138 (1991): 881–900; and D. M. Joel, "Resin ducts in the mango fruit: A defense system," *Journal of Experimental Botany* 31 (1984): 1707–18.

The role of bark surfaces in selecting or repelling organisms (epiphytes and lianas) is documented by the following articles: D. H. Benzing, "Bark surfaces and the origin and maintenance of diversity among angiosperm epiphytes: A hypothesis," *Selbyana* 5 (1981): 248–55; B. M. Boom and S. F. Mori, "Falsification of two hypotheses on liana exclusion from tropical trees possessing buttresses and smooth bark," *Bulletin of the Torrey Botanical Club* 109 (1982): 447–55; C. Boudreault, S. Gauthier, and Y. Bergeron, "Epiphytic lichens and bryophytes on *Populus tremuloides* along a chronosequence in the southwestern boreal forest of Quebec, Canada," *Bryologist* 103 (2000): 725–38; E. J. F. Campbell and D. Newberry, "Ecological relationships between lianas and trees in lowland rain forest in Sabah, East Malaysia," *Journal of Tropical Ecology* 9 (1993): 469–90; S. Eversman, "Epiphytic lichens of a ponderosa pine forest in southeastern Montana," *Bryologist* 85 (1982): 204–13; E. E. Hegarty, "Vine-host interactions," in *The Biology of Vines*, ed. F. E. Putz and H. A. Mooney (New York: Cambridge University Press, 1991), 357–76; J. M. H. Knops, T. H. Nash, and W. H. Schlesinger, "The influence of epiphytic lichens on the nutrient cycling of an oak woodland," *Ecological Monographs* 66 (1996): 159–79; S. Muthuramkumar and N. Parthasarathy, "Tree-liana relationships in a tropical evergreen forest at Varagalaiar, Anamalais, Western Ghats, India," *Journal of Tropical Ecology* 17 (2001): 395–409; J. Nabe-Nielsen, "Diversity and distribution of lianas in a Neotropical rain forest, Yasuni National Park, Ecuador," *Journal of Tropical Ecology* 17 (2001): 1–19; S. C. Sillett and T. R. Rambo, "Vertical distribution of dominant epiphytes in Douglas-fir forests of the Central Oregon Cascades," *Northwest Science* 74 (2000): 44–49; A. P. Smith, "Buttressing of tropical trees in relation to bark thickness in Dominica, B.W.I.," *Biotropica* 11 (1979): 159–60; G. C. Stevens, "Lianas as structural parasites: The *Bursera simaruba* example," *Ecology* 68 (1987): 77–81; and P. A. Wolsely, C. Moncrieff, and B. Aguirre-Hudson, "Lichens as indicators of environmental stability and change in the tropical forests of Thailand," *Global Ecology and Biogeography Letters* 4 (1996): 116–23.

Common names of these plants were mentioned in the chapter: frankincense = *Boswellia sacra* Flueckiger (Burseraceae); marking nut = *Semecarpus anacardium* L. f. (Anacardiaceae); purpleheart = *Swartzia cubensis* (Britton & P. Wilson) Standl. (Fabaceae); sweet potato = *Ipomoea batatus* (L.) Lam. (Convolvulaceae); banyan = *Ficus benghalensis* L. (Moraceae).

Chapter Ten: Iridescent Plants

The two quotations at the beginning of the chapter are from D. Young, "The portable earthlamp," in *The Planet on the Desk: Selected and New Poems, 1960–1990* (Middletown, CT: Wesleyan University Press, 1991), 138; and R. Hooke, *Micrographia, or, some Physiological Descriptions of Minute Bodies Made by Magnifying Glasses. With Observations and Inquiries thereupon* (London: Jo. Marfyn and Ja. Allstry, 1665). This chapter was partly adapted from a general article I wrote: "Iridescent blue plants," *American Scientist* 85 (1997): 56–63.

For a general description of the physical phenomena that produce this color, see references for chapter 3, plus a review of structural color in animals: D. L. Fox, *Animal Biochromes and Structural Colours* (Berkeley and Los Angeles: University of California Press, 1976). General descriptions of this phenomenon in plants are in P. W. Richards, *The Tropical Rainforest*, 2nd ed. (Cambridge: Cambridge University Press, 1990); and G. E. Stahl, "Über bunte Laublatter: Ein Beitrag zur Pflanzenbiologie II," *Annales de la Jardin Botanique de Buitenzorg* 13 (1896): 137–216.

For descriptions of structural color in vascular plants, see K. S. Gould and D. W. Lee, "Physical and ultrastructural basis of blue leaf iridescence in four Malaysian understory plants," *American Journal of Botany* 83 (1995): 45–50; R. M. Graham, D. W. Lee, and K. Norstog, "Physical and ultrastructural basis of blue leaf iridescence in two Neotropical ferns," *American Journal of Botany* 80 (1993): 198–203; C. W. Hébant and D. W. Lee, "Ultrastructural basis and developmental control of blue iridescence in *Selaginella* leaves," *American Journal of Botany* 71 (1984): 216–19; A. Nasrulhaq-Boyce and J. G. Duckett, "Dimorphic epidermal cell chloroplasts in the mesophyll-less leaves of an extreme shade tropical fern, *Teratophyllum rotundifoliatum* (R. Bonap.) Holtt.: A light and electron microscope study," *New Phytologist* 119 (1991): 433–44; D. W. Lee, "On iridescent plants," *Gardens Bulletin* (Singapore) 30 (1977): 21–29; and D. W. Lee and J. B. Lowry, "Physical basis and ecological significance of iridescence in blue plants," *Nature* 254 (1975): 50–51.

These references describe structural color in algae: G. Feldman, "Sur l'ultra-structure de l'appareil irisant du *Gastroclonium clavagum* (Roth.) Ardissone (Rhodophyceae)," *Comptes Rendues de l'Academie Sciences* (Paris) 270 (1970): 1244–46; L. M. Patrone, S. T. Broadwater, and J. L. Scott, "Ultra-structure of vegetative and dividing cells of the unicellular red algae *Rhodelia violacea* and *Rhodelia maculata*," *Journal of Phycology* 27 (1991): 742–53; and W. H. Gerwick and N. J. Lang, "Structural color: Chemical and ecological studies on iridescence in *Iridaea* (Rhodophyta)," *Journal of Phycology* 13 (1977): 121–27.

Structural colors in fruits were described in D. W. Lee, "Ultrastructural basis and function of iridescent blue colour of fruits in *Elaeocarpus*," *Nature* 349 (1991): 260–62; and D. W. Lee, G. T. Taylor, and A. K. Irvine, "Structural coloration in *Delarbrea michieana* (Araliaceae)," *International Journal of Plant Sciences* 161 (2000): 297–300.

Helicoidal structures accounting for iridescent color have also been described in animals (A. C. Neville and S. C. Caveney, "Scarabaeid beetle exocuticle as an optical analogue of cholesteric liquid crystal," *Biological Reviews* 44 [1969]: 531–52), and in plants (A. C. Neville and S. Levy, "The helicoidal concept in plant cell ultrastructure and morphogenesis," in *The Biochemistry of Plant Cell Walls*, ed. C. T. Brett and J. R. Hillman [Cambridge: Cambridge University Press, 1985]), 91–124.

References concerning the possible functions of iridescence in plants include K. S. Gould, D. N. Kuhn, D. W. Lee, and S. F. Oberbauer, "Why leaves are red," *Nature* 360 (1995): 40–41; and D. W. Lee, "Unusual strategies of light absorption in rain forest herbs," in *On the Economy of Plant Form and Function*, ed. T. Givnish (New York: Cambridge University Press, 1986), 105–31.

Chapter Eleven: Why Leaves Turn Red

Quotations at the beginning of the chapter are from A. Camus, "The Misunderstanding," in *Caligula and Three Other Plays*, trans. Justin O'Brien (New York: Alfred A. Knopf, 1958), 104; and H. Borland, *Sundial of the Seasons* (Philadelphia: J. P. Lippincott, 1964). The Thoreau quotation early in the chapter is from H. D. Thoreau, "Autumnal tints" (1862), in *The Writings of Henry David Thoreau*, vol. 5 (New York: Houghton Mifflin, 1906), 269.

This chapter was revised from an article: D. W. Lee and K. S. Gould, "Why leaves turn red," *American Scientist* 90 (2002): 524–31. Much of the supporting literature for this chapter is also available in a volume edited by those authors: *Anthocyanins in Leaves*, Advances in Botanical Research 37 (London: Academic Press, 2002). A dated but fascinating book on the history of this subject is M. Wheldale, *The Anthocyanin Pigments of Plants* (Cambridge: Cambridge University Press, 1916).

References on the distribution of anthocyanins in plants used in this chapter include D. W. Lee and T. M. Collins, "Phylogenetic and ontogenetic influences on the distribution of anthocyanins and betacyanins in leaves of tropical plants," *International Journal of Plant Sciences* 162 (2001): 1141–53; J. Sanger, "Quantitative investigations of leaf pigments from their inception in buds through autumn coloration to decomposition in falling leaves," *Ecology* 52 (1971): 1075–89; D. W. Lee, "Anthocyanins in leaves: Distribution, phylogeny, and development," in *Anthocyanins in Leaves*, ed. Gould and Lee, 37–53; and D. W. Lee, J. O'Keefe, N. M. Holbrook, and T. S. Feild, "Pigment dynamics and autumn leaf senescence in a New England deciduous forest, eastern USA," *Ecological Research* 18 (2003): 677–94.

Evidence for the protective function of anthocyanins is in the following articles: L. Chalker-Scott, "Environmental significance of anthocyanins in plant stress responses," *Photochemistry and Photobiology* 70 (1999): 1–9; D. C. Close and C. L. Beadle, "The ecophysiology of foliar anthocyanin," *Botanical Review* 69 (2003): 149–61; D. W. Lee and J. B. Lowry, "Young leaf anthocyanin and solar ultraviolet," *Biotropica* 12 (1980): 75–76; K. S. Gould, D. N. Kuhn, D. W. Lee, and S. F. Oberbauer, "Why leaves are sometimes red," *Nature* 378 (1995): 242–43; A. Post, "Photoprotective pigment as an adaptive strategy in the Antarctic moss *Ceratodon purpureus*," *Polar Biology* 10 (1990): 241–46; and D. W. Lee, S. Brammeier, and A. P. Smith, "The selective advantage of anthocyanins in developing leaves of mango and cacao," *Biotropica* 19 (1987): 40–49.

Evidence for antioxidant activity is taken from S. O. Neill and K. S. Gould, "Anthocyanins in leaves: Light attenuators or antioxidants?" *Functional Plant Biology* 30 (2003): 865–73; T. Tsuda, M. Watanabe, K. Ohshima, S. Norinobu, S. W. Choi, S. Kawakishi, and T. Osawa, "Antioxidant activity of the anthocyanin pigments cyanidin 3-O-β-D-glucoside and cyanidin," *Journal of Agricultural and Food Chemistry* 42 (1994): 2407–10; H. Yamasaki, "A function of colour," *Trends in Plant Science* 2 (1997): 7–8; and K. S. Gould, J. McKelvie, and K. R. Markham, "Do anthocyanins function as antioxidants in leaves? Imaging of H_2O_2 in red and green leaves after mechanical injury," *Plant, Cell, and Environment* 25 (2002): 1261–69.

Hypotheses concerning the biological functions of anthocyanins in leaves are included in P. D. Coley and T. M. Aide, "Red coloration of tropical leaves: A possible antifungal defense?" *Journal of Tropical Ecology* 5 (1989): 293–300; B. Juniper, "Flamboyant flushes: A reinterpretation of non-green flush colours in leaves," *International Dendrological Society Yearbook* (1993): 49–57; S. Lev-Yadun, "Weapon (thorn) automimicry and mimicry of aposematic colorful thorns in plants," *Journal of Theoretical Biology* 224 (2003): 183–88; and S. Lev-Yadun, A. Dafni, M. A. Flaishman, M. Inbar, I. Izhaki, G. Katzir, and G. Ne'eman, "Plant coloration undermines herbivorous insect camouflage," *BioEssays* 26 (2004): 1126–30.

Production and functions of other red pigments in leaves are discussed in J. Escribano, M. A. Pedreño, F. Gárcia-Carmona, and R. Muñoz, "Characterization of the anti-radical activity of betalains from *Beta vulgaris*," *Phytochemical Analysis* 9 (1998): 124–27; S. Kunz, G. Burkhardt, and H. Becker, "Riccionidins A and B: Anthocyanidins from the cell walls of the liverwort *Ricciocarpos natans*," *Phytochemistry* 35 (1994): 233–35; and K. Ida, K. Masamoto, T. Maoka, Y. Fujiwara, S. Takeda, and E. Hasegawa, "The leaves of the common box, *Buxus sempervirens* (Buxaceae), become red as the level of a red carotenoid anhydroeschscholzxanthin, increases," *Journal of Plant Research* 108 (1995): 369–76. Some unusual pigments are produced during autumn senescence: K. G. Chang, G. H. Fechner, and H. A. Schroeder, "Anthocyanins in autumn leaves of quaking aspen in Colorado," *Forest Science* 35 (1989): 229–36; and P. Matile, "Fluorescent idioblasts in autumn leaves of *Ginkgo biloba*," *Botanica Helvetica* 104 (1994): 87–92.

The discussion of anthocyanin function during the autumn looks at ideas and results from the following authors: M. Archetti and S. P. Brown, "The coevolution theory of autumn colours," *Proceedings of the Royal Society of London* B 271 (2004): 1219–23; W. D. Hamilton and S. P. Brown, "Autumn tree colours as a handicap signal," *Proceedings of the Royal Society of London* B 268 (2001): 1489–93; W. A. Hoch, E. L. Zeldin, and B. H. McCown, "Physiological significance of anthocyanins during autumnal leaf senescence," *Tree Physiology* 21 (2001): 1–8; W. A. Hoch, E. L. Singsaas, and B. H. McGown, "Resorption protection: Anthocyanins facilitate nutrient recovery in autumn by shielding leaves from potentially damaging light levels," *Plant Physiology* 133 (2003): 1416–25; N. P. A. Huner, G. Öquist, and F. Sarhan, "Energy balance and acclimation to light and cold," *Trends in Plant Science* 3 (1998): 224–30; S. B. Hagen,

I. Folstad, and S. W. Jakobsen, "Autumn colouration and herbivore resistance in mountain birch (*Betula pubescens*)," *Ecology Letters* 6 (2003): 807–11; H. M. Schaefer and D. M. Wilkinson, "Red leaves, insects, and coevolution: A red herring?" *Trends in Ecology and Evolution* 19 (2004): 616–18; P. G. Schaberg, A. K. Van den Berg, P. F. Murakami, J. B. Shane, and J. R. Donnelly, "Factors influencing red expression in autumn foliage of sugar maple trees," *Tree Physiology* 23 (2003): 325–33; T. S. Feild, D. W. Lee, and N. M. Holbrook, "Why leaves turn red in autumn: The role of anthocyanins in senescing leaves of red-osier dogwood," *Plant Physiology* 127 (2001): 566–74; H. M. Schaefer and G. Rolshausen, "Plants on red alert: Do insects pay attention?" *BioEssays* 28 (2006): 65–71; H. J. Ougham, P. Morris, and H. Thomas, "The colors of autumn leaves as symptoms of cellular recycling and defenses against environmental stresses," *Current Topics in Developmental Biology* 66 (2005): 135–60; P. Karageorgou and Y. Manetas, "The importance of being red when young: Anthocyanins and the protection of young leaves of *Quercus coccifera* from insect herbivory and excess light," *Tree Physiology* 26 (2006): 613–21; A. Sinkkonen, "Do autumn leaf colors serve as a reproductive insurance against sucking herbivores?" *Oikos* 113 (2006): 557–62; and the article by Lee et al. "Pigment dynamics and autumn leaf senescence," cited earlier in these notes. Check the Web site on autumn foliage at the Harvard Forest for more information on and photographs of this phenomenon: http://www.harvardforest.fas.harvard.edu/research/leaves/autumn_leaves.html. Jim Erdman, an ecologist retired from the U.S. Geological Survey, has made some interesting observations concerning the incidence of red coloration in aspens and the mineral content of the soil, mentioned briefly near the end of the chapter.

The common name of this plant was mentioned in the chapter: sugar maple = *Acer saccharum* Marshall (Aceraceae).

Chapter Twelve: Chlorophilia

The title of the chapter, "Chlorophilia," is certainly not my invention; it can be found on the Internet as an artists' cooperative and the title of a popular song. I do not know its origin. The quotations at the beginning of the chapter are from e. e. cummings, *Complete Poems, 1904–1962* (New York: Liveright, 1994); and D. Thomas, *The Poems of Dylan Thomas* (New York: New Directions, 2003). The quotation on the Ituri Pygmies is from C. M. Turnbull, *The Forest People: A Study of Pygmies of the Congo* (New York: Simon and Schuster, 1961). The quotations by Paul Shepard are from *Man in the Landscape* (New York: Alfred A. Knopf, 1967) and *Traces of an Omnivore* (Washington, DC: Island Press, 1996).

The origin of the word *jungle* is analyzed by M. R. Dove, "The dialectical history of 'jungle' in Pakistan: An examination of the relationship between nature and culture," *Journal of Anthropological Research* 48 (1992): 231–51. The ceremonial importance of giving flower versus animal parts as gifts is from F. Hallé, *In Praise of Plants* (Portland, OR: Timber Press, 2002). The idea of levels of influence of nature in our lives was adopted from F. H. Mahnke, *Color, Environment, and Human Response* (New York: Van Nostrand Reinhold, 1996).

The literature on color preferences, across age and culture, includes R. R. Crane and B. I. Levy, "Color scales in responses to emotionally laden situations," *Journal of Consulting Psychology* 26 (1962): 515–19; J. P. Guilford, "The affective value of color as a function of hue, tint, and chroma," *Journal of Experimental Psychology* 17 (1934): 342–70; M. Hemphill, "A note on adults' color-emotion associations," *Journal of Genetic Psychology* 157 (1996): 275–80; V. J. Jones-Molfese, "Responses of neonates to colored stimuli," *Child Development* 48 (1977): 1092–95; Y. Tatibana, "Colour feeling of the Japanese: I, The inherent emotional effects of colours," *Tohuku Psychologia Folia* 5 (1937): 21–46; and P. Valdez and A. Mehrabian, "Effects of color on emotions," *Journal of Experimental Psychology* 123 (1994): 394–409.

The following articles highlight the controversy concerning the absoluteness of color categories, in addition to those articles used for chapter 2: H. C. Conklin, "Hanunóo color categories," *Southwestern Journal of Anthropology* 11 (1955): 339–44; H. C. Conklin, "Color categorization," *American Anthropologist* 75 (1973): 931–42; P. Kay, B. Berlin, and W. Merrifield, "Biocultural implications of systems of color naming," *Journal of Linguistic Anthropology* 1 (1991): 12–25; M. Haspelmuth, E. König, W. Oesterreicher, and W. Raible, eds., *Language Typology and Language Universals: An International Handbook* (New York: Walter de Gruyter, 2001); B. A. C. Saunders and J. van Brakel, "Are there nontrivial constraints on colour categorization?" *Behavioral and Brain Sciences* 20 (1997): 167–228; K. Uchikawa and R. M. Boynton, "Categorical color perception of Japanese observers: Comparison with that of Americans," *Vision Research* 27 (1987): 1825–33; and A. Wierzbicka, "The meaning of color terms: Semantics, culture, and cognition," *Cognitive Linguistics* 1 (1990): 99–150. For views on nature in Miami and tropical south Florida, see the numerous articles in an excellent volume of the *Journal of Decorative and Propaganda Arts*, no. 13, "Florida: Land of sunshine and happiness" (1998).

The following articles document the writing on natural light environments and human responses: J. A. Endler, "The color of light in forests and its implications," *Ecological Monographs* 63 (1993): 1–25; S. J. C. Gaulin and D. H. McBurney, *Evolutionary Psychology*, 2nd ed. (Upper Saddle River, NJ: Pearson / Prentice Hall, 2003); D. W. Lee, "Canopy dynamics and light climates in a tropical moist deciduous forest," *Journal of Tropical Ecology* 5 (1989): 65–79; D. W. Lee, K. Baskaran, M. Mansor, H. Mohamad, and S. K. Yap, "Light intensity and spectral quality effects on Asian tropical rainforest tree seedling development," *Ecology* 77 (1996): 568–80; I. R. Gilbert, P. G. Jarvis, and H. Smith, "Proximity signal and shade avoidance differences between early and late successional trees," *Nature* 411 (2001): 792–95; and N. J. Dominy and P. W. Lucas, "Ecological importance of trichromatic vision to primates," *Nature* 410 (2001): 363–66. On nonvisual animal and human responses to light, see D. M. Berson, "Strange vision: Ganglion cells as circadian photoreceptors," *Trends in Neurosciences* 26 (2003): 314–20; G. C. Brainard, J. P. Hanefin, J. M. Greeson, B. Byrne, G. Glickman, E. Gerner, and M. D. Rollag, "Action spectrum for melatonin regulation in humans: Evidence for a novel circadian photoreceptor," *Journal of Neuroscience* 21 (2001): 6405–21; S. S. Campbell and P. J. Murphy, "Extra-ocular circadian phototransduction in humans," *Science* 279 (1998): 396–99; D.-J. Dijk and S. W. Lockley, "Functional genomics of sleep and circadian rhythm," *Journal of Applied Physiology* 92 (2002): 852–62; I. Provencio, I. R. Rodriquez, G. Jiang, W. P. Hayes, E. F. Moreira, and M. D. Rollag, "Photoreceptive net in the mammalian retina," *Nature* 415 (2002): 493–97; C. Cajochen, M. Münch, S. Kobialka, K. Kräuchi, R. Steiner, P. Oelhafen, S. Orgül, and A. Wirz-Justice, "High sensitivity of human melatonin, alertness, thermoregulation, and heart rate to short wavelength light," *Journal of Clinical Endocrinology and Metabolism* 90 (2004): 1311–16; R. A. Renema, F. E. Robinson, J. J. R. Feddes, G. M. Fasenko, and M. J. Zuidhof, "Effects of light intensity from photostimulation in four strains of commercial egg layers: 2, Egg production parameters," *Poultry Science* 80 (2001): 1121–31; I. Rosenbom, I. Biran, Y. Chaiseha, S. Yahav, A. Rosenstrauch, D. Sklan, and O. Halevy, "The effect of a green and blue monochromatic light combination on broiler growth and development," *Poultry Science* 83 (2004): 842–45; and C. K. Levenick and A. T. Leighton Jr., "Effects of photoperiod and filtered light on growth, reproduction, and mating behavior of turkeys: 1, Growth performance of two lines of males and females," *Poultry Science* 67 (1988): 1505–13. For a concise review on applications of nonvisual color reception in humans, I used J. Raloff, "Light impacts: Hue and timing determine whether rays are beneficial or detrimental," *Science News* 169 (2006): 330–32.

These articles were the sources of information about the visual preferences for natural environments: E. R. Gimblett, R. M. Itami, and J. E. Fitzgibben, "Mystery in an information processing model of landscape preference," *Landscape Journal* 4 (1985): 87–95; J. H. Heerwagen and G. H. Orians, "Adaptation to windowlessness: A study of the use of visual décor in windowed and windowless offices," *Environment and Behavior* 18 (1986): 623–39; R. B. Hull and G. R. B. Revell, "Cross-cultural comparison of landscape scenic beauty evaluations: A case study in Bali," *Journal of Environmental Psychology* 9 (1989): 177–91; R. Kaplan, "Nature at the doorstep: Residential satisfaction and the nearby environment," *Journal of Architectural Planning Research* 2 (1985): 115–27; S. Kaplan and R. Kaplan, *The Experience of Nature: A Psychological Perspective* (New York: Cambridge University Press, 1994); G. H. Orians, "Habitat selection: General theory and applications to human behavior," in *The Evolution of Human Social Behavior*, ed. J. S. Lockard (New York: Elsevier, 1980), 49–66; G. H. Orians, "Evolved responses to landscapes," in *The Adapted Mind: Evolutionary Psychology and the Generation of Culture*, ed. J. H. Barkow, L. Cosmides, and J. Tooby (New York: Oxford University Press, 1992), 555–79; J. F. Talbot and R. Kaplan, "Needs and fears: The response to trees and nature in the inner city," *Journal of Arboriculture* 10 (1984): 222–28; and T. D. White, G. WoldeGabriel, B. Asfaw, S. Ambrose, Y. Beyene, R. L. Bernor, J.-R. Boisserie, et al., "Asa Issie, Aramis, and the origin of

Australopithecus," *Nature* 440 (2006): 883–89. Anatomical evidence that *Australopithecus* climbed in trees is based on the skeleton of a child: Z. Alemseged, F. Spoor, W. H. Kimbel, R. Bobe, D. Geraads, D. Reed, and J. G. Wynn, "A juvenile early hominin skeleton from Dikika, Ethiopia," *Nature* 443 (2006): 296–301.

For the portrayal of landscapes in human mythology, see D. W. Lee, "The natural history of the Rāmāyaṇa," in *Hinduism and Ecology*, ed. C. K. Chapple and M. E. Tucker (Cambridge, MA: Harvard University Press, Center for the Study of World Religions, 2001), 245–68, which uses the quotations used in the chapter; and S. I. Pollock, *The Rāamāyaṇa of Vālmīki: An epic of ancient India*, vol. 2, *Ayodhyākanda* (Princeton, NJ: Princeton University Press, 1986). Also see J. Appleton, *The Symbolism of Habitat* (Seattle: University of Washington Press, 1990); and S. Schama, *Landscape and Memory* (New York: Random House, 1995). I add a little science fiction here: K. Kerr, *Snare* (New York: Tom Doherty and Associates, 2003), quoted as a vision of a planet with other colors; and a discussion of green (but plastic) trees: M. H. Krieger, "What's wrong with plastic trees," *Science* 179 (1973): 446–54.

For literature on the therapeutic value of plants, see F. E. Kuo, W. C. Sullivan, and R. L. Coley, "Fertile ground for community: Inner-city neighborhood common spaces," *American Journal of Community Psychology* 26 (1996): 823–51; F. E. Kuo and W. Sullivan, "Aggression and violence in the inner city: Effects of environment via mental fatigue," *Environment and Behavior* 33 (2001): 543–70; C. A. Lewis, *Green Nature / Human Nature* (Urbana: University of Illinois Press, 1996); J. Lollman, *Why We Garden* (New York: Henry Holt, 1994); E. O. Moore, "A prison environment's effect on health care service demands," *Journal of Environmental Systems* 11 (1982): 17–34; D. Relf, ed., *The Role of Horticulture in Human Well-Being and Social Development* (Portland, OR: Timber Press, 1992); H. W. Schroeder, "The felt sense of natural environments," in *Coming of Age: Proceedings of the Twenty-first Annual Conference of the Environmental Design Research Association*, ed. R. I. Selby, K. H. Anthony, J. Choi, and B. Orland (Oklahoma City: Environmental Design Research Association, 1990), 192–95; A. L. Turner, "The therapeutic value of nature," *Journal of Operational Psychiatry* 6 (1976): 64–74; R. S. Ulrich, R. F. Simons, B. D. Losito, E. Fiorito, M. A. Miles, and M. Zelson, "Stress recovery during exposure to natural and urban environments," *Journal of Environmental Psychology* 11 (1991): 201–30; R. Ulrich, "Natural versus urban scenes: Some psychophysiological effects," *Environment and Behavior* 13 (1981): 523–66; R. Ulrich, "Aesthetic and affective responses to natural environments," in *Behavior and the Natural Environment*, ed. I. Altman and J. F. Wohlwill (New York: Plenum, 1983), 86–125; R. Ulrich, "View through a window may influence recovery from surgery," *Science* 224 (1984): 420–21; J. F. Wohlwill, "The concept of nature: A psychologist's view," in *Behavior and the Natural Environment*, ed. Altman and Wohlwill, 5–37. For children's experiences in nature affecting their behavior and learning, I used N. M. Wells, "At home with nature: Effects of 'greenness' on children's cognitive functioning," *Environment and Behavior* 32 (2000): 775–95; and A. F. Tayler, F. E. Kuo, and W. C. Sullivan, "Coping with ADD: The surprising connection to green play settings," *Environment and Behavior* 33 (2001): 54–77. Rachel Kaplan discussed the attention-restoration theory in "The nature of the view from home: Psychological benefits," *Environment and Behavior* 33 (2001): 507–42. For a book-length discussion of the importance of nature experiences in child development, see Richard Louv, *Last Child in the Woods: Saving Our Children from Nature-Deficit Disorder* (Chapel Hill, NC: Algonquin Books, 2005). David Abram wrote a remarkable book on the perception of nature and authentic living: *The Spell of the Sensuous: Reception and Language in a More Than Human World* (New York: Vintage Books, 1996). In 1997, Robert Costanza and many co-authors published an article titled "The value of the world's ecosystem services and natural capital" (*Nature* 387:253–60), which provoked considerable controversy about the value of conducting such an analysis and playing the game of economic valuation, including a response by Stuart Pimm in the same issue. Although economic analysis has been valuable in developing techniques such as debt for nature swaps or using carbon credits to preserve forests, there is considerable concern, which I share, about playing the economic game. See D. J. McCauley, "Selling out on nature," *Nature* 443 (2006): 27–28. At the end, I cite Francis Hallé one more time, from *In Praise of Plants*.

ILLUSTRATION NOTES

U nless directly credited, the photographs in this book were taken by the author. Until 2002, I exclusively used Kodachrome films, primarily Kodachrome 64, and several 35-mm SLR cameras. After 2002, when Kodachrome was removed from the market, I used Ektachrome VS-100 film and a Canon EOS SLR. Most photographs from the past two years were taken with a digital camera, a Nikon Coolpix 4500, and since October 2006 with a Coolpix 8800.

All of the figures in this book were crafted by Stacy West. In many cases, the information was obtained from specific sources, which are given in the notes below. The molecular structures were produced by Roberto Roa, from a variety of sources. I particularly used Trevor Goodwin, ed., *Chemistry and Biochemistry of Plant Pigments* (London: Academic Press, 1965); B. B. Buchanan, W. Guissem, and R. L. Jones, *Biochemistry and Molecular Biology of Plants* (Rockville, MD: American Society of Plant Physiologists, 2000); and Trevor Robinson, *The Organic Constituents of Higher Plants*, 4th ed. (North Amherst, MA: Cordus Press, 1980). These structures are also available in the Chemical Abstracts database (Sci-Finder) and on the Web at a variety of sites. Transmission-electron-microscope (TEM) photographs were taken and samples prepared by George Taylor of the Electron Microscope Facility in the Department of Biological Sciences at Florida International University (FIU), unless an article is cited as the source of the photographs. Scanning-electron-microscope (SEM) photographs were taken at the FIU Florida Center for Analytical Electron Microscopy (FCAEM) by Barbara Maloney. I frequently consulted D. J. Mabberley, *The Plant Book*, 2nd ed. (Cambridge: Cambridge University Press, 1997), which provided much information on nomenclature.

Certain words will be repeated frequently in these notes, and I am substituting the following abbreviations for brevity:

FIU   Florida International University
FTBG   Fairchild Tropical Botanic Garden
JPM   Jardin des Plantes, Montpellier, France
LM   Light-microscope photograph
SEM   Scanning-electron-microscope photograph
TEM   Transmission-electron-microscope photograph
TS   Transverse section

Chapter One: Coloring Our Bodies with Plants

1.2. *Left*, *Muscari comosum* (L.) Mill. (Liliaceae), photographed in Montpellier, France, 1/1978. *Center*, *Achillea millefolium* L. (Asteraceae), photographed near Ephrata, WA, 5/2004. *Right*, *Althaea rosea* L. (Malvaceae), photographed in Ephrata, WA, 8/2003.

1.3. *Left*, flowers of *Bixa orellana* L. (Bixaceae), photographed at FTBG, 10/2003. *Right*, fruits and seeds of the same, photographed at FTBG, 2/1992.

1.4. *Left*, flowers of *Lawsonia inermis* L. (Lythraceae), FTBG, 10/2003. *Right*, designs on the hands of Neha Batra created in the holy town of Kurukshetra, India, 11/2003. Photograph used by permission of Neha and her mother Neema.

1.6. *Left*, flower of *Papaver rhoeas* L. (Papaveraceae), photographed near Montpellier, France, 5/1978. *Right*, *Nymphaea nouchali* Burm. f. (Nymphaeaceae), photographed at JPM, 5/1978.

1.7. Detail of an Indian miniature, nineteenth century, purchased by the author in Jodhpur, Rajasthan, in 1985. The painting depicts the night that Lord Krishna was born to Devaki and Vasudev, as their eighth child while they were imprisoned. The thousand-headed snake is the bed of Lord Vishnu and protects the baby from the storm.

1.8. Illustration originally drawn by William Hodges, one of the artists on Cook's voyage, then engraved by J. Hall for the published record of the voyage: J. Cook, *A Voyage Towards the South Pole and Around the World Performed by His Majesty's Ships the Resolution and Adventurer in the Years 1772, 1773, 1774, and 1775*, 2 vols. (London: W. Strahan and T. Cadell, 1777). Additional information on this illustration is available in R. Joppien, *The Art of Captain Cook's Voyages* (Melbourne: Oxford University Press, 1985–88).

1.9. Tsukioka Yoshitoshi (1839–92) was a pupil of Kuniyoshi, and became known late in life for his printmaking, particularly his portrayals of the Japanese underworld. This print was one of a series of 28 *Famous Murders with Verse* (1866).

1.10. Tattoo, photographed, 5/2003.

1.11. *Rubia tinctorum* L. (Rubiaceae), photographed in the Padua Botanical Garden, Italy, 9/1978.

1.12. Ikat weaving purchased by Carol Lee in 1975, photographed 5/2004.

1.13. *Top*, weaver in Ubud, Bali, photographed 3/1975. *Bottom*, Minangkabau head cloth (late nineteenth century), purchased in Bukit Tinggi, Sumatra, Indonesia, in 1975, photographed 6/2004.

1.14. *Left*, flowers of *Haematoxylum campechianum* L. (Fabaceae), photographed at FTBG, 3/1992. *Right*, wood of the same species, photographed at Walter Goldberg's residence, Miami, 10/2003.

1.16. Illustration from G. Rosetti, *The Plictho. Instructions in the Art of the Dyers Which Teaches the Dyeing of Woolen Cloths, Linens, Cottons, and Silk by the Great Art as Well as by the Common*, translated from the first edition of 1548 by S. M. Edelstein and H. C. Borghetty (Cambridge, MA: MIT Press, 1969).

Chapter Two: Light, Vision, and Color

2.1. Peacock feathers (*Pavo cristatus* L.), photographed 7/1980.

2.8. Diagrams of constructive interference were adapted from D. W. Lee, "Iridescent blue plants," *American Scientist* 85 (1997): 56–63.

2.10. This solar spectrum, above the atmosphere and at sea level, was adopted from J. P. Kerr, G. W. Thurtell, and C. B. Tanner, "An integrating pyranometer for climatological observation stations and meso-scale networks," *Journal of Applied Meterology* 6 (1967): 688–94.

2.11. Eye evolution diagram adapted from J. R. Cronly-Dillon, "Origin of invertebrate and vertebrate eyes," in *The Evolution of the Eye and Visual System*, ed. J. R. Cronly-Dillon and R. L. Gregory (Boca Raton, FL: CRC Press, 1991), 15–51.

2.12. Eye diagram was redrawn from one created by Helga Kolb, seen on her Web site (http://www.webvision.med.utah.edu), and used in H. Kolb, "How the retina works," *American Scientist* 91 (2003): 28–35.

2.13. Data on sensitivities of visual pigments of various animals was primarily taken from T. H. Goldsmith, "The evolution of visual pigments and colour vision," in *Vision and Visual Dysfunction*, ed. P. Gouras, 62–89 (Boca Raton, FL: CRC Press, 1991); and D. M. Chen and T. H. Goldsmith, "Four spectral classes of cone in the retinas of birds," *Journal of Comparative Physiology* A 159 (1986): 473–79; and that of the bat from Y. Winter, J. López, and O. von Helversen, "Ultraviolet vision in a bat," *Nature* 425 (2003): 612–14.

2.15. This version of the 1931 chromaticity diagram of the CIE (Commission Internationale de l'Eclairage) was produced by Dr. Hong Zhang, and a further explanation is available at www.biyee.net. The two axes allow the precise determination of the exact hue in the color space.

Chapter Three: The Palette

3.1. I collected these common wildflowers in the Swift River Preserve, Petersham, MA, 8/2004. They are identified by number on the accompanying line drawing: 1—*Lobelia cardinalis* L. (Lobeliaceae); 2—*Phlox paniculata* L. (Polemoniaceae); 3—*Impatiens capensis* Boj. ex Baker (Balsaminaceae); 4—*Eupatorium dubium* Poir.

(Asteraceae); 5—*Spiraea alba* Du Roi (Rosaceae); 6—*Aster divaricatus* L. (Asteraceae); 7—*Solidago altissima* L. (Asteraceae); 8—*Chelone glabra* L. (Scrophulariaceae); 9—*Aster prenanthoides* Muhl. ex Willd. (Asteraceae).

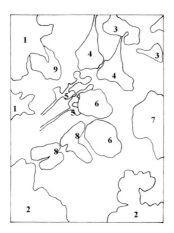

Key for wildflowers in figure 3.1.

3.5. *Bottom*, standard absorbance spectrum of chlorophyll *a* and *b*, adapted from W. G. Hopkins and N. P. A. Huner, *Introduction to Plant Physiology*, 3rd ed. (Hoboken, NJ: John Wiley and Sons, 2003).

3.6. *Left*, *Tricleocarpa cylindrica* (J. Ellis & Solander) Huisman & Borowiwtzka 1990, native to the Caribbean, and photographed by Diane and Mark Littler. Photograph is 3.5 cm across. For more information, see D. S. Littler and M. W. Littler, *Caribbean Reef Plants: An Identification Guide to the Reef Pants of the Caribbean, Bahamas, Florida, and Gulf of Mexico* (Washington, DC: Offshore Graphics, 2000). *Right, Nereocystis* sp. (Phaeophyta) was photographed near the Monterey Peninsula, CA, 10/1972. Stipes about 4 cm in diameter.

3.8. *Top, Balsamorhiza sagittata* (Pursh) Nutt. (Asteraceae), photographed at the Beezley Hills Preserve, a Nature Conservancy property in Grant County, WA, 5/2004. *Bottom*, absorbance spectrum obtained from reference in figure 3.5.

3.9. Absorbance ranges for various carotenoid pigments in this diagram were derived from R. Vetter, G. Englert, N. Rigassi, and U. Schwieter, "Spectroscopic methods," in *Carotenoids*, ed. O. Isler (Basel: Birkhäuser Verlag, 1971), 189–266.

3.10. The blue lobster (*Homarus americanus* H. Milne Edwards, 1836, class Crustacea) was photographed 9/2005 by Michael Tlusty, who is on the scientific staff of the New England Aquarium in Boston. This is a four-year-old male lobster fed on a diet low in carotenoid pigment. This diet produced a high portion of all astaxanthin pigment in the protein-bound form, and bright blue. After boiling, the lobster will turn bright red, when the astaxanthin pigment chromophore is freed from the protein.

3.13. *Left, Antirrhinum* 'Luminaire Harvest Red' (Scrophulariaceae) developed and photographed by Ball Horticultural Company. *Right, Butea monosperma* (Lam.) Kuntze (Fabaceae), photographed from a tree in my front yard, Miami, 3/2004.

3.14. These metabolic pathways are available in a variety of textbooks; this one was summarized from B. B. Buchanan, W. Guissem, and R. L. Jones, *Biochemistry and Molecular Biology of Plants* (Rockville, MD: American Society of Plant Physiologists, 2000).

3.15. Absorbance peaks and ranges of various flavonoids were adapted from T. Goodwin, ed., *Chemistry and Biochemistry of Plant Pigments* (London: Academic Press, 1976); and K. R. Markham, *Techniques of Flavonoid Identification* (London: Academic Press, 1982).

3.16. *Top left, Rosa gymnocarpa* Nutt. ex Torr. & Gray (Rosaceae), photographed at a spring west of Ephrata, WA, 6/1966. *Bottom,* the absorbance spectrum of cyanidin-3-glucoside is from J. B. Harborne, *Comparative Biochemistry of the Flavonoids* (London: Academic Press, 1967).

3.17. *Top left,* betanidin structure is from H. A. Stafford, "Anthocyanins and beta-lains: Evolution of the mutually exclusive pathways," *Plant Science* 101 (1994): 91–98. *Top right,* an unknown species of *Opuntia* (Cactaceae), photographed along a suburban street in Miami, 5/2004. *Bottom,* absorbance spectrum is from D. Strack and V. Wray, "Recent advances in betalain analysis," in *Caryophyllales: Evolution and Systematics,* ed. H.-D. Behnke and T. J. Mabry, 263–77 (Heidelberg: Springer-Verlag, 1994).

Chapter Four: The Canvas

4.1. *Left,* leaf undersurface pubescence of *Chrysophyllum oliviforme* L. (Sapotaceae) from a tree in my backyard in Miami, LM taken with both incident and transmitted illumination. Length of image 3.5 mm. *Right,* leaves and flowers of same plant, photographed 6/2005.

4.2. TEM of *Schefflera actinophylla* (Endl.) Harms (Araliaceae) leaf. Distance across cell is approximately 20 μm.

4.4. *Left,* leaves and inflorescences of *Salix caroliniana* Michx. (Salicaceae), photographed in the FIU Environmental Preserve, 1/2004. *Right,* undersurface detail is SEM. Width of field 120 μm.

4.5. *Capparis cynophallophora* L. (Capparaceae) is native to south Florida and was photographed in my yard. The top photograph includes the ripe fruit opened to reveal red pulp and black seeds. The leaf undersurface scales were photographed as in figure 4.1; the distance across the image is 3.5 mm.

4.7. This illustration was taken from G. E. Stahl, "Über bunte Laubblätter," *Annals of the Botanical Garden Buitenzorg* 13 (1896): 137–216. A, *Eranthemum cooperi* Hook. (Acanthaceae); B, *Piper porphyrophyllum* N. E. Br. (Piperaceae); C, *Begonia falcifolia* Hook. f. (Begoniaceae).

4.8. *Top, Peperomia rotundifolia* (L.) Kunth (Piperaceae), photographed on a branch fallen from a tropical rain-forest tree near Saül, French Guiana, 4/1978. Each leaf is about 6 mm in diameter. *Bottom, Lithops otzeniana* Nel (Aizoaceae) from South Africa; photographed in the greenhouse of JPM, 3/1978. Plants 2 cm across.

4.9. *Left, Ulva* sp. (Chlorophyta), photographed at the Kampong, on the shore of Biscayne Bay, Miami. Thallus is about 6 cm across. *Right,* paradermal LM of thallus; height of photograph about 480 μm.

4.10. The diagrams of the optical properties of sea lettuce are from unpublished data based on the methods in D. W. Lee and R. Graham, "Leaf optical properties of rainforest sun and extreme shade plants," *American Journal of Botany* 73 (1986): 1100–1108.

4.11. *Dieffenbachia ×picta,* probably a hybrid between *D. seguine* (Jacq.) Schott and *D. maculata* (Lodd.) G. Don f. (Araceae), photographed at a retail plant nursery in Miami, 2/2004. Leaves over 30 cm long; leaf TS 480 μm thick.

4.12. The leaf optics of dumb cane. Again, these are unpublished data based on the methods in the reference in figure 4.10.

4.13. TS of *Triolena hirsuta* Triana (Melastomataceae) collected at La Selva Research Station in Costa Rica. Leaf 160 μm thick. The microscope slide was used in D. W. Lee, R. A. Bone, S. L. Tarsis, and D. Storch, "Correlates of leaf optical properties in tropical forest sun and extreme-shade plants," *American Journal of Botany* 77 (1990): 370–80.

4.14. The pieces of dumb cane leaf (*Dieffenbachia ×picta*) are each about 4 cm across.

4.15. The photograph on the *right* is described in more detail in a review: C. Winefield, "The final steps in anthocyanin formation: A story of modification and sequestration," in *Anthocyanins in Leaves,* ed. K. S. Gould and D. W. Lee, Advances in Botanical Research 37 (London: Academic Press, 2002), 55–74. *Left, Eustoma* 'ABC 3-4 Purple', developed and photographed by Ball Horticultural Company.

4.16. These details of a leaf TS of *Triolena hirsuta* were derived from figure 4.13. Chloroplasts were drawn in to simulate their positions after movement.

4.17. This figure describing the Willstätter-Stoll model of leaf optics was taken from their classic work: R. Willstätter and A. Stoll, *Untersuchungen über die Assimilation der Kohlensäure* (Berlin: Springer-Verlag, 1918).

Chapter Five: Patterns

5.1. *Episcia cupreata* Hanst. (Gesneriaceae), photographed in the conservatory at FTBG, 9/2003. This plant is native to the understory of tropical rain forests in Latin America, and many color forms have been selected.

5.3. This diagram was redrawn from an illustration in the textbook W. K. Purves, D. Sadava, G. H. Orians, and H. C. Heller, *Life: The Science of Biology* (Sunderland, MA: Sinauer and Associates, 2003).

5.4. The diagram was adapted from T. A. Steeves and I. M. Sussex, *Patterns in Plant Development,* 2nd ed. (Cambridge: Cambridge University Press, 1989). *Upper right*, LM of radial section of stem apex of *Plectranthus scutellarioides* (L.) R. Br. (Lamiaceae). Image 900 μm wide. *Lower left*, LM radial image of shoot apex of *Ricinus communis* L. (Euphorbiaceae). Image width 1,020 μm. *Lower right*, SEM of partially dissected stem apex of *Artabotrys hexapetalus* (L. f.) Bhandari (Annonaceae). See U. Pozsluszny and J. B. Fisher, "Thorn and hook ontogeny in *Artabotrys hexapetalus* Annonaceae)," *American Journal of Botany* 87 (2000): 1651–70. Image 300 μm wide.

5.5. *Left*, *Cordyline fruticosa* (L.) Goeppert (Agavaceae), popular as a landscape plant in Miami and photographed in a nursery there, 1/2003. *Right*, *Hosta* sp. (Agavaceae), probably a complex hybrid cultivar, photographed in a residential garden in Cambridge, MA, 9/2004.

5.6. *Top*, *Fittonia verschaffeltii* (Lem.) Van Houte (Acanthaceae), tropical rainforest understory plant in Peru, photographed in the JPM conservatory, 10/1978. *Bottom*, hybrid vandaceous orchid *Ascocenda* 'Princess Mikasa, variety Indigo', photographed at RF Orchids south of Miami, 12/2003.

5.7. Broken tulip cultivar (*Tulipa* ×, complex hybrid [Liliaceae]), photographed by Allan Meerow, Poland.

5.8. *Left*, variegated form of *Pieris japonica* D. Don ex D. Don (Ericaceae), photographed in Bellevue, WA, 5/2004. *Right*, variegated variety of *Duranta erecta* L. (Verbenaceae), photographed at FTBG, 10/2003.

5.9. *Left*, flower of *Delonix regia* (Hook.) Raf. (Fabaceae), photographed in Miami, 6/2004. *Right*, variegated leaves of *Plectranthus* 'Kong Red Coleus' (Lamiaceae), developed and photographed by Ball Horticultural Company.

5.10. *Top*, variegated form of *Sonerila margaritacea* Lindl. (Melastomataceae), photographed at the Gardens of Marnier La Postelle, near Nice, France, 2/1978. *Bottom*, juvenile leaves of *Aframomum longipetiolatum* Koechl. (Zingiberaceae) are always variegated with round anthocyanic spots; photographed in La Makande, Gabon, 2/1999.

5.11. *Top*, *Aglaonema commutatum* Schott (Araceae). *Bottom*, *A. commutatum* 'Pewter'. Both photographed in a landscape nursery in Miami, 1/2003.

5.12. Photographs of *Zea mays* L. (Poaceae) grains. *Top*, 'Harvest' maize, crosses of Native American races. *Bottom*, mutant with transposable element, courtesy of Virginia Walbot. In the aleurone layer of these grains, removal of the *Mu1* transposable element from the *bronze2::mu1* allele produces the purple dots on the pink background, where the anthocyanins are concentrated in the cell vacuole. Otherwise they accumulate in the less acidic cytoplasm, producing the pinkish color. See V. L. Chandler and K. J. Hardeman, "The *Mu* elements of *Zea mays*," *Advances in Genetics* 30 (1992): 77–122; and G. N. Rudenko and V. Walbot, "Expression and post-transcriptional regulation of maize transposable element *MuDR* and its derivatives," *Plant Cell* 13 (2001): 553–70.

5.13. Cultivars of the flowers whose genetic controls are the best understood. *Left*, *Petunia* 'Wave Purple' (Solanaceae). *Right*, *Antirrhinum* 'Luminaire Sugarplum' (Scrophulariaceae). Both developed and photographed by Ball Horticultural Company.

5.14. Mutant of *Antirrhinum majus* L. (Scrophulariaceae). *Left*, the *pallida* mutant expressed in flowers. *Right*, petal where the *pallida* mutant was excised by switching to cool temperature. See A.-G. Rolland-Lagan, J. A. Bangham, and E. Coen, "Growth dynamics underlying petal shape and asymmetry," *Nature* 422 (2003): 161–65.

Chapter Six: Leaves

6.1. Leaves of *Schefflera actinophylla* (Endl.) Harms (Araliaceae), photographed in my neighborhood in Miami, 7/2003.

6.2. *Left*, *Schefflera actinophylla*, photographed in my neighborhood, 7/2003. *Right*, LM of TS of fresh leaf. Thickness 360 μm.

6.3. TS of *Schefflera actinophylla* leaf. Thickness 360 μm. This section was used in research cited in figure 4.13.

6.4. *Left*, TEM of palisade parenchyma cell of *Schefflera actinophylla* leaf. Cell approximately 20 μm across. *Right*, SEM of the undersurface of *S. actinophylla* leaf, showing the high density of stomata. Photographed at FIU, FCAEM. Image 200 μm wide.

6.5. Optical properties of *Schefflera actinophylla* leaf. Data from D. W. Lee, R. A. Bone, S. L. Tarsis, and D. Storch, "Correlates of leaf optical properties in tropical forest sun and extreme-shade plants," *American Journal of Botany* 77 (1990): 370–80.

6.6. *Left*, leaves of *Conocarpus erectus* L. (Combretaceae), photographed in my front yard in Miami, 6/2005. Leaves about 4 cm long. *Right*, SEM of the leaf surface of the silver form. Width 1.2 mm.

6.7. Leaves of *Cecropia peltata* L. (Cecropiaceae). *Top left*, large leaves, over 30 cm across, photographed at FTBG, 8/2003. *Top right*, juvenile leaves, photographed in the St. Elie Forest, French Guiana, 2/1978. *Bottom*, SEM of pubescence of adult leaf. Width of image 270 μm.

6.8. *Tillandsia recurvata* Baker (Bromeliaceae), photographed in my yard in Miami, 6/2005. Leaves about 5 cm long. *Right*, an SEM of the surface of these leaves, revealing the elegant scales. Image width 1.2 mm.

6.9. *Coccothrinax argentata* (Jacq.) L. H. Bailey (Arecaceae), native to the Florida Keys and Cuba, photographed on the FIU campus, 5/2004. Fronds about 1 m in diameter. *Right*, SEM detail of waxes on undersurface. Image 500 μm across.

6.10. Northrop Canyon is an ecosystem preserve in Grant County, WA. *Top*, view looks east toward the head of the canyon. *Bottom left*, *Artemisia tridentata* Nutt. (Asteraceae), photographed 8/2003. *Bottom right*, *Chrysothamnus nauseosus* (Pall.) Britton (Asteraceae), photographed 9/2005.

6.11. The forests on slopes beneath the summit plateau of Mount Kinabalu, Sabah, Borneo, at an elevation of 3,500 m; all photos taken 3/1974. *Top right*, tomentose form of *Leptospermum recurvum* Hook. f. (Myrtaceae). Leaves around 4 mm long. *Bottom right*, detail of hairs taken in dissecting microscope. *Bottom left*, glaucous blue leaf tips of *Schima wallichii* (DC.) Korth. (Theaceae). Leaves about 3 cm long. *Bottom center*, flowers of same species, 4 cm in diameter.

6.12. *Larrea divaricata* Cav. subsp. *tridentata* (DC.) Felger & Lowe (Zygophyllaceae), photographed at the Philip Boyd Deep Canyon Research Center, Palm Desert, CA, by Tom Philippi, 3/2004.

6.13. *Albizia lebbeck* (L.) Benth. (Fabaceae) of the Asian tropics, naturalized in south Florida and photographed in my neighbor's yard, 9/2003. *Left*, photographed at 3 p.m. *Right*, photographed at 10 p.m. Flowers 5 cm across.

6.14. *Calathea lutea* G. F. W. Met. (Marantaceae), photographed in pasture in Bocas del Toro, Panama, 6/2003. Leaves 2 m long. Unpublished data for the optical properties of the leaves use the methods of D. W. Lee and R. Graham, "Leaf optical properties of rainforest sun and extreme shade plants," *American Journal of Botany* 73 (1986): 1100–1108.

6.15. *Left, Begonia mazae* Ziesenh. (Begoniaceae), native to Amazonian region, photographed at FTBG, 8/1988. Leaves approximately 8 cm in diameter. *Right*, LM of TS of fresh leaf, photographed 8/1988. Thickness approximately 420 µm.

6.16. *Codiaeum variegatum* (L.) Blume (Euphorbiaceae). *Left*, leaf, photographed at FTBG, 4/2004. *Right*, fresh hand TS, photographed from same leaf.

6.17. Variegated leaves. *Top left*, prayer plant, *Maranta leuconeura* C. J. Morren (Marantaceae), understory plant of Amazonian rain forests, photograph by Francis Hallé in the Botanical Garden at Lyon, France. Leaf about 10 cm long. *Top right*, close-up of leaf surface of *M. leuconeura*, LM partially illuminated by incident light, from plants at JPM, 9/1978. Distance across photograph 1.0 mm. *Bottom left, Erythronium americanum* Ker Gawler (Liliaceae), photographed in the Gahanna Woods, outside of Columbus, OH, 4/1972. Leaves approximately 8 cm long. *Bottom right, Pilea cadierei* Gagnepain & Guillaumin (Urticaceae), native to Indochina, photographed at FTBG, 9/2003. Leaves 5 cm long.

6.18. *Byttneria aculeata* (Jacq.) Jacq. (Malvaceae), photographed in small forest gap at Barro Colorado Island, Panama, 5/1983. Leaves approximately 12 cm long.

6.19. *Top, Psychotria ulviformis* Steyerm. (Rubiaceae), photographed in understory of primary rain forest near Saül, French Guiana, 2/1978. *Bottom,* fresh leaf TS photographed from plants established in JPM, 9/1978. Leaf thickness 130 µm.

6.20. *Top left, Scindapsus pictus* Hassk. variety 'Argyraeus' (Araceae), small liana in tropical rain forests of Southeast Asia, photographed in Montpellier, France, 6/1978. Leaves 8 cm long. *Top right*, fresh leaf TS of same plant. Leaf 900 µm thick. *Center*, paradermal section of epidermal cells. Diameters about 70 µm. *Bottom*, SEM of leaves, provided by George Taylor, FIU.

6.21. Leaf optical properties of *Anthurium warocqueanum* Moore (Araceae). *Top left*, photographed in JPM, 9/1978. Portion of leaf shown is 15 cm across. *Bottom left*, leaf refraction by paradermal section. Width of image 1.1 mm. *Right*, model of cell optical properties from R. A. Bone, D. W. Lee, and J. N. Norman, "Epidermal cells functioning as lenses in leaves of tropical rainforest shade plants," *Applied Optics* 24 (1985): 1408–12.

6.22. *Monophyllaea patens* Ridl. (Gesneriaceae) and *Cyathodium foetidissimum* Schffn. (Marchantiaceae), photographed on the edge of an opening in the main sanctuary of Batu Cave, near Kuala Lumpur, Malaysia, 10/2005. Large leaf in the photograph is 4 cm long.

6.23. Popular houseplant, *Philodendron oxycardium* Schott (Araceae), purchased in a local nursery in Miami, photographed before and after 30 minutes of exposure to sun, 2/2004.

6.24. *Top,* these two photographs of an unknown species of *Saintpaulia* (Gesneriaceae) were sent to me in 1987 by B. L. Burtt, an expert on the gesneriads, particularly the understory herbs from the Old World tropics. *Bottom,* I took these photographs of *Selaginella serpens* (Desv. ex Poir.) Spring (Selaginellaceae) from plants in the FIU conservatory at 12 noon and 12 midnight. The leaves are about 2 mm long.

6.25. *Top left, Heliconia collinsiana* Griggs (Musaceae), photographed at FTBG, 1/2003. *Top right,* unknown species of *Agave* (Agavaceae), photographed at FTBG, 12/2002. *Bottom left, Warszewiczia coccinea* (Vahl) Klotzsch (Rubiaceae), photographed at FTBG, 7/2004. Bracts 5 cm long. *Bottom right, Musa coccinea* Andrews (Musaceae), photographed at the Fruit and Spice Park, south of Miami, 12/2003. These plants were provided by Chad Husby.

6.26. *Sarracenia leucophylla* Raf. (Sarraceniaceae), photographed at Weeks Reserve, near Mobile, AL, 7/2003. Leaves are approximately 25 cm tall.

6.27. This aerial photograph was taken by Richard Grotefendt for research conducted by Stephanie Bohlman, then a Ph.D. student at the University of Washington, 2003. Copyright Richard A. Grotefendt, Grotefendt Photogrammetric Services, Inc., P.O. Box 1794, North Bend, WA 98045, rich@envirophotos.com. The grid marks in the photograph are produced by this special camera to allow precise rectification of the image.

6.28. This northern Everglades satellite image map was produced by John Thomas Jones and Gregory Jean-Claude Desmond of the U.S. Geological Survey, Reston, VA. It is a composite image of bands 3 (red) and 4 (near-infrared) and a panchromatic band (green to near-infrared) acquired by the Landsat 7 ETM satellite on 2/05/2000.

Chapter Seven: Flowers

7.1. Closeup of flower lip of Blc 'Hillary Rodham Clinton', photographed at RF Nursery, south of Miami, 12/2003. Flower provided by Robert Fuchs.

7.2. *Top left, Solandra maxima* (Sessé & Mociño) P. Green (Solanaceae), photographed at FTBG, 2/2004. Flower is 15 cm long. *Top right, Bauhinia purpurea* L. (Fabaceae), photographed on FIU campus, 2/2004. Flowers 9 cm wide. *Bottom, Alstroemeria* sp., hybrid cultivar (Liliaceae), photographed at Miami flower vendor, 3/2004. Flowers 4 cm wide.

7.3. Drawing by Priscilla Fawcett, used by permission of FTBG, from L. Van der Pijl and C. Dodson, *Orchid Flowers: Their Pollination and Evolution* (Coral Gables, FL: Fairchild Tropical Garden / University of Miami Press, 1966).

7.4. Orchid flower anatomy and pigment distribution, of Blc 'Hillary Rodham Clinton'. *Top right*, petals, 4 cm across. *Middle right*, paradermal view, 1 mm across. *Bottom right*, LM of TS of petal, 450 μm across. *Bottom left*, lip TS, 500 μm across.

7.5. Orchid flowers (Orchidaceae). *Left*, Norman's Orchids earned first place for exhibits of 400 square feet at the 2004 Miami Orchid Show. *Top right*, *Cattleya aurantiaca* P. N. Don, photographed in our garden in Miami, 4/2004. Flowers 5 cm wide. *Bottom center*, *Orchis rotundifolia* Banks ex Pursh, photographed in Jasper National Park, Alberta, Canada, 8/1971. Flowers 1.5 cm across. *Bottom right*, *Ophrys apifera* Huds., photographed at Antibes, Côte d'Azur, France, 1/1978. Flowers about 1.5 cm across.

7.6. Three species of orchids involved in the production of 'Hillary'. *Top*, *Laelia purpurata* Lindl. & Paxton, photographed at Alpha Orchids, Miami, 10/2003. Flowers 9 cm across. *Bottom left*, *Cattleya labiata* Lindl., photographed by Tom Fennell II at Orchid Jungle in Homestead, FL, provided to me by Bill Peters (saved from the wreckage of Hurricane Andrew, 8/1992). *Bottom right*, *Rhyncholaelia digbyana* (Lindl.) Schltr., photographed at Whimsy Orchids, Homestead, with the help of Bill Peters, 4/2004. Flower 11 cm wide.

7.7. *Rosa* 'Double Delight', photographed in my mother's garden, Ephrata, WA, 8/2003. Flower 9 cm across.

7.8. *Pelargonium* ×*hybridum* (L.) L'Hérit (Geraniaceae) 'Designer Red Geranium' developed and photographed by Ball Horticultural Company. Individual flowers 4 cm in diameter.

7.9. *Pavonia multiflora* Juss. (Malvaceae), photographed at FTBG, 1/2004. Flowers 4 cm across. Sepals are the showy part of flower.

7.10. *Paeonia officinalis* L. (Paeoniaceae), photographed in JPM, 4/1978. Flowers 10 cm in diameter.

7.11. *Delphinium nuttallianum* Pritz. ex Walp. (Ranunculaceae), photographed at the Beezley Hills Reserve, Grant County, WA, 4/2004. Flowers about 2 cm across.

7.12. Anthocyanin absorbances for this diagram are from J. B. Harborne, "Functions of flavonoids in plants," in *Chemistry and Biochemistry of Plant Pigments*, ed. T. Goodwin (London: Academic Press, 1965), 736–78.

7.13. Betalain-containing flowers. *Left*, cultivars of *Bougainvillea* ×*spectabilis* (Nyctaginaceae), photographed at FTBG, 5/2004. Flowers 3.5 cm across. *Right*, Miller's pincushion cactus, *Pediocactus simpsonii* Britton & Rose, Beezley Hills Reserve, Grant County, WA, 4/2004. Flowers 3 cm in diameter.

7.14. *Left*, *Helianthus annuus* L. (Asteraceae), floricultural variety photographed at a street florist's display in Miami, 12/2002. *Right*, TS of ray flower, photographed 5/2004. Ray 250 μm thick.

7.15. *Eschscholzia californica* Cham. (Papaveraceae), photographed in home garden, Bellevue, WA, 9/2005. Flowers 3.5 cm in diameter.

7.16. *Cassia fistula* L. (Fabaceae), native to tropical deciduous forests of Southeast Asia. *Left*, street tree photographed in Miami, 6/2006. *Right*, flowers photographed from the same tree, 2 cm in diameter.

7.17. *Top left, Tagetes* 'Durango Bee' (Asteraceae), developed and photographed by Ball Horticultural Company. Flowers 4 cm across. *Top right*, paradermal view of ray flower, showing contribution of anthocyanins and carotenoids to color. Width of image 420 μm. *Bottom left*, TS of ray flower. Convex epidermal cells contain chromoplasts in cytoplasm and anthocyanins in central vacuole. Width of image 260 μm. *Bottom right*, TEM of tip of epidermal cell showing chromoplasts in cytoplasm and vacuole below. Width of image is 12 μm.

7.18. *Delonix regia* (Hook.) Raf. (Fabaceae). *Top left*, street tree in Miami, 5/2004. Tree about 5 m tall. *Top right*, detail of flag petal, approximately 1 cm across, 260 μm thick. *Bottom*, LM of TS (*left*) and paradermal section (*right*) of petal, revealing anthocyanins and chromoplasts below, photographed 5/2004. Distance across images is 200 μm.

7.19. Diagram depicting pigment interactions was adapted from T. Toto and T. Kondo, "Structure and molecular stacking of anthocyanins: Flower color variation," *Angewandte Chemie* 30 (1991): 17–33.

7.20. Flowers in which blue colors are caused by the modification of anthocyanins. *Top left, Hydrangea macrophylla* (Thunb.) Ser. (Hydrangeaceae), photographed in garden in Cambridge, MA, 9/2004. Flowers 2 cm in diameter. *Top right, Commelina communis* L. (Commelinaceae), photographed in garden in Petersham, MA, 9/2004. Flowers 2 cm wide. *Bottom left, Plumbago auriculata* Lam. (Plumbaginaceae), photographed from plants in my neighborhood, 6/2004. Flowers 1.5 cm in diameter. *Bottom right, Centaurea cyanus* L. (Asteraceae), photographed in garden, Ephrata, WA, 9/2005. Flowers 2 cm wide.

7.21. *Left* and *top right, Strongylodon macrobotrys* A. Gray (Fabaceae), photographed at FTBG, 3/2003. Flowers about 4 cm long. *Bottom right, Dendrobium* 'Peng Seng' (Orchidaceae), photographed at FTBG, 5/2006. Flowers about 3.5 cm across.

7.22. *Top left, Hibiscus tiliaceus* L. (Malvaceae), photographed on FIU campus, 12/3. Flowers 7 cm in diameter. *Top right*, paradermal LM of petals of old flower. Width of photograph 360 μm. *Bottom, Brunfelsia australis* Benth. (Solanaceae), photographed in neighbor's yard, Miami, 11/2003. Flowers 3 cm across.

7.23. *Top left, Anemone coronaria* L. (Ranunculaceae), photographed in JPM, 4/1978. Flowers 3 cm across. *Top right, Crocus sativus* L. (Iridaceae), photograph courtesy of Philippe Latour, www.aromatique.com. *Bottom left, Echinocereus engelmannii* (Engelm.) Lem. (Cactaceae), photographed in Sonoran Desert 70 km north of Phoenix, AZ, 5/2006. Flowers about 7 cm across. *Bottom right, Medinilla* sp. (unidentified, one of some 400 species in the Old World tropics [Melastomataceae]),photographed at FTBG, 5/2006. Individual flowers about 2 cm across.

7.24. LM of TS of flower petals. *Top left, Tagetes patula* L. (Asteraceae), photographed 11/2003. Ray thickness is about 250 μm. *Top right, Bauhinia purpurea* L. (Fabaceae), photographed 3/2004. Petal thickness is 350 μm. *Bottom left, Ascocenda*

'Princess Mikasa' (Orchidaceae), photographed 12/2003. Petal thickness is 170 μm. *Bottom right, Butea monosperma* (Lam.) Kuntze (Fabaceae), photographed 3/2004. Petal thickness is 170 μm.

7.25. *Top left, Durio zibethinus* Murray (Malvaceae), photographed in Ulu Langat, Malaysia, 1/1974. Flowers 4 cm across. *Bottom left, Eonycteris spelea*, photographed 1974, photographer unknown. *Right, Kigelia pinnata* L. (Bignoniaceae), photographed at FTBG, 5/2004. Flowers 7 cm long.

7.26. *Left,* bee pollination of *Cattleya,* illustrated by Priscilla Fawcett, used by permission of FTBG, from L. Van der Pijl and C. Dodson, *Orchid Flowers: Their Pollination and Evolution* (Coral Gables, FL: Fairchild Tropical Garden / University of Miami Press, 1966). *Right, Cattleya bowringiana* Veitch (Orchidaceae), photographed at Alpha Orchids, Miami, 10/2003. Flower 7 cm across.

7.27. *Top, Stapelia gigantea* N. E. Br. (Asclepiadaceae), photographed at FTBG, 5/2004. Flowers 7 cm across. *Bottom, Amorphophallus paeoniifolius* (Dennst.) Nicolson (Araceae), photographed in my front yard in Miami, 5/2004. Inflorescence 25 cm wide.

7.28. Ultraviolet reflectance by flowers. *Top, Hedychium coronarium* J. König (Zingiberaceae). Flowers 5 cm across. *Bottom, Acmella pilosa* R. K. Jansen (Asteraceae). Flower heads 2 cm wide. All photographs by Scott Zona at FTBG in 2003. He used a Minolta Dimage 7Hi digital camera with a standard Kodak 18-A filter, giving a sensitivity range of 350–400 nm.

7.29. *Left, Costus malortieanus* var. *amazonicus* Loes. *Right, Costus barbatus* Suesseng. (Zingiberaceae). Both photographed at FTBG, 5/2006. Flowers approximately 4 and 3 cm long, respectively.

7.30. Pollination and flower color. *Top left, Ipomopsis aggregata* (Pursh) V. E. Grant (Polemoniaceae), photographed by Nanci Morin at the Arboretum at Flagstaff, near Flagstaff, AZ, 5/2004. *Top right, Raphanus sativus* L. (Brassicaceae), photographed by Jeff Abbas in San Joaquin County, CA. *Bottom left,* small field of color forms of *R. sativus,* photographed near Davis, CA, by Sharon Strauss (see S. Y. Strauss, R. E. Irwin, and V. M. Lambrix, "Optimal defence theory and flower petal colour predict variation in the secondary chemistry of wild radish," *Journal of Ecology* 92 [2004]: 132–41). *Bottom right, Mimulus luteus* L. (Scrophulariaceae), photographed by Rodrigo Medel at Los Pelumbres, east of Salamanca, Chile (3,730 m), 2000 (see details in R. Medel, C. Botto-Mahan, and M. Kalin-Arroyo, "Pollinator-mediated selection on the nectar guide phenotype in the Andean monkey flower, *Mimulus luteus,*" *Ecology* 84 [2003]: 1721–32).

7.31. Pollination syndromes in monkey flowers (Scrophulariaceae) in the Sierra Nevada of California. *Top left, Mimulus cardinalis* Dougl. ex Benth., photograph by Jeff Abbas, Horseshoe Bend Recreation Area, Mariposa County, CA, 9/2001. *Top right,* pollination by hummingbird, photograph by Doug Schemske. *Bottom right, Mimulus lewisii* Pursh, photograph by Jeff Abbas, Alpine County, CA, 7/2001. *Bottom left,* pollination by bumblebee, photograph by Doug Schemske. For photo documentation by Abbas see http://elib.cs.berkeley.edu/photos/flora/. For documentation by

Schemske, see H. D. Bradshaw, S. M. Wilbert, K. G. Otto, and D. W. Schemske, "Genetic mapping of floral traits associated with reproductive isolation in monkey-flowers (*Mimulus*)," *Nature* 376 (1995): 762–65; H. D. Bradshaw, K. G. Otto, B. E. Frewen, J. K. McKay, and D. W. Schemske, "Quantitative trait loci affecting differences in floral morphology between two species of monkey flower (*Mimulus*)," *Genetics* 149 (1998): 367–92; and D. W. Schemske, "Pollinator preference and the evolution of floral traits in monkeyflowers (*Mimulus*)," *Proceedings of the National Academy of Sciences U.S.* 96 (1999): 11910–15.

Chapter Eight: Fruits and Seeds

8.1. This apple box label was loaned to me by my cousin Dale Easley, who grew up in Okanogan, WA, an apple-producing area, and who has developed an appreciation for the beauty and historical value of these labels. They document a vanished era, of which we both have early memories.

8.2. These antique apple varieties were collected by the Worcester Horticultural Society, now located at the Tower Hill Botanic Garden, Boylston, MA. I have added a commercial grocery fruit of 'Red Delicious' for contrast. 1, 'Roxbury'—MA, 1630; 2—'Lady', France, 1628; 3—'Pomme Grise', ?, 1800; 4—'Tolman Sweet', MA, 1822; 5—'Winter Banana', IN, 1876; 6—'Palouse', WA, 1879; 7—'Red Delicious', IA, 1888; 8—'Jonathan', NY, 1826; 9—'Oliver', AR, 1831; 10—'Opalescent', OH, 1899; 11—'Red Canada', US, 1822; 12—'Grimes', WV, 1804; 13—'Porter', MA, 1800; 14—'Rhode Island Greening', RI, 1650; 15—'Malinda', VT, 1860; 16—'Yellow Bellflower', NJ, 1800; 17—'Winter Pearmain', England, 1600s. These were photographed during harvest, 10/2004. Note that a 'Yellow Bellflower' was the source tree for the origin of the 'Red Delicious'. Seeds from varieties that originated prior to the nineteenth century would have been collected by Johnny Appleseed in his voyage west.

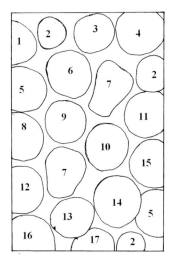

Key for apples in figure 8.2.

8.3. The size of each of these images is approximately 1.5 mm from top to bottom.

8.4. *Top,* a *Fragaria* ×*ananassa* variety (hybrid between *F. chiloensis* Duchesne and *F. virginiana* Duchesne [Rosaceae]), from local grocery, 8/2004. "Fruits" up to 3.5 cm long. *Bottom,* an individual achene on surface of receptacle, LM using incidental illumination. Achene less than 1 mm long.

8.5. *Capsicum annuum* L. (Solanaceae), from local grocer, photographed 1/2004. *Top right,* LM of paradermal section of skin of green bell pepper. Image is 400 μm across. *Bottom right,* LM of paradermal section of red bell pepper. Image 150 μm across.

8.6. *Left, Solanum lycopersicum* L. (Solanaceae). *Right,* paradermal LM view through tomato skin, photographed 9/2004, digital image. Image 260 μm across.

8.7. *Solanum melongena* L. (Solanaceae). *Top right,* paradermal LM through skin. Image is 600 μm across. *Bottom right,* LM of TS through skin and cortex of fruit, photographed Miami, 3/2004. Images 800 μm across.

8.8. *Citrus reticulata* Blanco (Rutaceae). *Left,* fruits on tree in Miami, 1/2004. Fruit 4 cm across. *Right,* LM of TS of fruit skin. Image 380 μm across.

8.9. Squashes. Numbers on outline of figure indicate the varieties present, belonging to the species *Cucurbita maxima* Duchesne and *C. moschata* Duchesne. This photograph, and the information on varietal identity, was provided by Deena Decker-Walters. For more information on identifying fruits see www.cucurbit.org. Varietal identification: *C. maxima:* 4—Hubbard; 5—'Boston Marrow'; 7—'Turk's Turban'; 8—buttercup; 9—'Golden Delicious'. *C. moschata:* 1—'Vegetable Spaghetti'; 2—field pumpkin; 3—'Cream of the Crop Hybrid'; 6—crookneck squash; 10—'Bicolor Pear'; 11—'Striped Pear'; 12—acorn; 13—'Table Gold'; 14—zucchini. *Bottom,* LM of TS through rind of summer squash. Image is 1.04 mm across.

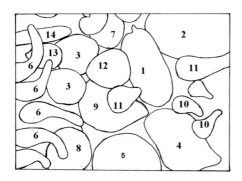

Key for squashes in figure 8.9.

8.10. *Top, Erythrina herbacea* L. (Fabaceae), photographed at the Gifford Arboretum, University of Miami, 2/2004. Flowers 4 cm long. *Bottom: left, Abrus precatorius* L. (Fabaceae); *center, Ricinus communis* L. (Euphorbiaceae); *right, Adenanthera pavonina* L. (Fabaceae), photographed in Miami, 11/2003, from seeds collected locally. Distance bottom to top is 7 cm.

8.11. *Top*, inflorescence and seeds of *Heliconia stricta* Huber (Musaceae), photographed at FTBG, 5/2005. Red bracts about 8 cm long. *Bottom left*, *Heliconia mariae* Hook. f., photographed at La Selva Research Station, Costa Rica, 5/1990. Seeds are 7 mm long. *Bottom right*, *Zamia furfuracea* L. f. (Zamiaceae), photographed in my yard in Miami, 12/2002. Seeds 1.3 cm long.

8.12. *Guaiacum officinale* L. (Zygophyllaceae), digital photograph taken at FTBG, 3/2003. Flowers 2.3 cm across.

8.13. *Top left*, *Pithecellobium keyense* Britton (Fabaceae), digital photograph taken at FTBG, 4/2004. Seeds about 7 mm across. *Top right*, *Acacia auriculiformis* A. Cunn. ex Benth. (Fabaceae), digital photograph of pods collected in my neighborhood, Miami, 4/2004. Seeds 6 mm across. *Bottom left*, *Myristica fragrans* Houtt. (Myristicaceae), photographed by J. E. Armstrong in Kerala, India, 1982. Fruit is 4 cm long. *Bottom right*, *Ravenala madagascariensis* Sonn. (Strelitziaceae), fruits obtained from neighbor's yard, Miami, digital photograph, 6/2003. Distance across aril-covered seeds is 6 mm.

8.14. *Top, Phaseolus vulgaris* L. (Fabaceae). *Bottom left*, Native American seeds obtained from nativeseeds.org: *top* from *left* to *right*, 'Tarahumara Ojo de Cabra'; 'Zuni Shalako'; 'Appaloosa'; *bottom* from *left* to *right*, 'Mountain Pima Vapai Bavi'; 'Hopi Purple String Bean'; 'Hopi Black Pinto'. *Bottom right*, standard cultivars: *top* from *left* to *right*, white; kidney; black; *bottom* from *left* to *right*, pinto; navy.

8.15. *Left*, *Mangifera indica* L. (Anacardiaceae), variety 'Glen', photographed in my yard in Miami, 5/2004. Fruits 11 cm long. *Top right*, *Pouteria sapota* (Jacq.) H. Moore & Stearn (Sapotaceae), photographed from neighbor's tree, Miami, 6/2004. Fruits 12 cm long. *Bottom right*, *Durio zibethinus* Murray (Malvaceae), photographed in Ulu Langat, Malaysia, 7/1975. Fruit 20 cm long.

8.16. *Top left*, *Schinus terebinthifolius* Raddi (Anacardiaceae), photographed at FIU Environmental Preserve, 12/2002. Fruits about 3 mm across. *Top right*, *Lasianthus attenuatus* Jack (Rubiaceae), Menglun Reserve, Yunnan, China, fruit 6 mm across. *Bottom left*, *Rosa gymnocarpa* Nutt. ex Torr. & Gray (Rosaceae), photographed near San Poil River, WA, 8/2003. Hips 1.3 cm long. *Bottom right*, *Sorbus cashmiriana* Hedl. (Rosaceae), photographed at the Bloedel Reserve, Bainbridge Island, WA, 11/1994. Fruits 6 mm across.

8.17. *Top left*, *Tabernaemontana orientalis* G. Don (Apocynaceae), photographed at FTBG, 12/2002. Flowers 2.5 cm across. *Top right*, *Aechmea woronowii* Harms (Bromeliaceae), photographed at FTBG, 11/2003. Fruits 1.2 cm across. *Center left*, *Chamaedorea seifrizii* Burret (Arecaceae), photographed at FTBG, 9/2003. Fruits 4 mm across. *Center right*, *Cola digitata* Mast. (Malvaceae), photographed by Francis Hallé, Ivory Coast, 1963. *Bottom left*, fruit of simpoh air (*Dillenia suffruticosa* [Griff.] Martelli [Dilleniaceae]), seeds quickly dispersed by birds after the fruits explode open, photographed at Rimba Ilmu, Univeriti Malaya, Petaling Jaya, Malaysia, 10/2005. *Bottom right*, *Polygala oreotrophes* B. L. Burtt (Polygalaceae) in understory of upper montane forest, photographed at Gunong Ulu Kali, 2,200 meters, Selangor, Malaysia, 11/2005.

8.18. *Left*, drawing by Francis Hallé of fruits of a *Manotes* sp. (Connaraceae) from living material in the Ivory Coast, 1965. *Top right*, *Harpullia pendula* Planchon ex F. Muell. (Sapindaceae), photographed at FTBG, 12/2002. Fruits 3.5 cm long. *Bottom right*, *Blighia sapida* König (Sapindaceae), cultivated in West Indies but native to West Africa, photographed from neighbor's yard, Miami, 9/2003. Fruits 6.5 cm wide.

8.19. *Top*, *Clerodendron minahassae* Teijsm. & Binn. (Verbenaceae), photographed at FTBG, 7/2004. Fruits 5 mm across. *Bottom*, pigments separated on filter paper.

8.20. *Left*, *Actaea pachypoda* Elliott (Ranunculaceae), Petersham, MA, 9/2004. Fruits 6 mm across. *Right*, *Sassafras albidum* (Nutt.) Nees (Lauraceae), photographed by Ken Robertson of the Illinois Natural History Survey, Coles County, IL, 9/1987. Fruits 7 mm across.

Chapter Nine: Stems and Roots

9.1. *Pinus ponderosa* Douglas ex Lawson & P. Lawson (Pinaceae), photographed along San Poil River south of Republic, WA, 8/2003. Bark 20 cm across.

9.2. *Left*, *Eucalyptus deglupta* Blume (Myrtaceae), photographed at FTBG, 12/2002. Bark 17 cm across. *Right*, *E. torelliana* F. Muell., photographed at FTBG, 12/2002. Bark 20 cm across.

9.3 *Top left*, *Ricinus communis* L. (Euphorbiaceae), wasteland in Miami, 1/2004. *Top right*, LM of TS of young stem with primary growth, 700 μm across. *Bottom left*, old stem, 10 cm in diameter. *Bottom right*, LM of TS of older stem with secondary growth, 2.4 mm across. Microscope slides from Turtox.

9.4. *Left*, *Amorphophallus paeoniifolius* (Dennst.) Nicolson (Araceae), photographed at FTBG, 1/2003. Petiole 5 cm in diameter. *Right*, *Didierea trollii* Capuron & Rauh (Didiereaceae), photographed at FTBG, 12/2002. Stem 6 cm in diameter.

9.5. *Left*, *Gigantochloa atroviolacea* E. A. Widjaja (Poaceae), photographed in my yard in Miami, 12/2002. Culm 5 cm in diameter. *Right*, *Cyrtostachys renda* Blume (Arecaceae), photographed at FTBG, 8/2003. Stems 5 cm in diameter.

9.6. Palm trunks (Arecaceae). *Top left*, *Phoenix canariensis* Hort. ex Chabaud, photographed at FTBG, 12/2002. Image 35 cm across. *Top right*, *Roystonea regia* (Kunth) Cook, photographed at FTBG, 9/2003. Image 20 cm across. *Bottom left*, *Pseudo-phoenix vinifera* (Mart.) Becc., photographed at FTBG, 1/2003. Image 16 cm across. *Bottom right*, *Daemonorops ochrolepis* Becc., FIU conservatory, 1/2003. Culm 4.5 cm in diameter.

9.7. *Bursera simaruba* (L.) Sarg. (Burseraceae). *Top left*, trunk 40 cm wide, photographed at FIU campus, 12/2002. *Top right*, dark-field LM of TS of young stem, image 2.5 mm wide; microscope slide prepared by David Storch. *Bottom left*, LM of TS of living stem, image 500 μm wide, 11/2003. *Bottom right*, bark detail 5 cm across, photographed at FIU campus, 12/2002.

9.8. *Top left*, *Ceiba insignis* (Kunth) Gibbs & Semir (Malvaceae), photographed at FTBG, 12/2002. Image 15 cm across. *Top right*, *Lysiloma sabicu* Benth. (Fabaceae), photographed at FTBG, 12/2003. Image 25 cm across. *Middle right*, *Melaleuca quinquenervia* (Cav.) S. T. Blake (Myrtaceae), photographed in Miami neighborhood, 10/2003. Image 25 cm across. *Bottom left, Quercus virginiana* L. (Fagaceae) with *Polypodium polypodioides* (L.) Watt. (Polypodiaceae) as epiphyte, photographed at FTBG, 12/2002. Image 22 cm across. *Bottom right, Arbutus menziezii* Pursh (Ericaceae), photographed at Bellevue, WA, 5/2004. Image 20 cm across.

9.9. *Top,* Crestone, CO, with the Sangre de Cristo Mountains in background. *Bottom left,* trunk of *Populus tremuloides* Michx. (Salicaceae). Trunk 18 cm across. *Bottom right,* trunk of *Populus trichocarpa* Torr. & A. Gray ex Hook. (Salicaceae). Image 15 cm across. All photos 3/2004.

9.10. *Tilia americana* L. (Tiliaceae). *Top left,* trunk of young tree with buds, photographed at Denver Botanical Garden, CO, 3/2004. *Right,* dark-field LM of TS of young stem. Distance across image 2.4 mm. *Bottom left,* TS of young stem. Distance across image 690 µm. Both microscope slides from Turtox.

9.11. *Left,* trunks of (*left*) *Thuja plicata* Donn ex D. Don (Cupressaceae) and (*right*) *Pseudotsuga menziesii* (Mirbel) Franco (Pinaceae), photographed at Deception Falls Preserve, WA, 5/2004. *Top right, P. menziesii* bark, photographed at Icicle Creek, Leavenworth, WA, 8/2003. Trunk 1 m in diameter. *Bottom right,* detail of *T. plicata* bark, photographed at Deception Falls, 5/2004. Image 15 cm wide. I learned much of the habits of wolf lichen (*Letharia vulpina* [L.] Hue) during a field trip in July 2006 accompanied by Susanne Altermann, who was studying the habitat preferences and genetic variation of the lichen for her Ph.D. dissertation at the University of California, Santa Cruz.

9.12. *Left, Ficus microcarpa* L. f. (Moraceae), photographed at FIU campus, 9/2003. Image 10 cm wide. *Top right, Clusia rosea* Jacq. (Clusiaceae), photographed at FIU campus, 9/2003. Image 20 cm wide. *Bottom right, Hevea brasiliensis* L. (Euphorbiaceae), photographed in Negeri Sembilan, Malaysia, 12/1974.

9.13. *Left, Staudtia gabonensis* Warb. (Myristicaceae), photographed at La Makande, Gabon, 2/1999. Image 20 cm wide. *Center, Dracaena draco* (L.) L. (Agavaceae), FTBG, 10/2003. Stem 10 cm across. *Right, Sanguinaria canadensis* L. (Papaveraceae), digital photograph from Ion Exchange Native Seed and Plant Nursery, Allamakee County, IA, by Jeff Abbas. Flower 5 cm across.

9.14. Diagram of resin ducts reprinted from D. M. Joel, "Resin ducts in the mango fruit: A defence system," *Journal of Experimental Botany* 31 (1980): 1707–18.

9.15. *Left, Metopium toxiferum* (L.) Krug & Urban (Anacardiaceae), photographed at Montgomery Botanical Center, Miami, 12/2003. Distance across photo 20 cm. *Right,* antique Japanese lacquer bowl, purchased by my wife's uncle Ed Rotsinger during the occupation after World War II.

9.16. Hardwoods, photographed from the FTBG wood collection, 5/2004. *Left, Tectona grandis* L. f. (Verbenaceae). *Center, Dalbergia sissoo* Roxb. ex DC (Fabaceae). *Right, Juglans nigra* L. (Juglandaceae). Distance across photo 16 cm.

9.17. Root vegetables from local grocer, photographed 1/2004. *Top left, Solanum tuberosum* L. (Solanaceae). *Top right*, paradermal LM of purple heritage potato skin. Image 480 μm across. *Bottom left*, *Beta vulgaris* L. (Chenopodiaceae). *Bottom right*, *Daucus carota* L. (Apiaceae).

9.18. *Left,* processing rhizomes at a turmeric trader's in Cochin, Kerala, India, 3/1989. *Right*, close-up of *Curcuma longa* L. (Zingiberaceae).

9.19. *Top*, *Sequoia sempervirens* (D. Don) Endl. (Taxodiaceae), photographed at Redwood National Park, CA, 8/1969. Trunk 3 m in diameter. *Bottom, Pinus ponderosa* Douglas ex Lawson & P. Lawson (Pinaceae), photographed at Icicle Creek, WA, 8/2003. Base 2 m across.

Chapter Ten: Iridescent Plants

10.1. *Top, Trogonoptera brookiana* Wallace, photographed at a butterfly farm, Cameron Highlands, Malaysia, 7/1993. *Bottom,* butterfly scales photographed by LM at unknown magnification, Malaysia, 1975.

10.2. *Selaginella willdenowii* (Desv.) Bak. (Selaginellaceae), photographed at FTBG, 10/2003. Anole is 6 cm long, including tail.

10.3. *Top left, Selaginella willdenowii*, with drop of water. Photograph about 5 cm wide. *Top right*, TS by TEM of leaf, by Charles Hébant (see C. W. Hébant and D. W. Lee, "Ultrastructural basis and developmental control of blue iridescence in *Selaginella* leaves," *American Journal of Botany* 71 [1984]: 216–19). TS about 140 μm thick. *Bottom left*, surface of leaf, LM with incident illumination, photographed in Malaysia, 1976. Image about 3 mm across. *Bottom right*, TS of outer cell wall by TEM, 1976. Two layers near surface each are about 70 nm thick.

10.4. *Left, Picea engelmannii* Parry (Pinaceae), photographed in yard, Ephrata, WA, 5/2004. *Right*, SEM of needles from this tree. Image 800 μm wide.

10.5. *Danaea nodosa* (L.) Small (Marattiaceae). *Top*, blue juvenile frond and larger adult frond of plant, photographed at La Selva Research Station, Costa Rica, 5/1984. *Bottom left*, TEM of TS of outer cell wall and cytoplasm of epidermis. Image is 6 μm wide. *Bottom right*, TEM of TS of cell wall. Image is 2.9 μm top to bottom.

10.6. Diagram of helicoidal thickening adapted from D. W. Lee, "Iridescent blue plants," *American Scientist* 85 (1997): 56–63.

10.7. *Diplazium tomentosum* Blume (Athyriaceae); *Phyllagathis rotundifolia* Blume (Melastomataceae). Collected together in the Bukit Lanjang Forest Reserve, near Kepong, Malaysia, photographed 7/1993. *Bottom left*, LM of TS of leaf of *Diplazium tomentosum*. About 130 μm thick. *Right*, detail of helicoidal thickening in epidermal cell wall. Vertical distance is 2 μm. For these photographs see K. S. Gould and D. W. Lee, "Physical and ultrastructural basis of blue leaf iridescence in four Malaysian understory plants," *American Journal of Botany* 83 (1995): 45–50.

10.8. *Top, Dictyota mertensii* (Martins) Kützing (Phaeophyta); *bottom, Ochtodes secundiramea* (Montague) M. Howe (Rhodophyta); both iridescent blue algae native to the Caribbean, photographed by Diane and Mark Littler. Photographs are 9 and

38 cm across, respectively. For more information, see D. S. Littler and M. W. Littler, *Caribbean Reef Plants: An Identification Guide to the Reef Plants of the Caribbean, Bahamas, Florida, and Gulf of Mexico* (Washington, DC: Offshore Graphics, 2000).

10.9. *Top, Begonia pavonina* Ridl. (Begoniaceae), photographed with *Selaginella willdenowii* and *Diplazium tomentosum*, in the Bukit Lanjang Forest Reserve, Malaysia, 1976. *Middle left*, LM of TS of leaf 220 μm thick. *Middle right*, epidermal surface with convex curvature and flecks of blue color coming from within. Cells are about 60 μm in diameter. *Bottom left*, TEM of iridoplast; 1.6 μm top to bottom. *Bottom right*, LM of TS of fresh leaf section, showing chloroplasts in palisade cells and iridoplasts in epidermal cells. Leaf about 230 μm thick. For technical details of micrograph, see article cited in note for figure 10.7.

10.10. Ultrastructure of *Phyllagathis rotundifolia*. *Top left*, LM of leaf TS showing positions of iridoplasts in epidermal cells. Leaf 230 μm thick. *Top right*, TEM detail of iridoplast showing thylakoids pressed together; 1 μm top to bottom. *Bottom*, TS of entire iridoplast by TEM. Structure is 1.5 μm thick. For technical details of micrographs, see article cited in note for figure 10.7.

10.11. *Trichomanes elegans* Rich. (Hymenophyllaceae). *Top left*, entire plant, about 50 cm high; *bottom left*, leaf detail, photographed near Saül, French Guiana, 2/1978. Microscopic details of the living plant, photographed at La Selva Research Station, Costa Rica, 5/1982: *Top right*, surface of frond, showing iridescence by chloroplasts; 270 μm across image. *Bottom center*, paradermal view of cells with transmitted illumination; 170 μm across image. *Bottom right*, detail of frond; 2 mm across.

10.12. Ultrastructural details of *Trichomanes elegans*. *Left*, outer epidermal wall with modified chloroplasts adjacent to it. Height is 9.8 μm. *Right*, detail of chloroplast. Height is 2.7 μm. For details, see R. M. Graham, D. W. Lee, and K. Norstog, "Physical and ultrastructural basis of blue leaf iridescence in two Neotropical ferns," *American Journal of Botany* 80 (1993): 198–203.

10.13. *Elaphoglossum herminieri* (Bory ex Fée) T. Moore (Lomariopsidaceae), collected from Saül, French Guiana, 2/1978. *Left*, entire frond, 25 cm long. *Center*, detail of fronds, approximately 2 cm wide. *Right*, LM of TS of fresh hand section, revealing chloroplasts and structures in epidermal cells (iridoplasts?). Image height is 260 μm.

10.14. *Elaeocarpus angustifolius* Blume (Elaeocarpaceae), collected and photographed at Chapman Field USDA Station in Miami, 10/1989. *Top left*, branch with fragrant flowers and tiny juvenile fruits. Flowers about 1 cm long. *Top right*, Indian sadhu outside of Samadhi Shrine of Swami Nityananda, Ganeshpuri, Maharashtra, India, 7/1989. *Bottom left*, ripe iridescent blue fruits, photographed by George Valcarce and Michael Upright, 7/1990. Fruit about 2.2 cm in diameter. *Bottom right*, light reflected from fruit surface, 7/1990. Image 3.5 mm wide.

10.15. Details of *Elaeocarpus angustifolius* fruit. *Left*, LM of fresh TS of outer fruit wall. Height of image about 150 μm. *Right*, TEM of TS of epidermal cell, reveal-

ing iridescent structure with multiple layers; 30 μm top to bottom. For details, see
D. W. Lee, "Ultrastructural basis and function of iridescent blue colour of fruits in
Elaeocarpus," *Nature* 349 (1991): 260–62.

10.16. Iridescent blue fruits of *Delarbrea michieana* (F. Muell.) F. Muell. (Araliaceae) from Queensland, Australia. *Top left*, whole fruits, 1.7 cm long, photographed
5/2000. *Top right*, LM of TS of fruit skin, showing approximate position of iridosomes; 300 μm thick. *Bottom left*, TEM of iridosome and nearby chloroplasts; 6 μm
top to bottom. *Bottom right*, detail of iridosome, showing interference layers; top to
bottom 500 nm. For more details of these images, see D. W. Lee, G. T. Taylor, and
A. K. Irvine, "Structural coloration in *Delarbrea michieana* (Araliaceae)," *International
Journal of Plant Sciences* 161 (2000): 297–300.

10.17. This figure was adapted from D. W. Lee, "Iridescent blue plants," *American
Scientist* 85 (1997): 56–63.

Chapter Eleven: Why Leaves Turn Red

11.1. *Left*, *Adiantum* sp., photographed near Saül, French Guiana, 2/1978. *Right*,
the young leaves of *Zamia skinneri* Warsz. (Zamiaceae), photographed at FTBG,
5/1994.

11.2. *Top*, *Sphagnum* sp., growing on summit of Mount Marcy, NY, photographed
9/1971. *Bottom*, sporophyte growing from red foliose gametophytes of the liverwort
Lepidolaena taylorii (Gott.) Trev., at Swanson University Reserve, Auckland, New Zealand, photograph by Cortwa Hooijmaijers, 9/2002.

11.3. LM of paradermal section of young leaf of *Strongylodon macrobotrys* A. Gray
(Fabaceae), from FTBG; 2.4 mm top to bottom.

11.4. *Top*, *Elaeocarpus angustifolius* Blume (Elaeocarpaceae), Chapman Field USDA
Station, Miami, 2/1988. *Bottom*, *Terminalia catappa* L. (Combretaceae), collected in
Miami neighborhood, 2/2004.

11.5. Red leaf flush in tropical trees. *Top*, *Mesua ferrea* L. (Clusiaceae), photographed in my Miami yard, 3/2004. Leaves about 6 cm long. *Bottom*, young (ca. 5 cm
long) leaves of *Terminalia catappa* L. (Combretaceae), FIU campus, 2/2004.

11.6. *Top left*, small tree of *Theobroma cacao* L. (Malvaceae), Saül, French Guiana,
2/1978. Tree height 3 m. *Top right*, young translucent leaves, photographed from
plants in Montpellier, France, 6/1978. Distance across image 10 cm. *Bottom*, LM of
paradermal section of young leaf, showing anthocyanins distributed in parenchyma
cells around veins, JPM, 6/1978. Distance across image 1 mm.

11.7. *Top*, young leaves of *Mangifera indica* L. (Anacardiaceae) var. 'Glen', 12 cm
long, photographed in my yard, 3/2003. *Bottom*, fresh LM of TS of young leaf,
3/2004. Leaf 330 μm thick.

11.8. *Top left*, *Calathea* hybrid leaves, photographed in Miami nursery, 1/2003. *Top
right*, LM of TS of leaf, 6/1999; 200 μm thick. *Bottom*, LM of paradermal section of
leaf undersurface, 6/1999; 1 mm across image.

11.9. *Top, Tradescantia zebrina* Loudon (Commelinaceae), JPM, 3/1978. Leaves 6 cm long. *Bottom*, paradermal LM view of leaf undersurface of same plant. Image length 1 mm.

11.10. *Top left*, plant of *Nymphaea odorata* Willd. (Nymphaeaceae), photographed in Everglades National Park, FL, 6/1998. Leaf 14 cm across. *Top right*, undersurface detail of same plant. *Bottom*, paradermal LM view of leaf undersurface, photographed 11/2003. Width of image 270 μm.

11.11. Floating leaves of *Victoria amazonica* (Poeppig) Sowerby (Nymphaeaceae), FTBG, 7/2004. *Top*, entire plant, leaves about 1 m across. *Bottom*, undersurface leaf detail, photograph 13 cm across.

11.12. *Top*, young red leaves of *Bougainvillea spectabilis* Willd. (Nyctaginaceae), photographed at FIU, 6/2006. Leaves 4 cm long. *Bottom*, LM of TS of young leaf, 6/1999. Leaf 270 μm thick.

11.13. Young leaves of temperate plants. *Top left, Photinia glabra* (Thunb.) Maxim. (Rosaceae), photographed at my cousin Dale Easley's house in Bellevue, WA, 5/2004. Leaves 6 cm long. *Top right, Mahonia aquifolium* (Pursh) Nutt. (Berberidaceae), photographed in Tumwater Canyon, Wenatchee River, WA, 5/2004. Mature leaves 5 cm long. *Bottom, Rosa* 'Paradise', photographed in my mother's garden, Ephrata, WA, 4/2004. Leaves 4 cm long.

11.14. *Top*, flushing red leaves and small flowers of *Simarouba glauca* DC. (Simaroubaceae), Gifford Arboretum, University of Miami, FL, 2/2004. Adult leaflets 5 cm long. *Bottom*, leaf LM of TS, 7/1984. Thickness 450 μm.

11.15. This plate was taken from Édouard Morren, "Dissertation sur les feuilles vertes et colorées," C. annoot-Braeckman, 1858, Gand, Belgium (Ph.D. dissertation).

11.16. Photograph taken 10/2004.

11.17. Collage of leaves collected at the Harvard Forest, in central Massachusetts, 10/1998. Species can be located by numbers on the outlined drawing: 1—*Hamamelis virginiana* L.; 2—*Viburnum alnifolium* H. Marsh.; 3—*Viburnum cassinoides* L.; 4—*Acer saccharum* H. Marsh.; 5—*Acer rubrum* L.; 6—*Fraxinus americana* L.; 7—*Prunus serotina* J. F. Ehrh.; 8—*Vaccinium corymbosum* L.; 9—*Betula alleghaniensis* Britton; 10—*Fagus grandifolia* J. F. Ehrh.

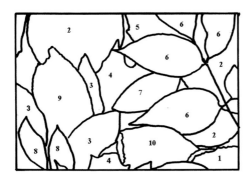

Key for leaves in figure 11.17.

11.18. *Top left*, senescing yellow leaves of *Hamamelis virginiana* L. (Hamamelidaceae). Leaves about 8 cm long. *Top right*, LM of fresh TS of leaf, showing carotenoid pigments in plastids. Leaf thickness 310 μm. Both photographs at the Harvard Forest, MA, 10/1998. *Bottom*, data methods described in D. W. Lee and R. Graham, "Leaf optical properties of rainforest sun and extreme shade plants," *American Journal of Botany* 73 (1986): 1100–1108.

11.19. *Top left*, senescing red leaf of *Quercus rubra* L. (Fagaceae). *Top right*, detail of senescing leaf surface, 4 mm across. *Bottom left*, see note for figure 11.18 for methods. *Bottom right*, LM of TS of leaf. Thickness 320 μm. Photographed at Harvard Forest, Petersham, MA, 10/1998.

11.20. *Top*, senescing leaves and berries of *Cornus stolonifera* Michx. (Cornaceae), Delaware Water Gap, NJ, 10/1996. *Bottom left*, LM of TS of leaf, Harvard Forest, MA, 11/1998. Thickness 330 μm. *Bottom right*, detail of leaf undersurface. About 5 mm across image.

11.21. For details on methods and results summarized in this diagram, see T. S. Feild, D. W. Lee, and N. M. Holbrook, "Why leaves turn red in autumn: The role of anthocyanins in senescing leaves of red-osier dogwood," *Plant Physiology* 127 (2001): 566–74.

11.22. *Top*, *Pseudowintera colorata* (Raoul) Dandy (Winteraceae), photographed by Kevin Gould in the Waitakeri Range, North Island, New Zealand, 2001. Detailed information for images of lesions and fluorescence appears in K. S. Gould, J. McKelvie, and K. R. Markham, "Do anthocyanins function as antioxidants in leaves? Imaging of H_2O_2 in red and green leaves after mechanical injury," *Plant, Cell, and Environment* 25 (2002): 1261–69.

11.23. *Top*, *Ginkgo biloba* L. (Ginkgoaceae). Leaves were collected and photographed from an old tree on the Radcliffe College campus, Cambridge, MA, 10/2004. Leaf blades are 3–4 cm long. *Bottom*, *Citharexylum fruticosum* L. (Verbenaceae), photographed on FIU campus, 3/2003. Leaf about 8 cm long.

11.24 *Top*, *Larix potaninii* Batalin (Pinaceae) tree needles turn gold in the autumn before falling, Jade Snow Mountain Park, Lijiang, Yunnan, China, 11/2006. *Bottom*, a stand of *Populus tremuloides* Michx. (Salicaceae), growing on mountain slopes near Poncha Springs Pass, at the head of the San Luis Valley in Colorado, showing yellow and red forms during the autumn; photograph by Catherine Kleier, 9/2003.

Chapter Twelve: Chlorophilia

12.6. Photographs taken in 1991. Spectral diagram from D. W. Lee, "Canopy dynamics and light climates in a tropical moist deciduous forest," *Journal of Tropical Ecology* 5 (1989): 65–79.

INDEX

···

Page numbers in italic refer to figures.